Designing Pigeonpea Genome for a Future-Ready Crop

Grown primarily by the smallholder farmers in tropical and subtropical regions, pigeonpea is pivotal to subsistence agriculture in Asia, Africa, Latin America, and the Caribbean. The last two decades have witnessed several research breakthroughs in pigeonpea. Following soybean, pigeonpea became the second food legume crop for which the whole genome sequence made available for use in genetic research and breeding. Since then, there have been tremendous efforts to develop and implement cutting-edge technologies for improving pigeonpea. These include genome sequencing of the wider germplasm sets, development of pangenome, mapping of economically important traits, multi-omics studies, and gene editing, among others. Given this, our book ***Designing Pigeonpea Genome for a Future-Ready Crop*** explores the current trends in pigeonpea research together with underscoring potential future directions and the challenges that lie ahead.

Features

- Presents the achievements of genomics research and molecular breeding of pigeonpea
- Provides current status of using wild relatives for pigeonpea improvement and optimized strategies to accelerate identification and incorporation of new genetic variation in breeding programs
- Presents the significance of mutation breeding in pigeonpea varietal improvement and future opportunities for integrating mutation-based techniques with cost-effective high-throughput approaches
- Discusses growing access to muti-omics datasets, pangenome diversity and genomic prediction studies and their relevance to future varietal development
- Demonstrates applications of generation advancement techniques such as speed breeding to accelerate breeding decisions and genetic gain in pigeonpea
- Follows the scope of brining heritable genetic changes for improving pigeonpea's response to biotic and abiotic stresses through implementation of modern gene-editing methods

The different chapters in this book have been contributed by the experts in the field. The book will serve as a comprehensive and reliable source for researchers and academicians and inform policy decisions.

Depiction and Designing of Plant Genomes
Series Editor: Chittaranjan Kole and Phullara Kole

The Cannabis Genome: A New Era in Genetics, Genomics, and Breeding
Edited by Angel Fernandez i Marti and Jaime Prohens

Designing Sorghum Genome for NextGen Agriculture
Edited by Dinesh Kumar Saini and S. V. Krishna Jagadish

The Turmeric Genome
Edited by Prasath Duraisamy, Aarthi Sounderarajan, and Chittarajan Kole

The Aloe Genome
Edited by Vinod Chhokar and Alka Jangra

The Melon Genome: Designing for NextGen Agriculture
Edited by Rakesh Kumar and Chittaranjan Kole

Designing Pigeonpea Genome for a Future-Ready Crop
Edited by Abhishek Bohra, Ashok Kumar Parihar, and Girish Prasad Dixit

For more information, please visit our website: www.routledge.com/Depiction-and-Designing-of-Plant-Genomes/book-series/DDPG

Designing Pigeonpea Genome for a Future-Ready Crop

Edited by
Abhishek Bohra
Ashok Kumar Parihar
Girish Prasad Dixit

CRC Press
Taylor & Francis Group
Boca Raton London New York

CRC Press is an imprint of the
Taylor & Francis Group, an **informa** business

Designed cover image: Cover Design by Chittaranjan Kole; Image courtesy of Abhishek Bohra

First edition published 2026
by CRC Press
2385 NW Executive Center Drive, Suite 320, Boca Raton FL 33431

and by CRC Press
4 Park Square, Milton Park, Abingdon, Oxon, OX14 4RN

CRC Press is an imprint of Taylor & Francis Group, LLC

ISBN: 978-1-032-99240-2 (hbk)
ISBN: 978-1-032-99242-6 (pbk)
ISBN: 978-1-003-60309-2 (ebk)

DOI: 10.1201/9781003603092

Typeset in Times
by SPi Technologies India Pvt Ltd (Straive)

Dedication

This book series is dedicated to our brothers/brothers-in-law and sisters/sisters-in-law –

Bansi Badan Kole
Niranjan Kole
Bharati Nandi Nee Kole
Arati Kundu Nee Kole
Ram Ranjan Kole

Contents

Preface to the Series

Genome sequencing of the model plant Arabidopsis in 2000 followed by the crop plant rice in 2002 flagged of genome sequencing in an array of plants. As of March 5, 2024, 4,604 genomes of 1,482 different plant species have been sequenced, and the number has been continuing to grow exponentially. The list includes several crop plants, as well as some model plants. Sequencing of the nuclear genomes was followed by sequencing of extranuclear genomes, such as mitochondria and chloroplast and by several studies on functional genomics. However, detailed reports on them are scattered in the literature including research articles and reviews. Therefore, a comprehensive compilation of deliberations on the sequences with enumeration of the sequences, annotation, assembly, gene families and comparative genomics of these useful genomes and elucidation of functional genomics, including transcriptomics, metabolomics, and proteomics, in these higher plants is timely.

A number of higher plants provide the basic F5—food, feed, fuel, fiber, and furniture—needs; and also, nutraceuticals, beverages, luxury, and aesthetic items. Genetic improvement in these plants with regard to their utilities were being done in the last century mainly by traditional breeding. Advent of molecular breeding and genetic engineering in the last two decades of the last century strengthened the breeding strategies in the economic plants. Advances in plant genomics has ultimately revolutionized the concepts and strategies of crop improvement. Employment of high-throughput genotyping and phenotyping together with other genomics assisted breeding tools like gene discovery, allele mining, genome-wide association studies, genomic selection, and genome-wide prediction, etc. has facilitated accelerated and precise plant breeding. Recent emergence of the genome editing technique has enormously supplemented the molecular and genomics tools, making it possible to design and tailor the genomes according to the needs. However, success stories of next-generation breeding techniques need to be compiled for easy access to the readers.

This present book series entitled, Depiction and Designing Plant Genomes, includes a number of books each dedicated to the deliberations on genomic attributes and/or genetic improvement of a higher plant and hopefully will serve as handy compendium for readers.

The books of this series have been edited by highly reputed scientists, and I remain thankful for their invaluable academic efforts and cooperation. We are also thankful to the publisher and specifically to its Senior Commissioning Editor (Chemistry/Agricultural Sciences), Dr. Renu Upadhyay, and her Editorial Assistant, Tisha Sharma for their outstanding and timely help and cordial cooperation since inception of this book series.

Finally, we must express my thanks to our children, Sourav and Devleena, for assisting us in editing the books of this series and also bearing with us for our channelizing the time that should have been kept for them toward editing this book series.

Kolkata, August 3, 2024

Chittaranjan Kole
Phullara Kole

Preface

Pigeonpea is an important food legume crop of rain-fed agriculture. According to the FAO data, pigeonpea is cultivated worldwide in 5.38 mha area producing 4.58 mt annually. More than 80% of the global pigeonpea area is in India, followed by Myanmar (7.6%), Kenya (5.5%), Malawi (4.75%), and Tanzania (4.48%). The pigeonpea crop ensures the food security and livelihood generation of the vulnerable populations in tropical and subtropical regions. The high level of carbohydrates, proteins, minerals, vital amino acids, and vitamins in pigeonpea help complement the traditional cereal-based diets that are prevalent in low-income countries. The health benefits and therapeutic uses of various parts of the pigeonpea plant, including seeds, leaves, and roots, are well documented.

Traditional breeding efforts directed at enhancing yield, quality and resistance performance of pigeonpea have led to development and release of several varieties in Asia and Africa. Despite the growing availability of improved pigeonpea varieties, the productivity of the crop has remained stagnated at around 850–900 kg/ha. A number of production constraints limit the potential of pigeonpea, and the crop is sensitive to a variety of biotic stresses, including diseases (e.g., blight, wilt, and sterility mosaic) and insect pests, mainly pod borers and pod fly. Among abiotic constraints, water logging, salinity and cold (frost) have been reported to deteriorate crop performance.

Recent years have witnessed the development of cutting-edge genomic tools and modern breeding methods to augment pigeonpea improvement programs. Notable in the context are decoding of the pigeonpea reference genome in 2011 accompanied by the large-scale genome resequencing of germplasm collections and elucidation of pangenome diversity. Growing access to genome-wide nucleotide sequence information and large-scale phenotyping data open new opportunities to tailor future pigeonpea cultivars to climate change requirements.

Genome-wide analyses of various economically important traits of pigeonpea have provided breeders with the low-cost and high-throughput molecular assays to reach rapid and accurate breeding decisions. Recent research on increasing generational turnover of pigeonpea have resulted in optimization of rapid generation advancement methods, speed breeding protocols, and genomic selection models to curtail the lengthy breeding cycles and enhancing genetic gain in pigeonpea improvement programs. Likewise, modern gene-editing methods introduce precise heritable changes in the genome in rapid and cost-effective manner. In parallel to varietal improvement, exploitation of heterosis has proven to be a promising alternative to overcome the exiting yield barriers in pigeonpea. In this context, discovery and application of male sterility systems in pigeonpea have deepen our understanding of the genetic, biochemical, and physiological mechanisms underlying male sterility and fertility restoration. Recent advances in cytoplasmic male sterility (CMS) in pigeonpea has opened new opportunities to harness the potential of hybrid vigor for pigeonpea improvement. Our present book is well-timed with the growing demand for innovative and sustainable solutions to address the impact of climate change on global production of important food crops. The book has 12 chapters from 73 authors comprehensively exploring the latest trends, opportunities and challenges that lie ahead in pigeonpea research and breeding. We believe that the book will be a great resource for researchers, university faculties, personnel from seed industry and policy makers. The new insights offered from the book will facilitate a better understanding for developing efficient strategies for accelerated pigeonpea improvement.

The editors are also thankful to Dr. Mangi Lal Jat (Director General, ICAR and Secretary DARE, India) and Dr. Devendra Kumar Yadava (Deputy Director General-Crop Science, ICAR, India) for their support. The editors thank Prof. C. Kole, Series Editor, for his invitation and help in editing this volume. We gratefully acknowledge the support received from Tisha Sharma and Renu Upadhyay from CRC Press/Taylor and Francis Group. The editors express their gratitude to their family members for all cooperation and moral support received during the development of the book.

Abhishek Bohra
Ashok Kumar Parihar
Girish Prasad Dixit

About the Editors

Abhishek Bohra is currently working as Senior Scientist at ICAR-Indian Institute of Pulses Research (IIPR), Kanpur, India. He has served as Principal Investigator for a national hybrid breeding program for nearly a decade period. He has contributed toward the development of various pigeonpea hybrids and varieties. He has made contributions to develop new germplasm for pollen sterility trait for application in hybrid development and identification of promising genetic stocks that serve as donors for enhancing key breeding traits. Considering genomics-assisted breeding approach, he is associated with projects on developing novel experimental populations, identifying important genes/DNA markers associated with economically important traits and marker-assisted selection. His research work has resulted in publication of more than 150 research papers in journals of international repute. He is contributing to the scientific community and society by serving as Editor (*Functional and Integrative Genomics*, *PLoS One*, *Crop Science*, *Heliyon*, *Plant Breeding*, *Functional Plant Biology and Plant Methods*) and Reviewer to various international journals. He has a h-index of 44, with cumulative citation of over 7500. He is a recipient of the Young Scientist Medal of Indian National Science Academy (INSA), India and Associate Fellow of National Academy of Agricultural Sciences (NAAS), India.

Ashok Kumar Parihar obtained his Ph.D. in plant breeding and genetics from Maharana Pratap University of Agriculture and Technology, Udaipur, Rajasthan, India. He is currently working as Senior Scientist in the Crop Improvement Division of ICAR, Kanpur, and also associated as Principal Investigator of plant breeding program in All India Coordinated Research Project on Rabi Pulses. During his research career of more than 14 years, he has developed 12 high yielding multiple disease resistance field pea varieties suitable for different agro-climatic conditions. He has so far published more than 75 research papers in different national and international journals. He has also published 6 books, 35 book chapters, >100 popular articles and 10 technical bulletins. He has been honored with many prestigious awards, including Fellow of NAAS, India and Fellow of Indian Society of Genetics and Plant Breeding (ISGPB), India and Indian Society of Pulse Research and Development (ISPRD), India.

Girish Prasad Dixit is an eminent agricultural scientist with over 30 years of experience in pulses research and development. He received a Ph.D. in 1993 from Banaras Hindu University, Varanasi. He is currently serving as Director at ICAR – IIPR, Kanpur, India. He has led an Indian national chickpea program as Project Coordinator for about eight years, focusing on integrated breeding approaches for improving yield potential, improved tolerance to stresses, market-preferred grain traits, suitability to machine harvesting, and modernization of programs through digitalization. He has led several international and national projects. He has developed 25 high yielding varieties of pulses. He has published more than 100 research papers and 25 book chapters, and edited 5 books.

Contributors

Hubert Adoukonou-Sagbadja
Department of Genetic and Biotechnology
University of Abomey-Calavi
Cotonou, Republic of Benin

Ronaldine Arlienne Ahouanmagnagahou
Department of Genetic and Biotechnology
University of Abomey-Calavi
Cotonou, Republic of Benin

Amandeep
Department of Plant Sciences
North Dakota State University
Fargo, North Dakota, USA

Ruchi Asati
Department of Genetics and Plant Breeding
Rani Lakshmi Bai Central Agricultural University (RLBCAU)
Jhansi, India

Ulrich Ayikpon
Department of Genetic and Biotechnology
University of Abomey-Calavi
Cotonou, Republic of Benin

Harmeet Singh Bakala
University of Missouri
Fisher Delta Research, Extension and Education Center
Portageville, Missouri, USA

Shruthi H. Belliappa
Department of Genetics
International Crops Research Institute for the Semi-Arid Tropics (ICRISAT)
Hyderabad, India

Jyotsna Bharti
Nutritional improvement of Crops Group
International Centre for Genetic Engineering and Biotechnology (ICGEB)
New Delhi, India

Abhishek Bohra
Division of Crop Improvement
ICAR – Indian Institute of Pulses Research (IIPR)
Kanpur, India

Naresh Bomma
Pigeonpea breeding
International Crops Research Institute for the Semi-Arid Tropics (ICRISAT)
Hyderabad, India

Angelo Bossikponnon
Faculty of Sciences and Techniques
University of Abomey-Calavi
Cotonou, Republic of Benin

Puja Chakraborty
Nutritional Improvement of Crops Group
International Centre for Genetic Engineering and Biotechnology (ICGEB)
New Delhi, India

Subhash Chand
Crop Improvement Division
ICAR-Indian Grassland and Fodder Research Institute (IGFRI)
Jhansi, India

Anup Chandra
Division of Crop Protection
ICAR – Indian Institute of Pulses Research (IIPR)
Kanpur Uttar Pradesh, India

S. K. Chaturvedi
Department of Genetics and Plant Breeding
Rani Lakshmi Bai Central Agricultural University (RLBCAU)
Jhansi, India

Rakesh Choudhary
Department of Genetics and Plant Breeding
Rani Lakshmi Bai Central Agricultural University (RLBCAU)
Jhansi, India

Sunita Choudhary
Center of Excellence in Genomics and Systems Biology (CEGSB) and Center for Pre-Breeding Research (CPBR),
International Crops Research Institute for the Semi-Arid Tropics (ICRISAT)
Patancheru, Hyderabad, Telangana, India

Vijay A. Dalvi
DCM Shriram Ltd
Gurugram, India

Antara Das
Division of Plant Biotechnology
ICAR – Indian Institute of Pulses Research (IIPR)
Kanpur, India

Dibendu Datta
Regional Station, Phanda, Bhopal
ICAR – Indian Institute of Pulses Research (IIPR)
Kanpur, India

K Dharani Sree
Pigeonpea breeding
International Crops Research Institute for the Semi-Arid Tropics (ICRISAT)
Hyderabad, , India

Girish Prasad Dixit
Division of Crop Improvement
ICAR – Indian Institute of Pulses Research (IIPR)
Kanpur, Pradesh, India

Prakash Gangashetty
Pigeonpea Breeding,
International Crops Research Institute for the Semi-Arid Tropics (ICRISAT)
Hyderabad, India

Sunil Gangurde
Center of Excellence in Genomics and Systems Biology (CEGSB) and Center for Pre-Breeding Research (CPBR),
International Crops Research Institute for the Semi-Arid Tropics (ICRISAT)
Hyderabad, India

Surendra Ghritlahre
Regional Station, Phanda, Bhopal
ICAR – Indian Institute of Pulses Research (IIPR)
Kanpur, India

Isha Gupta
Nutritional Improvement of Crops Group
International Centre for Genetic Engineering and Biotechnology (ICGEB)
New Delhi, India

Harshavardan J. Hilli
School of Agriculture
Sanskriti University, Mathura
Uttar Pradesh, Mathura, India

Arshad Husain
Division of Plant Biotechnology
ICAR – Indian Institute of Pulses Research (IIPR)
Kanpur, India

Jagdish Jaba
Center of Excellence in Genomics and Systems Biology (CEGSB) and Center for Pre-Breeding Research (CPBR),
International Crops Research Institute for the Semi-Arid Tropics (ICRISAT)
Hyderabad, India

Souframanien Jegadeesan
Nuclear Agriculture and Bio Technology Division
Bhabha Atomic Research Centre, Trombay
Mumbai, India

Uday Chand Jha
Division of Crop Improvement
ICAR – Indian Institute of Pulses Research (IIPR)
Kanpur, India

Rashmi Kaul
Nutritional Improvement of Crops Group
International Centre for Genetic Engineering and Biotechnology (ICGEB)
New Delhi, India

Tanushri Kaul
Nutritional Improvement of Crops Group
International Centre for Genetic Engineering and Biotechnology (ICGEB)
New Delhi, India

Mehakpreet Kaur
Department of Biological Sciences
North Dakota State University
Fargo, North Dakota, USA

Dhanapati Keerthana
Department of Genetics and Plant Breeding
Rani Lakshmi Bai Central Agricultural University (RLBCAU)
Jhansi, India

Kadam Kiran
Department of Genetics and Plant Breeding
Rani Lakshmi Bai Central Agricultural University (RLBCAU)
Jhansi, India

Nidhi Kumari
Regional Station, Phanda, Bhopal
ICAR – Indian Institute of Pulses Research (IIPR)
Kanpur, India

T R Likhithashree
Department of Genetics and Plant Breeding
University of Agricultural Sciences
Bengaluru, India

R. Shubhra Maithreyi
Nutritional Improvement of Crops Group
International Centre for Genetic Engineering and Biotechnology (ICGEB)
New Delhi, India

Alok Kumar Maurya
Division of Crop Improvement
ICAR – Indian Institute of Pulses Research (IIPR)
Kanpur, India

Sean Mayes
Center of Excellence in Genomics and Systems Biology (CEGSB) and Center for Pre-Breeding Research (CPBR),
International Crops Research Institute for the Semi-Arid Tropics (ICRISAT)
Hyderabad, India

Abhinav Moghiya
Center of Excellence in Genomics and Systems Biology (CEGSB) and Center for Pre-Breeding Research (CPBR),

International Crops Research Institute for the Semi-Arid Tropics (ICRISAT)
Hyderabad, India

Nainika Nagar
Regional Station, Phanda, Bhopal
ICAR – Indian Institute of Pulses Research (IIPR)
Kanpur, India

Manish K Pandey
Center of Excellence in Genomics and Systems Biology (CEGSB) and Center for Pre-Breeding Research (CPBR),
International Crops Research Institute for the Semi-Arid Tropics (ICRISAT)
Hyderabad, India

Ashok Kumar Parihar
Division of Crop Improvement
ICAR – Indian Institute of Pulses Research (IIPR)
Kanpur, India

Sugyan Preet
Nutritional Improvement of Crops Group
International Centre for Genetic Engineering and Biotechnology (ICGEB)
New Delhi, India

Mahesh Pujar
Pigeonpea breeding
International Crops Research Institute for the Semi-Arid Tropics (ICRISAT)
Hyderabad, India

Dhanasekar Punniyamoorthy
Nuclear Agriculture and Biotechnology Division
Bhabha Atomic Research Centre (BARC)
Mumbai, India

Sagar K Rangari
Center of Excellence in Genomics and Systems Biology (CEGSB) and Center for Pre-Breeding Research (CPBR),
International Crops Research Institute for the Semi-Arid Tropics (ICRISAT)
Hyderabad, India

Asish K. Rout
ICAR IIPR – RS Bhubaneswar
ICAR – Indian Institute of Pulses Research (IIPR)
Kanpur, India

A. K. Roy
ICAR – Emeritus Professor
Rani Lakshmi Bai Central Agricultural University (RLBCAU)
Jhansi, India

Rajib Roychowdhury
Center of Excellence in Genomics and Systems Biology (CEGSB) and Center for Pre-Breeding Research (CPBR),
International Crops Research Institute for the Semi-Arid Tropics (ICRISAT)
Hyderabad, India

Gitanjali Sahay
Regional Station, Phanda, Bhopal
ICAR – Indian Institute of Pulses Research (IIPR)
Kanpur, Uttar Pradesh, India

S J Satheesh Naik
Division of Crop Improvement
ICAR – Indian Institute of Pulses Research (IIPR)
Kanpur, India

K B Saxena
Al Ain City, UAE

Rachit K. Saxena
ICAR – Indian Agricultural Research Institute (IARI)
Gogamukh, India

M Senthil Kumar
Division of Basic Science
ICAR – Indian Institute of Pulses Research (IIPR)
Kanpur, India

PS Shanmugavadivel
Division of Plant Biotechnology
ICAR – Indian Institute of Pulses Research (IIPR)
Kanpur, India

Ira Sharma
Pigeonpea breeding
International Crops Research Institute for the Semi-Arid Tropics (ICRISAT)
Hyderabad, India

Mamta Sharma
Center of Excellence in Genomics and Systems Biology (CEGSB) and Center for Pre-Breeding Research (CPBR),
International Crops Research Institute for the Semi-Arid Tropics (ICRISAT)
Hyderabad, India

Aalok Shiv
Division of Crop Improvement
ICAR – Indian Institute of Sugarcane Research (IISR)
Lucknow, India

Navjot Singh Sidhu
Department of Genetics and Plant Breeding
University of Agricultural Sciences
Bengaluru, India

Gurjeet Singh
Department of Soil and Crop Sciences
Texas A&M University, AgriLife Research Center,
Beaumont, Texas, USA

Inderjit Singh
Department of Plant Breeding and Genetics
Punjab Agricultural University
Ludhiana, India

Pallavi Sinha
International Rice Research Institute (IRRI), ICRISAT Campus
Patancheru, India

GK Sujayanand
Division of Crop Protection
ICAR – Indian Institute of Pulses Research (IIPR)
Kanpur, India

R Sultana
School of Drought Stress Management
ICAR-National Institute of Abiotic Stress Management (NIASM)
Baramati, India

Kunal Tanwar
Nutritional Improvement of Crops Group
International Centre for Genetic Engineering and Biotechnology (ICGEB)
New Delhi, India

Arulprakash Thangaraj
Nutritional Improvement of Crops Group
International Centre for Genetic Engineering and Biotechnology (ICGEB)
New Delhi, India

Abha Tiwari
School of Crop Resistance System Research
ICAR-National Institute of Biotic Stress Management (NIBSM)
Raipur Chhattisgarh, India

Rachana Verma
Nutritional Improvement of Crops Group
International Centre for Genetic Engineering and Biotechnology (ICGEB)
New Delhi, India

Suruchi Vij
Department of Plant Breeding and Genetics
Punjab Agricultural University
Ludhiana, India

K Yaksha
International Crops Research Institute for the Semi-Arid Tropics (ICRISAT)
Hyderabad, India

Fiacre Gildas Zavinon
Faculty of Sciences and Techniques
University of Abomey-Calavi
Cotonou, Republic of Benin

1 Improving Pigeonpea for Climate Resilience

An African Perspective

Fiacre Gildas Zavinon, Ulrich Ayikpon, Angelo Bossikponnon, Ronaldine Arlienne Ahouanmagnagahou, and Hubert Adoukonou-Sagbadja

1.1 INTRODUCTION

Pigeonpea (*Cajanus cajan* L.), a versatile and resilient legume, plays a pivotal role in African agriculture. Predominantly cultivated by smallholder farmers, this crop is a crucial source of protein for millions across sub-Saharan Africa and contributes significantly to food security and rural incomes (Arora et al. 2017). In many rural areas, pigeonpea is a lifeline during periods of food scarcity, offering a reliable source of calories and protein when other crops fail (Ayenan et al. 2017). Its ability to thrive in marginal soils and harsh climates makes it particularly valuable for resource-poor farmers who lack access to irrigation or advanced agricultural inputs (Bohra et al. 2020a; Saxena et al. 2021a). Moreover, pigeonpea's compatibility with intercropping systems enhances its appeal, as it allows farmers to diversify their harvests and reduce risks associated with monocropping. Beyond its nutritional value, pigeonpea enriches the soil with nitrogen, reduces erosion, and enhances soil fertility, making it an integral component of sustainable farming systems (Saxena et al. 2023). Its adaptability to arid and semi-arid conditions has earned it recognition as a strategic crop for regions prone to recurrent droughts. Pigeonpea's extensive taproot system enables it to access underground water reserves, allowing it to endure prolonged dry spells – a trait increasingly valuable in the face of climate change. It is considered one of Africa's traditional drought-tolerant crops with high potential for mitigating the impacts of climate change on the continent (Odeny 2007).

Despite its numerous advantages, pigeonpea cultivation faces significant challenges, many of which are exacerbated by climate change (Mishra et al. 2017). Erratic rainfall patterns, rising temperatures, and extreme weather events such as heatwaves, prolonged droughts, and flooding disrupt its growth cycle, affecting

DOI: 10.1201/9781003603092-1

flowering, pod development, and overall yield stability. As a C3 plant, pigeonpea is particularly vulnerable to high temperatures (Slim & Omanga 2001), which accelerate evapotranspiration and lead to soil moisture deficits, while excessive rainfall increases the risk of waterlogging and fungal diseases like Fusarium wilt. Additionally, warmer conditions favor the proliferation of pests such as pod borers and aphids, further threatening productivity. These climate-induced stresses not only reduce yields but also jeopardize the livelihoods of smallholder farmers who depend on pigeonpea for food security and economic stability. Furthermore, limited access to improved seeds, inadequate knowledge of climate-smart agronomic practices, weak infrastructure for processing and marketing pigeonpea products compound these challenges (Kaoneka et al. 2016). Africa remains the most susceptible to climate change due to its vulnerability and limited capacity to cope with the physical, human, and socioeconomic consequences of climate extremes. As these extremes become more frequent and intense, their impact on key crops like pigeonpea threatens food security, rural livelihoods, and economic stability. Gaining a deeper insight into local climate variability is essential for developing adaptation strategies that are both effective and context-specific (Yeleliere et al. 2023). Developing resilient pigeonpea varieties, promoting sustainable farming techniques, and strengthening value chains are therefore urgent priorities to safeguard the crop's role in Africa's food systems and enhance agricultural resilience in the face of a changing climate.

In response to these challenges, numerous research efforts highlight innovative strategies to enhance pigeonpea's resilience to climate variability. These include adopting climate-smart agricultural practices, developing improved varieties with greater stress tolerance, and implementing sustainable resource management techniques to mitigate the adverse effects of climate change (Musokwa & Mafongoya 2021). Crop rotation with species having varying water requirements, for example, has emerged as a promising approach to mitigate drought-related stress in vulnerable regions (Choudhary et al. 2011). Furthermore, advancements in understanding the bio-physiological mechanisms of pigeonpea under stress conditions provide valuable insights into fostering its resilience. One of the most promising avenues for improving pigeonpea resilience lies in breeding programs aimed at developing climate-adapted varieties (Bohra et al. 2019; Lohithaswa et al. 2025). Advances in genomics have revolutionized pigeonpea breeding, enabling researchers to identify and incorporate genes associated with drought tolerance, pest resistance, and higher yields (Bohra et al. 2020a). Speed breeding protocols, pioneered by institutions like ICRISAT, have significantly reduced the time required to develop new pigeonpea lines, allowing for faster dissemination of improved varieties to farmers (Simion et al. 2022, Aggarwal et al. 2025). These innovations not only address immediate challenges but also lay the groundwork for long-term sustainability.

Sustainable agricultural practices play a crucial role in enhancing pigeonpea resilience, particularly in the face of increasing climatic stressors. Conservation agriculture – characterized by minimal soil disturbance, permanent soil cover, and crop rotation – has proven effective in maintaining soil health and moisture levels, which are critical for sustaining yields amid frequent droughts across Africa (Khaniya, 2024). Implementing integrated pest management (IPM) strategies is equally vital to mitigate the growing threat of pests and diseases exacerbated by climate change (Volp et al. 2025). IPM combines biological controls, such as natural predators and biopesticides, with cultural practices like crop rotation and intercropping to suppress pest populations (Saxena et al. 2023). For instance, intercropping pigeonpea with maize or sorghum disrupts the life cycles of major pests like pod borers, reducing infestations without excessive reliance on synthetic pesticides (Yohane et al. 2020). These holistic approaches not only enhance pigeonpea productivity but also contribute to environmental sustainability and long-term food security.

This chapter aims to delve into these strategies while offering an African perspective on research and development efforts. By addressing local realities, agro-climatic challenges, and opportunities for participatory innovation, we seek to highlight practical solutions tailored to the continent's needs (Chanda et al. 2019). Recent pre-extension demonstrations of newly developed pigeonpea varieties, such as those conducted in Ethiopia and Benin (Zavinon et al. 2022), underscore the potential of improved cultivars to boost productivity and farmer resilience. Additionally, integrating climate-smart practices into pigeonpea production systems has been identified as a cornerstone for building smallholder resilience against global climate impacts (Saxena et al. 2023).

Ultimately, this work aspires to serve as a practical and scientific roadmap for transforming pigeonpea into an even more robust pillar of African agriculture amidst a changing climate. By analyzing current challenges and exploring actionable solutions, we aim to inspire tangible interventions that support farmers, preserve ecosystems, and secure a sustainable future for generations to come. The integration of advanced breeding techniques, sustainable farming practices, and robust pest management strategies offers a comprehensive framework for addressing the multifaceted impacts of climate change on pigeonpea production.

1.2 UNDERSTANDING PIGEONPEA AND ITS ROLE IN AFRICAN AGRICULTURE

1.2.1 Pigeonpea Production and Botanical Description

1.2.1.1 Pigeonpea Production in Africa

The pigeonpea (*Cajanus cajan* (L.) Millsp.), a legume with a debated origin between Africa and the Indian subcontinent, is widely cultivated across sub-Saharan Africa, where its adaptability to drought and poor soils makes it a crucial crop for smallholder farmers. Africa accounts for approximately 14% of global pigeonpea production (Zavinon et al. 2018; Yohane et al. 2020), with Malawi and Tanzania leading the continent's output, contributing over 70% of total production. Other significant producers include Kenya, Uganda, Mozambique, and Ethiopia, while smaller-scale cultivation is also reported in Nigeria, Benin, Ghana, Togo, and Mali (Odeny 2007; Zavinon et al. 2018).

Despite its potential, pigeonpea remains an underutilized and "orphan" crop in many African countries (Varshney et al. 2011; Kaoneka et al. 2016), often grown on marginal lands by smallholder farmers with limited support in terms of research, extension services, and market access. However, it plays a dual role as both a food and cash crop, providing an affordable protein source while generating income for rural households. Additionally, its nitrogen-fixing ability enhances soil fertility, making it an essential component of sustainable farming systems.

Yield levels, however, remain relatively low across Africa, averaging between 600 and 800 kg per hectare, compared to potential yields exceeding 2,000 kg per hectare under improved agronomic practices. This yield gap is primarily attributed to climate variability, pest and disease pressures, and limited access to improved seed varieties and modern farming techniques. Addressing these constraints through research, breeding programs, and improved agronomic practices could significantly enhance pigeonpea productivity and its contribution to food security and climate resilience in Africa.

1.2.1.2 Botanical and Agronomic Characteristics of Pigeonpea

Pigeonpea (*C cajan* (L.) Millsp.) is a perennial legume often cultivated as an annual crop, belonging to the Fabaceae family. It is a hardy woody shrub with an extensive adaptation to diverse agroecological conditions, making it an essential crop in many African farming systems (Odeny 2007). Pigeonpea is characterized by its deep tap-root system, which can extend up to two meters into the soil, allowing it to access moisture stored in deep layers. This feature makes it particularly resilient to prolonged droughts, a crucial trait for semi-arid regions. The plant displays a wide range of growth habits, from determinate and compact to indeterminate and tall types, with heights varying between 0.5 and 4 meters. The leaves are trifoliate, with lanceolate leaflets covered in fine hairs to minimize water loss. Pigeonpea flowers are typically yellow or red, arranged in racemes or panicles, and follow a predominantly self-pollinating system, though cross-pollination can reach an average rate of 20%. The fruit is an elongated pod (5–9 cm), containing 2 to 9 seeds that vary in color, shape, and size, depending on the cultivar (Venkata et al. 2019).

Pigeonpea thrives in a range of environmental conditions, performing well in areas with annual rainfall between 600 and 1,200 mm. Although it prefers mean temperatures between 20°C and 30°C, it can tolerate short-term extremes from 10°C to 40°C. Its optimal soil pH ranges from 5.5 to 7.5, allowing it to grow in slightly acidic to neutral soils. The crop's nitrogen-fixing ability enhances soil fertility, reducing reliance on synthetic fertilizers and making it an essential component of sustainable farming systems. It is widely intercropped with cereals like maize, sorghum, and millet, as well as other legumes such as cowpea and black gram, helping to improve overall farm productivity. Pigeonpea cultivars are classified into extra-early (60–90 days), early (90–120 days), medium (120–150 days), and late-maturing (150–250 days) types (Upadhyaya et al. 2013), offering flexibility for different farming systems. Long-duration varieties are often grown in agroforestry systems, providing shade and ecosystem benefits, while short-duration varieties are preferred for field cultivation.

1.2.2 Nutritional and Socioeconomic Importance of Pigeonpea

1.2.2.1 Nutritional Importance

Pigeonpea (*C cajan* (L.) Millsp.) is a highly nutritious legume that plays a crucial role in enhancing food security, balanced diets, and poverty alleviation in Africa. It is a primary source of protein for many rural communities, particularly in regions where animal protein is scarce or expensive. Globally, about 20% of the population relies on pigeonpea as a major protein source. The seeds contain approximately 20–22% protein, making them a vital dietary component in cereal-based diets common across Africa. Pigeonpea protein includes essential amino acids such as lysine, methionine, and tryptophan, which improve the nutritional balance of traditional diets and help combat malnutrition. In addition to protein, pigeonpea seeds provide significant amounts of carbohydrates (57.3%), fat (1.5%), and ash (8.1%), making them an energy-rich food source (Saxena et al. 2002). The legume is consumed in various forms, including fresh green seeds, dried split peas (dhal), and flour. In Africa, whole dry seed is generally boiled and consumed, making it a staple in many households. Its protein composition includes two key globulins – cajanin and concajanin – accounting for 58% and 8%, respectively, of the total protein content. This rich nutritional profile makes pigeonpea a valuable crop for addressing food security challenges in Africa.

Beyond human consumption, pigeonpea also serves as high-quality fodder and feed for livestock. Its foliage contains approximately 23.7% crude protein and 35.7% fiber, making it an excellent supplement to low-quality forage (Khoury et al. 2015). Studies have shown that feeding livestock with pigeonpea fodder increases their intake of low-quality herbage, improving weight gain and overall productivity. The by-products of split and shriveled pigeonpea seeds are also used as an inexpensive alternative to high-cost animal feed ingredients such as soybean meal and fish meal. Research has demonstrated that pigeonpea seeds can effectively replace maize and groundnut cake in the diets of poultry, including broilers, pullet chicks, and layers, contributing to cost-effective livestock production. Pigeonpea's ability to provide both human and animal nutrition while thriving in harsh environments highlights its potential as a sustainable and multipurpose crop in Africa.

1.2.2.2 Socioeconomic Importance

Pigeonpea is not only a staple food but also a vital cash crop for smallholder farmers in Africa. Its resilience to drought and poor soils makes it a reliable source of income, particularly in semi-arid regions where other crops struggle to perform. The ability of pigeonpea to thrive in marginal environments allows farmers to cultivate it with minimal inputs, reducing production costs while ensuring household food security. The crop plays a significant role in rural livelihoods through local and international markets. Africa accounts for about 14% of global pigeonpea production, with major producers including Malawi, Tanzania, Kenya, Uganda, Mozambique, and Ethiopia. In countries like Malawi and Tanzania, pigeonpea exports – primarily to India, the world's largest consumer – generate substantial revenue for farmers and contribute to national economies (Kaoneka et al. 2016).

Furthermore, pigeonpea supports employment across the value chain, from production and processing to marketing and trade. Women, in particular, benefit from pigeonpea farming, as they are heavily involved in harvesting, processing, and selling the crop. The commercialization of pigeonpea-based products such as flour, split peas (dhal), and fortified food items has further increased its economic value, promoting agribusiness opportunities. Additionally, pigeonpea's nitrogen-fixing ability enhances soil fertility, benefiting crop rotation and intercropping systems. This improves overall farm productivity and sustainability, ensuring long-term agricultural resilience.

1.2.3 Current Cultivation Practices

Pigeonpea cultivation in Africa is shaped by diverse farming practices influenced by agroecological conditions, farmer preferences, and resource availability. Traditional knowledge, passed down through generations, is combined with modern agronomic techniques to optimize productivity. The crop is grown under both sole and intercropping systems, though intercropping remains the predominant practice. For example, in Malawi, Silim et al. (2005) reported that intercropping accounted for 90% of pigeonpea fields, while in Benin, 94% of farmers cultivate pigeonpea in association with other crops (Zavinon & Adoukonou-Sagbadja 2019). This practice, rooted in traditional farming methods, maximizes land use, enhances biodiversity, mitigates risks, and improves natural pest control. Common companion crops include maize, sorghum, millet, and cowpea. Despite the prevalence of intercropping, monocropping systems are also observed, particularly in Malawi, Tanzania, and Benin, where pigeonpea is sometimes cultivated in rotation with cereals such as maize or sorghum.

Land preparation for pigeonpea cultivation varies across regions but generally involves minimal tillage to preserve soil structure and maintain natural fertility. Farmers typically prepare the land manually using hand-hoes to clear and loosen the soil. Seeds are sown directly into the field, with planting methods differing based on regional practices and cropping systems. Farmers employ either broadcasting or direct seeding in rows, often placing seeds in shallow holes made with a stick or hoe. Planting densities depend on whether pigeonpea is grown as a sole crop or intercropped. In sole cropping systems, farmers generally space rows 75–100 cm apart, with 25–50 cm between plants. In intercropping systems, wider spacing is adopted to accommodate companion crops, with row spacing ranging from 100 to 150 cm and intra-row spacing between 50 and 75 cm. Farmers typically plant at the onset of the rainy season to optimize soil moisture availability.

Traditional landraces remain widely used due to their adaptability to local conditions, but research institutions have introduced improved varieties with higher yields, early maturity, and resistance to pests and diseases. However, access to quality seeds remains a challenge, as many farmers rely on saved seeds from previous harvests. Given that pigeonpea is primarily grown by smallholder farmers, the use of chemical inputs for weed and pest control is limited in Africa (Zavinon & Adoukonou-Sagbadja 2019). Pigeonpea is harvested when pods mature and turn brown. The process typically involves handpicking mature pods (Ayenan et al. 2017). Proper post-harvest handling, including drying and storage, is essential to minimize losses caused by pests and mold, ensuring both food security and marketability.

1.2.4 Challenges and Opportunities for Pigeonpea Cultivation in Africa

Pigeonpea cultivation in Africa is characterized by a unique combination of challenges and opportunities. While the crop holds immense potential to address food security, economic development, and environmental sustainability, several barriers hinder its widespread adoption and optimal productivity. This section explores these dynamics, drawing on recent studies and analyses to provide a comprehensive understanding of the current landscape.

1.2.4.1 Challenges Facing Pigeonpea Cultivation

One of the primary challenges facing pigeonpea production in Africa is the limited adoption of high-yielding cultivars (Kaoneka et al. 2016). Despite advancements in breeding programs, many farmers continue to rely on traditional landraces that are often low-yielding and susceptible to pests and diseases. This reluctance to adopt improved varieties can be attributed to factors such as inadequate access to seeds, lack of awareness about their benefits, and insufficient extension services to support farmers in transitioning to new practices (Fatokimi & Tanimonure 2021).

Diseases and pests further compound these challenges. Pigeonpea is particularly vulnerable to pod borers, Fusarium wilt, and sterility mosaic disease, which are exacerbated by climate change (Yohane et al. 2022; Volp et al. 2025). Rising temperatures and humidity create favorable conditions for these pathogens, leading to significant yield losses. For instance, pod borers alone can cause up to 80% damage in severely infested fields, undermining the economic viability of pigeonpea farming (Sharma, 2016). Besides the two most significant pests of pigeonpea (lepidopteran pod borers – *Helicoverpa armigera* and *Maruca vitrata*), a previously unreported stem borer has been observed in pigeonpea fields in Benin (Zavinon & Adoukonou-Sagbadja 2019) This insect burrows into the stems, disrupting nutrient transport and causing plant wilting, which can lead to total crop loss in heavily infested fields. To date, no similar pest has been documented in other pigeonpea-growing regions, highlighting the need for further research to understand its biology, distribution, and potential management strategies.

Another critical constraint is the lack of financial resources and market infrastructure. Many smallholder farmers face inadequate capital to invest in improved inputs, mechanization, or post-harvest processing technologies. Additionally, pricing issues and inconsistent market demand discourage farmers from prioritizing pigeonpea over more lucrative crops like maize or common beans (Simion et al. 2022). These economic barriers limit the crop's potential contribution to local economies and trade.

1.2.4.2 Opportunities for Growth and Development

Despite these challenges, pigeonpea presents numerous opportunities for enhancing agricultural resilience and livelihoods in Africa (Odeny et al. 2007; Kaoneka et al. 2016). One promising avenue lies in its adaptability to marginal soils and harsh climates, making it an ideal crop for regions prone to drought and soil degradation. Its nitrogen-fixing capabilities improve soil fertility, reducing the need for chemical fertilizers and promoting sustainable farming practices.

Recent innovations in breeding have also opened new possibilities for pigeonpea cultivation. The development of drought-tolerant and pest-resistant varieties through genomic interventions offers practical solutions to mitigate the impacts of climate change. For example, pre-extension demonstrations conducted in Ethiopia in 2024 showcased the potential of improved pigeonpea cultivars to boost yields by up to 20%, even under challenging climatic conditions (Lemma et al. 2024). Such advancements not only enhance productivity but also empower farmers to adapt to evolving environmental realities.

Agroforestry and intercropping systems represent another opportunity to maximize pigeonpea's benefits. By integrating pigeonpea with trees or other crops, farmers can diversify their harvests, reduce pest pressure, and improve ecosystem services. Intercropping pigeonpea with cereals like maize or sorghum has been shown to stabilize yields during periods of climatic uncertainty, providing a safety net against crop failures. These practices align with broader efforts to promote conservation agriculture and sustainable resource management.

Furthermore, there is untapped potential for value addition and market expansion. Processed pigeonpea products – such as flour, snacks, and ready-to-eat meals – are gaining popularity, creating opportunities for entrepreneurship and export (Fatokimi & Tanimonure 2021). Strengthening market linkages and investing in processing facilities could unlock additional revenue streams for farmers while addressing nutritional needs at scale.

1.3 CLIMATE CHANGE AND ITS IMPACTS ON PIGEONPEA PRODUCTION IN AFRICA

1.3.1 Overview of Climate Variability and Its Effects On Agriculture

Climate change poses a growing threat to agriculture worldwide, endangering food security and livelihoods. Africa is particularly vulnerable due to its heavy reliance on rainfed agriculture and limited adaptive capacity. Its geographical position, with most landmass in the warming tropics, further exacerbates climate risks. Climate variability in Africa is marked by rising temperatures, erratic rainfall, and increased occurrences of droughts and floods, all of which significantly impact agricultural productivity.

Temperature increases across the continent have become more pronounced in recent decades. By mid-century, average temperatures are expected to rise by approximately 2°C, with some regions experiencing even greater spikes during critical growing seasons. Studies suggest that under current emissions trajectories, Africa's temperatures could rise by 2.7°C by the 2050s and 4.5°C by the 2080s (Girvetz et al. 2017: Legg 2021). This increase accelerates evapotranspiration, leading to soil moisture deficits and more frequent droughts. Heat stress poses serious challenges for crops like pigeonpea, which, despite its resilience, may suffer yield reductions under prolonged high temperatures. Additionally, sub-Saharan Africa is projected to face more extreme heat events, threatening over 40% of rangeland productivity at 2°C warming levels, with worse effects at 3.75°C.

Variations in precipitation further compound these challenges. Rainfall has become increasingly unpredictable, with some regions experiencing prolonged dry spells and others suffering from excessive rainfall and flooding. In East Africa, the March-May rainy season has declined, likely due to rapid warming of the Indian Ocean, which increases precipitation over the ocean but reduces it on land (Williams & Funk 2011). In West Africa and the Sahel, rainfall projections show mixed trends, with some coastal areas experiencing a reduction of 50–100 mm per year (e.g., Côte d'Ivoire, Ghana, Nigeria), while parts of the Sahel (e.g., Burkina Faso, Niger, Senegal) see increased rainfall. These fluctuations lead to soil erosion, waterlogging, and nutrient depletion, threatening crop growth. As rainfall patterns continue to shift, sustainable water management and climate-resilient agricultural practices will be essential to safeguarding food production in Africa.

1.3.2 Impact of Climate Change on Pigeonpea Production

Pigeonpea is a key legume in Africa, valued for its drought tolerance and role in food security, soil fertility improvement, and income generation. However, climate change is severely disrupting its production, threatening smallholder farmers who depend on it. Rising temperatures, erratic rainfall, weakening traditional climatic patterns, and intensifying pest and disease pressures are major challenges affecting its productivity.

Although pigeonpea is known for its resilience, extreme heat above 35°C during flowering and pod-setting stages causes pollen sterility, flower abortion, and poor seed set, leading to significant yield reductions (Ramakrishna et al. 2021; Devi et al. 2023). Prolonged heat stress accelerates evapotranspiration, reducing soil moisture and shortening the growing period, particularly in regions experiencing more frequent and prolonged heatwaves. A 2°C rise in temperature combined with an 11% increase in rainfall could lower pigeonpea yields by 8%, while a 20% drop in rainfall could result in losses exceeding 20% (Kwena et al. 2021). In addition to yield reductions, climate change affects the nutritional composition of pigeonpea. Elevated CO_2 levels, while enhancing photosynthesis, dilute the plant's protein and micronutrient content, diminishing its nutritional value for millions of consumers (Dimes et al. 2009).

Rainfall variability further complicates pigeonpea production. As a predominantly rainfed crop, it is highly sensitive to changes in precipitation patterns. Erratic rainfall disrupts planting schedules, reduces germination rates, and increases the risk of moisture stress during critical growth stages. Early cessation of rains limits grain filling, resulting in shriveled seeds and lower yields, while excessive rainfall at maturity increases pod shattering and post-harvest losses. In addition, waterlogging caused by unpredictable rainfall patterns promotes root rot and fungal diseases (Choudhary et al. 2013). Changing climatic conditions also intensify biotic stressors.

Pest outbreaks are becoming more frequent as warmer temperatures create ideal conditions for insect proliferation. Pod borers (*Helicoverpa armigera*), aphids, and storage pests like bruchids (*Callosobruchus spp.*) are increasingly damaging crops both in the field and during storage. These pressures not only decrease productivity but also increase production costs due to the need for pest and disease management. Soil degradation, exacerbated by erratic rainfall and extreme weather events, further limits productivity. Farmers are forced to cultivate on increasingly degraded lands, reducing soil fertility and crop yields.

Across Africa, the effects of climate change on pigeonpea production vary by region, reflecting the continent's diverse agroecological zones and climatic conditions. In East Africa, where pigeonpea is a major staple and export crop, shifting rainfall patterns have significantly disrupted traditional cropping systems. In Kenya and Tanzania, unreliable rainy seasons have led to frequent planting failures, forcing farmers to adopt early maturing and drought-tolerant pigeonpea varieties. However, even these improved varieties struggle under prolonged dry spells, as reduced soil moisture affects root development and nutrient uptake. In Ethiopia, where pigeonpea is integrated into mixed farming systems, unseasonal heavy rains have resulted in severe waterlogging, increasing incidences of fungal diseases such as Fusarium wilt. These conditions not only lower yields but also threaten the sustainability of pigeonpea-based farming systems that millions depend on for food and income (Mishra et al. 2017).

In West Africa, particularly in Nigeria and Benin, local farmers emphasize the critical role of the harmattan, a dry and cool seasonal wind, in pigeonpea productivity (Zavinon & Adoukonou-Sagbadja 2019). Traditionally, the harmattan has created favorable conditions for flowering and grain filling, particularly for local varieties. However, with climate change, this phenomenon is becoming less intense and increasingly rare, disrupting pigeonpea growth cycles. The weakening of the harmattan results in higher nighttime temperatures and prolonged humidity, which negatively impact flowering, increase disease incidence, and ultimately reduce yields. Farmers who once relied on the predictability of the harmattan now struggle with irregular weather patterns, forcing them to seek alternative strategies for sustaining pigeonpea production. Moreover, extreme weather events, such as droughts and storms, have led to crop failures, reducing farmers' resilience and increasing reliance on external food sources.

Without urgent interventions, pigeonpea productivity in Africa will continue to decline. Investments in climate-resilient varieties, sustainable farming practices, and improved extension services are crucial to mitigating these threats. Scaling up climate-smart technologies and strengthening market access for smallholder farmers will be key to ensuring the long-term sustainability of pigeonpea production.

1.4 STRATEGIES FOR IMPROVING PIGEONPEA RESILIENCE

Pigeonpea remains an underutilized crop in many African countries, particularly in West Africa, where improvement efforts began relatively late. Unlike Eastern and Southern Africa, where production is significantly higher, long-term breeding programs supported by ICRISAT have led to the development of several improved genotypes through intensive genetic enhancement initiatives. However, with the increasing threats posed by climate change, there is a growing urgency to develop and promote resilient pigeonpea varieties suited to diverse agroecological conditions. This section examines genetic improvement strategies and agronomic approaches aimed at enhancing pigeonpea's adaptability and productivity in the face of climate challenges.

1.4.1 Genetic Improvement of Pigeonpea for Climate Resilience

The primary objective of pigeonpea breeding in Africa is to develop varieties capable of withstanding extreme climatic conditions while maintaining high yields and nutritional quality. Although pigeonpea varietal selection is still in its early stages in several African countries, considerable efforts have been made to collect, characterize, and improve genetic resources to enhance climate resilience.

1.4.1.1 Collection, Characterization, and Improvement of Genetic Resources

Germplasm collection and characterization have played a critical role in identifying landraces and wild relatives with valuable traits such as drought tolerance, early maturity, and resistance to pests and diseases (Bohra et al. 2022). National research institutions, in collaboration with ICRISAT and local partners, have undertaken several collection missions across East, West, and Southern Africa. These efforts have led to the identification of promising accessions with superior adaptive traits. In Malawi, Kenya, Uganda, and Tanzania, genetic diversity studies have revealed significant variability among local landraces, particularly in terms of flowering time, seed size, and tolerance to moisture stress (Manyasa et al. 2008; Silim et al. 2005; Pandey et al. 2016). In West Africa, where pigeonpea remains a minor crop, initial characterization efforts in Nigeria, Ghana, and Benin have identified resilient accessions with remarkable drought tolerance and resistance to pests (Zavinon et al. 2019). Although local pigeonpea varieties grown in Africa exhibit traits of resilience to climate change and are adapted to local agroecological conditions, they remain unimproved varieties with many undesirable characteristics, such as a late maturity cycle and low grain yield (Zavinon et al. 2019b; Yohane et al. 2020). The mean performances of pigeonpea landraces cultivated in Benin, as presented in Table 1.1, highlight these limitations. The data reveal significant variability in key agronomic traits, with most landraces exhibiting prolonged growth durations exceeding 180 days, which delays harvesting and increases exposure to adverse climatic conditions.

To address these challenges, breeding programs in Eastern and Southern Africa, supported by ICRISAT, have successfully developed extra-early maturing pigeonpea varieties with growth cycles of less than 120 days, significantly reducing exposure to climatic risks while ensuring timely harvests. In an effort to introduce improved genetic material in West Africa, exotic cultivars with short-duration cycles and high yield potential have been introduced to Benin (Zavinon et al. 2022) and are currently undergoing evaluation for their adaptability and potential adoption by local farmers. These initiatives aim to enhance productivity and resilience in pigeonpea farming systems across the region.

TABLE 1.1
Agronomic Performances of a Collection of Pigeonpea Landraces from Benin

Traits	Range	Mean	SD	CV	P > (F)
Plant Height (m)	194.5–401	303.46	40.56	13.37	<0.000
No. of Primary Branches	14.5–58.5	38.27	8.9	23.34	0.003
No. of Secondary Branches	24.5–187.9	88.54	34.21	38.64	<0.000
Days to 50% flowering	132.5–228	191.99	16.22	8.45	<0.000
Pod Lenght (cm)	4.25–8.2	5.15	0.56	10.98	0.025
Pod Width (mm)	6.5–10.8	8.31	0.8	10.31	<0.000
No. of Seed per Pod	3.1–4.82	4.04	0.35	8.83	0.244
Pod Bearing Length (cm)	31.71–74.94	49.92	8.05	16.14	0.004
100 Seed Weight (g)	5.3–12.05	9.65	1.17	12.21	<0.000
Days of 75% Maturity	209.5–309.5	277.06	18.34	6.62	0.251
No. of Pods per Plant	105.3–691.1	380	135.9	35.76	0.002
Seed Yield per Plant (g)	28.73–133.9	83.54	25.81	30.89	0.083

Source: Zavinon et al. (2019).

1.4.1.2 Harnessing Wild Relatives and Addressing Emerging Challenges

Wild relatives of pigeonpea (*Cajanus spp.*) present an untapped genetic reservoir for breeding climate-resilient varieties (Bohra et al. 2010). Species such as *Cajanus scarabaeoides* and *Cajanus platycarpus* possess inherent tolerance to drought, pests, and soil salinity, making them valuable for introgression breeding. In Africa, interspecific hybridization initiatives are underway to transfer beneficial genes from wild species into cultivated pigeonpea varieties. However, in West Africa, the presence of only one wild species, *Cajanus kerstingii* Harms, endemic to the region, raises the possibility of a secondary center of origin for pigeonpea. Recent studies (Zavinon et al. 2024) have speculated on the potential role of *C. kerstingii* in reinforcing the genetic diversity of cultivated pigeonpea in Benin. Exploring its genetic contribution remains crucial, as crop wild relatives offer valuable opportunities for developing more productive and resilient varieties. Integrating *C. kerstingii* into breeding programs could broaden the genetic base of pigeonpea in West Africa and improve its adaptation to climatic stressors.

1.4.1.3 Genomic Tools and Molecular Breeding for Climate Resilience

Advances in genomics and molecular breeding have accelerated pigeonpea improvement in Africa. The development of genetic maps using molecular markers such as simple sequence repeats (SSRs) and single nucleotide polymorphisms (SNPs) has provided valuable insights into the genetic control of key agronomic traits. Next-generation sequencing technologies, including genotyping-by-sequencing (GBS), have been employed to discover novel genetic variations in African pigeonpea germplasm, facilitating marker-assisted selection (MAS) and breeding efficiency. Recent studies in Benin and Eastern and Southern Africa (ESA) have leveraged SSR and SNP markers to assess the genetic diversity of pigeonpea collections, revealing substantial variability within local landraces. In Benin, SSR as well SNP markers have been used to characterize indigenous germplasm, identifying distinct genetic clusters that could serve as promising sources for breeding climate-resilient varieties (Zavinon et al. 2020; Kinhoégbè et al. 2022). Similarly, genome-wide SNP analyses in ESA have enabled the identification of superior genotypes with enhanced drought tolerance, pest resistance, and early maturity, strengthening ongoing breeding initiatives in the region (Kimaro et al. 2020; Yohane et al. 2022).

These molecular tools are now playing a crucial role in developing resilient pigeonpea varieties across Africa, ensuring that future cultivars combine both local adaptation and improved agronomic performance. Genome-wide association studies (GWAS) have been instrumental in identifying quantitative trait loci (QTLs) linked to drought tolerance, early flowering, and resistance to pod borer (*Helicoverpa armigera*). For instance, an association mapping study conducted on a collection of 80 accessions using 954 SNP markers identified a total of 22 marker-trait associations (MTAs) for nine key agronomic traits (Zavinon, 2019). Such findings have enabled breeders to integrate genomic selection (GS) into breeding programs, improving the prediction of desirable traits, and accelerating the development of climate-resilient varieties (Saxena et al. 2021b). Linkage mapping studies have further refined the understanding of genetic architecture in pigeonpea, allowing for more precise introgression of adaptive traits.

One of the most pressing challenges in pigeonpea production is the increasing prevalence of Fusarium wilt (*Fusarium udum*) and pod borer (*Helicoverpa armigera*), particularly in ESA, where pigeonpea is a major export crop. In response, breeding programs in Kenya, Tanzania, and Malawi have prioritized the development of wilt-resistant genotypes. Through screening and selection of resistant germplasm, several improved varieties, such as *ICEAP 00554* and *ICEAP 00557*, have been released with enhanced tolerance to Fusarium wilt (Kaoneka et al. 2016). Marker-assisted breeding has further accelerated the introgression of resistance genes, with significant progress made in identifying QTLs linked to wilt resistance. Pod borer (*Helicoverpa armigera*) is another major constraint in ESA, where outbreaks have become more frequent due to changing climatic conditions. The pest causes severe damage by feeding on flowers and pods, leading to significant yield losses. Traditional breeding efforts have focused on selecting genotypes with pod toughness and high trichome density, which reduce larval feeding. Additionally, recent advances in genetic engineering, such as the development of Bt pigeonpea, hold promise for controlling *pod borer* infestations. However, regulatory challenges and biosafety concerns have limited the deployment of transgenic pigeonpea varieties in Africa.

Hybrid pigeonpea breeding has emerged as a promising strategy for improving yield stability under climate stress (Bohra et al. 2025). Hybrid varieties, developed through cytoplasmic male sterility (CMS) systems, have demonstrated superior performance in drought-prone regions, significantly increasing productivity compared to traditional landraces. Research efforts led by ICRISAT in ESA have successfully developed and released early maturing hybrid cultivars, which not only enhance yield potential but also provide resilience against erratic rainfall patterns. In West Africa, particularly in Benin, recent initiatives have aimed at identifying parental lines from local and exotic pigeonpea collections to initiate hybridization programs (Zavinon et al. 2024). Phylogenetic analysis revealed a high level of genetic divergence between African and Asian germplasm. This divergence may be explained by a possible dual domestication process, which provides a valuable evolutionary perspective on the crop's genetic structure (Zavinon et al. 2020). Such pronounced divergence offers strong potential for generating heterotic combinations through inter-pool hybridization, thereby achieving significant yield gains. Hybridization between African and Asian pigeonpea gene pools thus holds great promise for developing high-yielding, climate-resilient cultivars. These improved hybrids would not only enhance productivity but also strengthen farmer adoption, potentially transforming pigeonpea production systems in West Africa.

Despite these advancements, several challenges remain in the genetic improvement of pigeonpea in Africa. Limited breeding infrastructure, inadequate funding, and weak seed delivery systems hinder the widespread adoption of improved varieties. To address these constraints, collaborative breeding programs involving public and private-sector stakeholders must be strengthened. Furthermore, participatory breeding approaches, engaging farmers in the selection and evaluation of improved lines, are crucial for ensuring the development of varieties that meet local needs. As climate change continues to threaten pigeonpea production, integrating traditional breeding methods with modern genomic tools will be key to sustaining productivity and ensuring food security across Africa.

1.4.2 Agronomic and Crop Management Strategies

Agronomic and crop management strategies play a crucial role in enhancing pigeonpea resilience to climate change, particularly in Africa, where smallholder farmers rely on low-input systems. Unlike genetic improvement, agronomic interventions offer immediate and cost-effective solutions to mitigate climate stress, improve soil fertility, and enhance productivity (Musokwa & Mafongoya 2021). Given the diverse agroecological zones where pigeonpea is cultivated – from the semi-arid tropics of ESA to the humid and sub-humid regions of West Africa – tailored agronomic practices are essential. Research has focused on optimizing planting density, intercropping, soil fertility management, water conservation, and integrated pest control to enhance productivity under increasingly unpredictable climatic conditions.

Optimizing planting density is key to improving pigeonpea performance across regions. In ESA, where commercial production is more developed, adjusting plant populations significantly improves water-use efficiency and yield stability. Studies in Kenya and Tanzania show that spacing of 75 cm × 30 cm for medium-duration varieties and 90 cm × 50 cm for long-duration ones enhances pod formation (Rao et al. 2012). In Malawi, staggered planting of early- and medium-duration varieties helps mitigate risks associated with erratic rainfall. In contrast, in West Africa, pigeonpea is primarily intercropped, leading to lower plant densities. Research in Nigeria and Ghana suggests that row spacing of 80 cm × 40 cm in cereal-based intercropping systems optimizes pigeonpea growth without reducing companion crop yields.

Intercropping and rotational systems improve soil fertility, stabilize yields, and enhance climate resilience (Saxena et al. 2021a). In Kenya, maize-pigeonpea intercropping increases soil nitrogen while reducing the impact of drought on cereals. In Tanzania and Uganda, push-pull technology incorporating desmodium and Napier grass has effectively controlled *Helicoverpa armigera* (pod borer) and Striga weeds while improving pigeonpea resilience (Kaoneka et al. 2016). In West Africa, where intercropping is common, pigeonpea-based rotations are also promoted to restore soil fertility and break pest cycles (Zavinon & Adoukonou-Sagbadja 2019).

Pigeonpea's ability to fix atmospheric nitrogen makes it valuable for sustainable soil fertility management. However, declining soil fertility remains a challenge. In Malawi, incorporating pigeonpea residues improves soil organic matter and boosts subsequent maize yields by 20–30% (Musokwa & Mafongoya 2021). In Kenya, pigeonpea leaf litter enhances soil moisture retention and reduces dependency on synthetic fertilizers. Biochar and compost use in Ethiopia and northern Kenya is gaining traction in combating soil degradation (Miriti et al. 2015). In West Africa, integrating pigeonpea into livestock farming has been explored for soil fertility enhancement through green manure.

Rainfed pigeonpea cultivation necessitates improved water conservation strategies due to climate change-induced rainfall variability. In semi-arid Zimbabwe and Mozambique, conservation agriculture techniques such as tied ridges, mulching, and minimum tillage improve soil moisture retention. In Kenya, zai pits – shallow planting basins capturing runoff – enhance pigeonpea survival during dry spells (Saxena et al. 2023). In Ethiopia, supplemental irrigation during pod development increases yields by 15%–20%. While irrigation infrastructure is limited in many regions, small-scale irrigation expansion in Malawi and Tanzania offers opportunities to enhance resilience to erratic rainfall.

Climate change has intensified pest and disease pressures, requiring IPM strategies. In ESA, Fusarium wilt (*Fusarium udum*) and pod borer (*Helicoverpa armigera*) are major constraints. Research in Malawi and Mozambique shows that using resistant varieties alongside early planting and soil solarization reduces Fusarium wilt incidence (Kumar et al. 2024). In Tanzania, neem-based botanical pesticides provide a low-cost alternative for pod borer control. In West Africa, where storage pests such as *Callosobruchus* spp. (bruchids) are problematic, hermetic storage technologies like Purdue Improved Crop Storage (PICS) bags effectively reduce post-harvest losses.

Farmer-led adaptation strategies are essential for improving pigeonpea resilience. Participatory research programs have facilitated the promotion of locally adapted agronomic practices. In Uganda, farmer field schools (FFS) support knowledge exchange on climate-smart pigeonpea farming, leading to increased adoption of drought-tolerant varieties and sustainable agronomic techniques. In Kenya and Tanzania, mobile-based extension services provide real-time climate advisories and agronomic recommendations. Similarly, in Malawi and Ethiopia, community seed banks preserve climate-resilient landraces and ensure seed security.

Agronomic strategies are critical for enhancing pigeonpea resilience to climate change in Africa. Improved planting density, intercropping, soil fertility management, and water conservation techniques have proven effective, yet adoption remains limited due to socioeconomic constraints and inadequate extension services. Scaling up climate-smart pigeonpea production requires stronger research-extension-farmer linkages, greater investment in sustainable land management, and policies supporting smallholder adaptation to climate risks. Integrating traditional knowledge with modern agronomic innovations will be key to ensuring the long-term sustainability of pigeonpea farming across Africa.

1.5 STRENGTHENING PIGEONPEA VALUE CHAINS FOR CLIMATE RESILIENCE

1.5.1 Seed Systems and Access to Climate-Resilient Varieties

The foundation of a resilient value chain lies in the availability and accessibility of improved seed systems (Varshney et al. 2021). High-quality seeds are critical for ensuring that farmers can cultivate pigeonpea varieties capable of withstanding the adverse effects of climate change, such as drought, heat stress, and pest infestations (Sarr et al. 2024). However, despite significant advancements in breeding programs, many smallholder farmers across Africa continue to rely on traditional landraces or recycled seeds, which often lack the resilience and productivity required to thrive under changing climatic conditions. In most African countries, the informal seed system remains the primary source of legume seeds, including pigeonpea, for the vast majority of smallholder farmers. In this system, farmers save seeds from previous harvests, which they either use for their own planting or sell to fellow farmers, sometimes even as grain for consumption when there is excess. This reliance on the informal system, while ensuring seed availability at the community level, poses challenges in terms of seed quality, genetic purity, and resilience to evolving climatic and biotic stresses.

Improved seed systems play a dual role: they provide farmers with access to high-yielding, climate-resilient varieties while also fostering trust in modern agricultural technologies (Varshney et al. 2021). Recent studies have highlighted the potential of genomic interventions and speed breeding protocols to accelerate the development of pigeonpea cultivars tailored to specific agro-climatic zones (Jones et al. 2002). For instance, newly developed varieties like ICEAP 00557, which exhibit enhanced tolerance to moisture stress and pod borer resistance, have shown promising results in field trials conducted in Ethiopia during 2024 (Saxena et al. 2018). These innovations not only address immediate challenges but also lay the groundwork for long-term sustainability.

However, the success of improved seed systems hinges on their accessibility to farmers, particularly those in remote or underserved regions (Varshney et al. 2021). Challenges such as inadequate distribution networks, high seed costs, and limited awareness about the benefits of improved varieties often hinder adoption (Soni & Modi 2024). To overcome these barriers, governments and development partners must invest in strengthening seed supply chains, promoting farmer education, and subsidizing inputs to make them more affordable. Public-private partnerships also offer a viable pathway to scale up seed production and dissemination, ensuring that even resource-poor farmers can access climate-resilient pigeonpea seeds (Kumar et al. 2024).

Moreover, participatory approaches that involve farmers in the breeding and selection process can enhance the relevance and acceptance of new varieties. By incorporating local knowledge and preferences into breeding programs, researchers can develop cultivars that align with farmers' needs and agroecological contexts (Tefera & Abass 2012). This collaborative approach not only fosters innovation but also builds confidence among farmers, encouraging them to adopt improved practices and technologies.

1.5.2 Post-Harvest Handling and Processing Innovations to Reduce Losses

Post-harvest losses represent a significant bottleneck in the pigeonpea value chain, undermining both food security and economic returns for farmers. Studies estimate that up to 30% of pigeonpea production is lost due to poor handling, storage, and processing practices, particularly in regions with limited infrastructure and technological capacity (Soni & Modi 2024). Addressing these inefficiencies is essential for enhancing the resilience and profitability of pigeonpea cultivation.

One of the primary causes of post-harvest losses is improper drying and storage. High moisture content in harvested grains creates favorable conditions for fungal growth and insect infestations, leading to spoilage and reduced marketability. Innovations in drying technologies, such as solar dryers and moisture meters, offer practical solutions to mitigate these risks. Solar dryers, for example, enable farmers to dry pigeonpea grains efficiently, even in areas with unreliable electricity access, while moisture meters help ensure optimal drying levels to prevent spoilage (Sarr et al. 2024).

Processing innovations also play a crucial role in reducing losses and adding value to pigeonpea products. Traditional methods of threshing and milling are labor-intensive and often result in significant grain breakage. Mechanized threshers and mills, designed specifically for pigeonpea, can significantly improve efficiency and reduce waste. Additionally, advancements in packaging technologies, such as hermetic storage bags, protect grains from pests and moisture during storage, extending their shelf life and preserving quality.

Value addition through processing represents another avenue for enhancing the pigeonpea value chain. Transforming raw pigeonpea grains into processed products – such as flour, snacks, or ready-to-eat meals – not only reduces perishability but also increases market opportunities and consumer appeal. Recent initiatives in Kenya and Tanzania have demonstrated the potential of small-scale processing units to empower rural entrepreneurs and create jobs along the value chain (Kumar et al. 2024). By investing in processing infrastructure and capacity-building programs, stakeholders can unlock the full economic potential of pigeonpea while addressing nutritional needs at scale.

Furthermore, research has shown that post-harvest treatments significantly affect the nutrient composition of pigeonpea. For instance, retention of iron, zinc, total sugars, vitamin A, protein, and vitamin C were recorded at 60 days of storage, highlighting the importance of proper storage techniques to preserve nutritional quality (Soni & Modi 2024). These findings emphasize the need for innovative solutions to minimize nutrient loss and maximize the health benefits of pigeonpea consumption. Technologies like fermentation and extraction are also being explored to enhance the nutritional profile of pigeonpea-based products, making them more appealing to health-conscious consumers.

1.5.3 Strengthening Market Linkages and Policy Support

A third critical component of strengthening pigeonpea value chains is the development of robust market linkages and supportive policies. Despite its nutritional and economic significance, pigeonpea remains an underutilized crop in many African countries, largely due to inconsistent market demand, pricing volatility, and weak institutional frameworks (Saxena et al. 2018). Strengthening market systems and fostering policy support are essential for creating an enabling environment that encourages investment and innovation in pigeonpea production.

Market linkages play a pivotal role in connecting farmers to buyers, processors, and exporters, ensuring that surplus production translates into income generation. Farmer cooperatives and producer organizations can serve as effective intermediaries, aggregating produce, negotiating better prices, and providing access to larger markets (Soni & Modi 2024). Digital platforms and mobile applications are increasingly being used to facilitate direct transactions between farmers and buyers, reducing reliance on middlemen and improving transparency in pricing. For example, initiatives in Malawi have leveraged digital marketplaces to connect pigeonpea farmers with international buyers, boosting export revenues and enhancing livelihoods.

Policy support is equally critical for fostering a conducive environment for pigeonpea cultivation and trade. Governments can incentivize production by providing subsidies for inputs, improving irrigation infrastructure, and investing in research and development (Sarr et al. 2024). Policies that promote crop diversification and integrate pigeonpea into national food security strategies can also enhance its visibility and importance within agricultural systems. Furthermore, regional trade agreements and harmonization of standards can facilitate cross-border trade, opening new opportunities for pigeonpea producers in Africa.

1.6 CONCLUSION

Pigeonpea is a nutritionally, economically, and environmentally valuable crop with significant potential to address climate challenges in Africa. Enhancing its resilience requires a holistic approach integrating genetic improvement, agronomic innovations, and strengthened value chains.

Genetic advancements, including molecular breeding and MAS, have led to climate-resilient varieties with improved drought tolerance, early maturity, and pest resistance. However, adoption remains limited due to restricted seed access, weak extension services, and socioeconomic barriers. Strengthening seed systems through public-private partnerships and participatory breeding is crucial for ensuring smallholder farmers benefit from these innovations (Bohra et al. 2020b, Varshney et al.

2021). Agronomic strategies complement genetic progress by offering immediate solutions to climate stress. Optimized planting density, intercropping, soil fertility management, and water conservation techniques enhance productivity under erratic climatic conditions. Innovations such as conservation agriculture, solar dryers, and hermetic storage help reduce post-harvest losses and improve marketability. However, adoption is constrained by limited knowledge dissemination and weak institutional support. Strengthening research-extension-farmer linkages and increasing investment in sustainable land management will be essential for scaling up climate-smart pigeonpea production. A resilient value chain is critical to unlocking pigeonpea's full potential. Improved seed systems, efficient post-harvest handling, and better market linkages can reduce losses and create economic opportunities for rural communities. Policy support and public awareness campaigns must also highlight pigeonpea's role in climate adaptation and food security.

Despite progress, challenges such as funding constraints, inadequate infrastructure, and fragmented markets persist. Integrating traditional knowledge with modern scientific advancements through collaborative efforts involving farmers, researchers, and policymakers will be key to making pigeonpea a pillar of sustainable agriculture in Africa.

REFERENCES

Aggarwal A, Jeena AS, Mehra K et al. (2025) Speed-bred crops for food security and sustainable agriculture. *Planta* 262: 34.

Arora S, Mahato AK, Singh P et al. (2017) A high-density intraspecific SNP linkage map of pigeonpea (*Cajanus cajan* L. Millsp.). *PLoS One* 12(6): e0179747.

Ayenan MA, Dansi A, Ahoton LE et al. (2017) Pigeonpea (*Cajanus cajan* L. Millsp.) production system, farmers' preferred traits and implications for variety development and introduction in Benin. *Agric Food Secur* 6: 1–11.

Bohra A, Jha UC, Godwin I et al. (2020b) Genomic interventions for sustainable agriculture. *Plant Biotechnol J* 18: 2388–2405.

Bohra A, Mallikarjuna N, Saxena KB et al. (2010) Harnessing the potential of crop wild relatives through genomics tools for pigeonpea improvement. *J Plant Biol* 37: 83–98.

Bohra A, Pareek S, Jones M et al. (2019) Genomic interventions to improve resilience of pigeonpea in changing climate. In: *Genomic designing of climate-smart pulse crops*. Springer Nature, Switzerland, pp. 107–134. ISBN 978-3-319-96931-2.

Bohra A, Saxena KB, Varshney RK , et al. (2020a) Genomics-assisted breeding for pigeonpea improvement. *Theor Appl Genet* 133(5): 1721–1737.

Bohra A, Tiwari A, Kaur P et al. (2022) The key to the future lies in the past: Insights from grain legume domestication and improvement should inform future breeding strategies. *Plant Cell Physiol* 63(11): 1554–1572.

Bohra A, Tiwari A, Pareek S et al. (2025) Past and future of cytoplasmic male sterility and heterosis breeding in crop plants. *Plant Cell Rep* 22: 33.

Choudhary AK, Raje RS, Datta S et al. (2013) Conventional and molecular approaches towards genetic improvement in pigeonpea for insect resistance. *Am J Plant Sci* 4: 372–385.

Choudhary AK, Sultana R, Pratap A et al. (2011) Breeding for abiotic stresses in pigeonpea. *J Food Legum* 24(3): 165–174.

Devi J, Sagar V, Mishra GP et al. (2023) Heat stress tolerance in peas (*Pisum sativum* L.): Current status and way forward. *Front Plant Sci* 13: 1108276. doi: 10.3389/fpls.2022.1108276

Dimes J, Cooper P, Rao KPC. (2009) *Climate change impact on crop productivity in the semi-arid tropics of Zimbabwe in the 21st century*. In: Humphreys, E. et al. *Proceedings of the Workshop on Increasing the Productivity and Sustainability of Rainfed Cropping Systems of Poor, Smallholder Farmers, Tamale, Ghana*, 22–25 September 2008. Colombo, Sri Lanka: CGIAR Challenge Program on Water and Food. https://hdl.handle.net/10568/21250

Fatokimi EO, Tanimonure VA (2021) Analysis of the current situation and future outlooks for pigeonpea (*Cajanus cajan*) production in Oyo State, Nigeria: A Markov Chain model approach. *J Agric Food Res* 6: 100218.

Girvetz E, Corner-Dolloff C, Lamanna C, Rosenstock T (2017) CSA-Plan: Strategies to put climate-smart agriculture (CSA) into practice. *Agric Develop*, 30: 12–16.

Jones RB, Freeman HA, Lo Monaco G (2002). Improving the access of small farmers in Eastern and Southern Africa to global pigeonpea markets. *Agric Res Ext Net*, 120: 1–11.

Kaoneka SR, Saxena RK, Silim SN, Odeny DK, Ganga Rao NVPR, Shimelis HA, Siambi M, Varshney RK (2016) Pigeonpea breeding in eastern and southern Africa: Challenges and opportunities. *Plant Breed* 135(2): 148–154.

Khaniya J (2024). Revolutionizing Agriculture: Unveiling the Power of Pigeonpea (*Cajanus cajan* (L.) Mill Sp.) for Sustainable Farming and Food Security. *Agri Articles* 04(02): 37–42

Khoury CK, Castañeda-Alvarez NP, Achicanoy HA et al. (2015) Crop wild relatives of pigeonpea (*Cajanus cajan* L. Millsp.): Distributions, ex situ conservation status, and potential genetic resources for abiotic stress tolerance. *Biol Conserv* 184: 259–270.

Kimaro D, Melis R, Sibiya J, Shimelis H, Shayanowako A (2020) Analysis of genetic diversity and population structure of pigeonpea (*Cajanus cajan* L. Millsp.) accessions using SSR markers. *Plants* 9(12): 1643.

Kinhoégbè G, Djèdatin G, Saxena RK, Chitikineni A, Bajaj P, Molla J, Agbangla C, Dansi A, Varshney RK (2022) Genetic diversity and population structure of pigeonpea (*Cajanus cajan* L. Millsp.) landraces grown in Benin revealed by genotyping-by-sequencing. *PLoS One* 17(7): e0271565.

Kumar NV, Sawargaonkar G, Rani S et al. (2024) Harnessing the potential of pigeonpea and maize feedstock biochar for carbon sequestration, energy generation, and environmental sustainability. *Bioresour Bioprocess* 11(5): 1–19. doi: 10.1186/s40643-023-00719-3

Kwena KM, Karuku GN, Ayuke F, Esilaba A (2021) Impact of climate change on maize and pigeonpea yields in semi-arid Kenya. In: *Environmental issues and sustainable development*. IntechOpen. https://doi.org/10.5772/intechopen.93321

Legg S (2021) IPCC, 2021: Climate change 2021—the physical science basis. *Interact* 49(4): 44–45. https://search.informit.org/doi/10.3316/informit.315096509383738

Lemma LL, Wada AB, Bune ME . (2024) On-farm pre-extension demonstration of newly released pigeonpea (*Cajanus cajan*) technology with proper agronomic practices in Gamo and Konso zones of southern Ethiopia. *Glob J Agric Res Dev* 11: 1–11.

Lohithaswa HC, Shreekanth SM, Banakara S, Rajanna TL, Kashyap GSS, Hosakoti S, Gonal B, Mallikarjuna MG, Gangashetty P (2025) Breeding climate-resilient pigeonpea in climate-change era: Current breeding strategies and prospects. In: Pandey MK., Mallikarjuna MG, Lohithaswa HCS, Aski M, Gupta S (eds) *Breeding climate-resilient and future-ready pulse crops*. Springer, Singapore. https://doi.org/10.1007/978-981-96-0483-8_9

Manyasa EO, Silim SN, Saxena KB et al. (2008) Diversity in Tanzanian pigeonpea (*Cajanus cajan* L. Millsp.) landraces and their response to environments. *Genet Resour Crop Evol* 55: 379–387.

Miriti JM,Esilaba AO, Rao KPC, Onyango JW, Kimani SK, Njeru PM, Lekasi JK (2015) Adapting agriculture to climate change: An evaluation of yield potential of maize, sorghum, common bean, and pigeonpea varieties in a very cool-wet region of Nyandarua County, central Kenya. In: *Adapting African agriculture to climate change: Transforming rural livelihoods*, pp. 97–108.

Mishra S, Singh R, Kumar R, Kalia A, Panigrahy SR (2017) Impact of climate change on pigeonpea. *Econ Aff* 62(3): 455–457.

Musokwa M, Mafongoya P (2021) Pigeonpea yield and water-use efficiency: A savior under climate change-induced water stress. *Agron* 11(1): 5.

Odeny DA(2007) The potential of pigeonpea (*Cajanus cajan* L. Millsp.) in Africa. *Nat Res Forum* 31 : 297–305

Pandey P, Kumar R, Pandey VR (2016) Genetic association studies for yield and yield-related traits in pigeonpea (*Cajanus cajan* L. Millsp.). *Legum Res* 39(2): 189–193.

Ramakrishna G, Kaur P, Singh A et al. (2021) Comparative transcriptome analyses revealed different heat-stress responses in pigeonpea (*Cajanus cajan*) and its crop wild relatives. *Plant Cell Rep* 40: 881–898. doi: 10.1007/s00299-021-02686-5

Rao NVPRG, Silim SN, Simtowe F, Siambi M, Monyo ES, Lyimo S, Ubwe R, Mbando F, Mligo J, Kananji ADG, Maideni FW (2012). Enhancing the productivity and production of pigeonpea in eastern and southern Africa. In: Abate T (ed.). *Four Seasons of Learning and Engaging Smallholder Farmers: Progress of Phase 1*. Research Report. Nairobi, Kenya: International Crops Research Institute for the Semi-Arid Tropics. 258 pp

Sarr M, Majili Z, Khalili N, Matavel CE, Mbwana HA, Kaingo J, Löhr K and Rybak C (2024) Adoption of processing technologies and innovative food preservation techniques: Findings from smallholders in the Lindi Region in Tanzania. *Front Sustain Food Syst* 7: 1169578. https://doi.org/10.3389/fsufs.2023.1169578

Saxena KB, Bohra A, Choudhary AK, Sultana R, Pazhamala L, Saxena RK (2021a) The alternative breeding approaches for improving yield gains and stress response in pigeonpea (*Cajanus cajan* L. Millsp.). *Plant Breed* 140: 76–86

Saxena KB, Chauhan SY Kumar, CVS, Hingane AJ, Kumar RV, Saxena RK, Rao GVR (2018) Developing improved varieties of pigeonpea. In: Sivasankar, S. et al. (ed.), *Achieving sustainable cultivation of grain legumes Volume 2: Improving cultivation of particular grain legumes*, Burleigh Dodds Science Publishing, Cambridge, UK. ISBN: 978 1 78676 140 8; www.bdspublishing.com. http://dx.doi.org/10.19103/AS.2017.0023.33

Saxena KB, Kumar RV, Rao PV (2002) Pigeonpea nutrition and its improvement. *Int J Plant Prod* 5: 227–260.

Saxena KB, Reddy LJ, Saxena RK (2023) Breeding high-protein pigeonpea genotypes and their agronomic and biological assessments. *Plant Breed* 142(2): 129–139.

Saxena RK, Jiang Y, Khan A et al. (2021b) Characterization of heterosis and genomic prediction-based establishment of heterotic pattern for developing better hybrids in pigeonpea. *Plant Genome* 14: e20125.

Sharma HC (2016) Host plant resistance to insect pests in pigeonpea: potential and limitations. *Legume Perspect* (11): 24–29.

Silim SN, Bramel PJ, Akonaay HB, Mligo JK, Christiansen JL (2005) Cropping systems, uses, and primary *in situ* characterization of Tanzanian pigeonpea (*Cajanus cajan* L. Millsp.) landraces. *Genet Resour Crop Evol* 52: 645–654.

Simion T, Ersulo D, Fikre A (2022) Performance evaluation of pigeonpea (*Cajanus cajan* L. Millsp.) variety for registration in the lowland areas of Ethiopia. *Adv Agric* 2022(1): 7013602.

Slim SN, Omanga PA (2001) The response of short-duration pigeonpea lines to variation in temperature under field conditions in Kenya. *Field Crop Res* 72: 97–108.

Soni RK, Modi G (2024) Innovations and strategies in post-harvest handling and value addition of horticultural produce. *AGBIR* 40(4):1206–1209.

Tefera T, Abass A (2012) Improved postharvest technologies for promoting food storage, processing, and household nutrition in Tanzania, Ibadan, Nigeria. *International Institute of Tropical Agriculture*. https://hdl.handle.net/10568/24886

Upadhyaya HD, Sharma S, Reddy KN, Saxena R, Varshney RK Gowda CLL (2013) Pigeonpea. In: Mohar Singh, Hari D. Upadhyaya and Ishwari Singh Bisht (ed.). *Genetic and genomic resources of grain legume improvement*. Elsevier, Pages 181–202, ISBN 9780123979353, https://doi.org/10.1016/B978-0-12-397935-3.00008-6

Varshney RK, Bohra A, Roorkiwal M, Barmukh R, Cowling W, Chitikineni A, Lam HM, Hickey LT, Croser J, Edwards D, Farooq M, Crossa J, Weckwerth W, Millar AH, Kumar A, Bevan MW, Siddique KHM (2021) Rapid delivery systems for a food-secure future. *Nat Biotechnol* 39: 1179–1181. doi: 10.1038/s41587-021-01079-z. PMID: 34584247.

Varshney RK, Chen W, Li Y, Bharti AK, Saxena RK, Schlueter JA, Donoghue MT, Azam S, Fan G, Whaley AM, Farmer AD, Sheridan J, Iwata A, Tuteja R, Penmetsa RV, Wu W, Upadhyaya HD, Yang SP, Shah T, Saxena KB, Jackson SA (2011) Draft genome sequence of pigeonpea (Cajanus cajan), an orphan legume crop of resource-poor farmers. *Nat Biotechnol* 30(1): 83–89.

Venkata SK, Rama GR, Saxena RK et al. (2019) Pigeonpea improvement: An amalgam of breeding and genomic research. *Plant Breed* 138(4): 445–454.

Volp T, Jat BL, Jaba J, Zalucki MP, Furlong MJ. (2025) Integrated pest management in pigeonpea: Progress and prospects. *J Appl Entomol* 0: 1–21.

Williams AP, Funk CA, (2011) A westward extension of the warm pool leads to a westward extension of the Walker circulation, drying eastern Africa. *Clim Dyn* 37: 2417–2435.

Yeleliere E, Antwi-Agyei P, Guodaar L (2023) Farmers' response to climate variability and change in rainfed farming systems: Insight from lived experiences of farmers. *Heliyon* 9(9). e19656, ISSN 2405-8440, https://doi.org/10.1016/j.heliyon.2023.e19656

Yohane EN, Shimelis H, Laing M, Mattew I, Shayanowako A (2020) Phenotypic divergence analysis in pigeonpea (*Cajanus cajan* L. Millspaugh) germplasm accessions. *Agron* 10: 1682. doi:10.3390/agronomy10111682

Yohane EN, Shimelis H, Laing M Shayanowako A (2022) Genetic diversity and grouping of pigeonpea (*Cajanus cajan* Millspaugh) germplasm using SNP markers and agronomic traits. *PLoS One* 17(11): e0275060. https://doi.org/10.1371/journal.pone.0275060

Zavinon F, Adoukonou-Sagbadja H (2019) Pigeon pea [*Cajanus cajan* (L.) Millsp] cultivation, its major constraints and ethnobotanical status in Southern Benin. *J Agron Crop Res* 7(6): 95–105.

Zavinon F, Adoukonou-Sagbadja H, Ahoton L, Vodouhê R, Ahanhanzo C (2018) Quantitative analysis, distribution and traditional management of pigeonpea (*Cajanus cajan* L. Millsp.) landrace diversity in southern Benin. *Eur Sci J* 14(9): 184–211.

Zavinon F, Adoukonou-Sagbadja H, Bossikponnon A, Dossa H, Ahanhanzo C (2019) Phenotypic diversity for agro-morphological traits in pigeonpea landraces (*Cajanus cajan* L. Millsp.) cultivated in southern Benin. *Open Agric* 4(1): 487–499.

Zavinon F, Adoukonou-Sagbadja H, Keilwagen J, Lehnert H, Ordon F, Perovic D (2020) Genetic diversity and population structure in Beninese pigeonpea (*Cajanus cajan* L. Huth) landrace collection revealed by SSR and genome-wide SNP markers. *Genet Resour Crop Evol* 67: 191–208.

Zavinon F, Fonhan N, Atrokpo A, Djossou, Adoukonou-Sagbadja H (2022) Genotype × environment interaction and agronomic performance analysis in exotic pigeonpea (*Cajanus cajan* L. Millsp) cultivars in Benin. *Int J Appl Agric Sci* 8(6): 251–258.

Zavinon FG. 2019. Genetic diversity and association mapping of candidate genes/QTLs for agronomic traits in pigeonpea [Cajanus cajan (L.) Millsp.] collections from Benin. PhD dissertation. *Doctoral School of Life and Earth Sciences, University of Abomey-Calavi*, Benin. 198 p.

Zavinon FG, Djossou R, Gbéto M, Fonhan N, Kouke R, Adoukonou-Sagbadja H. (2024) SSR-marker-assisted evaluation of genetic diversity in local and exotic pigeonpea cultivars in Benin for parental genotype selection. *Plant Gene* 37: 2352–4073.

2 Harnessing the Potential of Pigeonpea Wild Relatives for Climate Change Resilience

Dibendu Datta, Nidhi Kumari, Abhishek Bohra, Surendra Ghritlahre, SJ Satheesh Naik, Uday Chand Jha, Gitanjali Sahay, Nainika Nagar, and Girish Prasad Dixit

2.1 INTRODUCTION

Pigeonpea (*Cajanus cajan* (L.) Millsp.) is the second most important pulses crop of India where it is known by several names such as red gram, arhar, tur, harad, rahat, kandi, tuvara. It also assumes global importance as a grain legume in the ranks of beans and broad beans, lentil, peas, and chickpeas. The cultivated pigeonpea is essentially a perennial shrub, normally survives for three to five years, but largely cultivated as an annual crop except for kitchen gardens, wasteland, bunds, and hilly tracts, wherein it is allowed to grow for years. Pigeonpea is grouped in fabaceae family under the genusCajanus, subtribeCajaninae, tribePhaseoleae, orderFabales. It can grow reasonably well in marginal lands, as well as under fairly low input conditions. It has been grown for long by the small and marginal farmers for subsistence farming owing to its hardy nature that provides a good source of protein with high nutritive value (rich in lysine, tyrosine, cystine and arginine, iron, and iodine), fodder, and fuel wood besides other traditional uses such as basket making.

Pigeonpea is cultivated in different parts of the world in Asia, Africa, and the Caribbean islands, but its distribution is asymmetric over the globe wherein India, Myanmar, Tanzania, Malawi, and Kenya are the major growers (Bohra et al., 2020). In Indian subcontinent, pigeonpea is also cultivated in Nepal, Bangladesh, and Sri Lanka. In India, pigeonpea occupied 4.82 mha area in geographically diverse regions and soil types (Ghritlahre et al., 2024; Anonymous, 2022, Annual PC Report, AICRP on pigeonpea 2021–2022). It is one of the most important crops of pulses-based cropping system either as a sole crop or as intercrop with soybean, cotton, and other kharif crops (Saxena et al., 2021a). The crop area has asymmetric distribution and is concentrated in a few states—namely, Karnataka, Maharashtra, Gujarat, Uttar Pradesh, Madhya Pradesh, Telangana, Andhra Pradesh, and Jharkhand despite wide

 DOI: 10.1201/9781003603092-2

adaptability of pigeonpea to grow in diverse agroclimatic zones. In addition, Chhatisgarh, Odisha, Bihar, and Rajasthan also contribute significant area. During a period of 11 years (2013–2014 to 2023–2024) the average production of pigeonpea varied from 2.561 mt in 2015–2016 to 4.873 mt in 2016–2017, and during 2022–2024, the production was staggering between 3.312 to 3.385 mt (Anonymous, 2024; DES-DAC&FW website, GOI). The trends of area, production, and productivity of pigeonpea in India during the last decade are represented graphically in Figure 2.1. Furthermore, as per the normal estimates if production data for five years (2017–2018 to 2021–2022), Maharashtra and Karnataka contributed about 54% of production (Anonymous, 2024, DES-DAC&FW). At the national level, total area under pigeonpea has increased in the last five decades, but the area has declined drastically in the North Eastern Plains Zone where long duration pigeonpea is traditionally cultivated.

The productivity of the majority of crops is being adversely affected by global warming, a phenomenon that is rising trend. Mishra et al. (2017) have studied the impact of temperature and rainfall on pigeonpea productivity over a period of 32 years in Gujrat. The regression analysis indicated that 1°C rise in mean maximum temperature will reduce the yield by 20.8%. Kwena et al. (2021) have analyzed the maize-pigeonpea cropping system in Kenya, which is projected to witness a rise by 2°C and 11% rainfall by 2100; and predicted that there will be a 4%–20% (depending on models used) decline in yield of short duration varieties of pigeonpea. Yadav et al. (2021) modeled the impact of climate change on short and long duration pigeonpea during a period of 85 years from 2010 to 2095. They predicted delayed anthesis and maturity and decreased grain yield (14%–66% for long and short duration genotypes, respectively, in reference varieties). The alterations in seasonality, intensity, and duration of precipitation; mean, maximum, and minimum temperatures; and

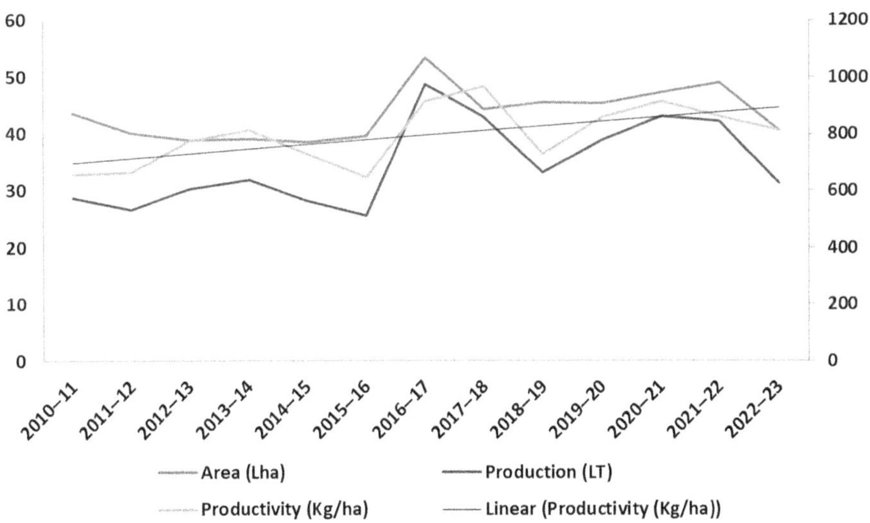

FIGURE 2.1 Trend in area, production, and productivity over the decade.

humidity and extreme weather phenomena affect crop growth, population of weedy species, and population of beneficial insects and microorganisms. These changes lead to a shift in insect-pest dynamics, besides the buildup of the population of harmful pathogens in the soil and the cropping system beyond the threshold level and, therefore, poses serious threat to crop health and productivity. Under the prevailing circumstances and forecasting, there is a need to develop climate-resilient varieties to faceoff the direct and indirect vagaries of climate.

2.2 PRODUCTIVITY LIMITING FACTORS: BIOTIC AND ABIOTIC STRESSES

Pigeonpea faces several biotic and abiotic stresses that contribute to significant yield losses worldwide (Rana et al., 2016). The major biotic stresses limiting the pigeonpea production worldwide include diseases, *viz.*, *Fusarium* wilt (FW; caused by *Fusarium udum* Butler), *Phytophthora* blight (*Phytophthora cajani*), and sterility mosaic virus (SMV), as well as insects, *viz.*, pod borer (*Helicoverpa armigera* and *Maruca vitrata*) and pod fly (*Melanagromyza chalcosoma*). FW and PSB are widespread fungal diseases prevalent across all pigeonpea-growing regions and are responsible for substantial yield losses (Ravikumara et al., 2022). FW incidences may vary from 30% to 60% at flowering and crop maturity stages, resulting even in complete crop failure (Sharma et al., 2016). Apart from FW, PSB is emerging as one of the deadliest diseases of pigeonpea under the changing climatic conditions. Intermittent rainfall, coupled with temperatures ranging between 27°C and 33°C and relative humidity exceeding 75%, has been conducive to the development of Phytophthora stem blight in pigeonpea and may result in 100% yield loss (Ganiga et al., 2020). Sterility mosaic disease (SMD), often referred to as the "green plague," is a major constraint in pigeonpea production, characterized by partial or complete suppression of flowering (Patil & Kumar, 2015). In India alone, SMD incurred estimated annual yield losses exceeding US$300 million (Patil & Kumar, 2015). The pod borer (*Helicoverpa armigera*)was recognized as the most destructive insect pest of pigeonpea, affecting cultivars across all maturity groups—including extra-early, early, medium, and late-maturing types. In the semiarid tropics alone, the annual yield loss attributed to this pest was estimated at approximately US$317 million. On a global scale, *H. armigera* causes losses exceeding US$2 billion each year (Sharma, 2016).

Among abiotic stresses, waterlogging during the initial growth stage, low temperature during the flowering stage, high temperature during the pod formation and drought are impediments in realizing the productivity potential of pigeonpea cultivars. Although pigeonpea has deep roots, yield losses due to these stresses are large and widespread, especially when they occur during critical seedling and reproductive stages. Pigeonpea is a predominantly rainfed crop, and less than 5% area is under assured irrigation. Adequate moisture in the soil is essential for seed germination. Inadequate moisture availability and poor soil fertility may lead to crop failure under rainfed conditions with mild to severe drought stress. Pigeonpea can tolerate drought to a certain degree; nonetheless, drought is one of the limiting factors for realizing high yield. Intermittent drought at flowering and pod-filing is very detrimental and reduces yield drastically even in the short duration varieties, which are otherwise capable of escaping terminal drought.

Pigeonpea is a quantitative short-day plant with the late-maturing types having a strict day-length requirement for induction of flowering. The phenological responses, influenced by photoperiod and temperature, have played a major role in the evolution of the various crop production systems. Pigeonpea is highly sensitive to salinity and poses a major constraint in raising the crop in salt-affected land. Salinity affects plant growth, development, and yield.

2.3 GENETIC BASE FOR CULTIVAR DEVELOPMENT

Genetic variability among breeding populations is the prerequisite for deriving the desirable allelic combinations to design the most suitable high yielding genotypes encompassing biotic and abiotic resistances for enhancing and stabilizing the productivity of pigeonpea; which is why the success of breeding programs depends on the genetic base of parents and trait donors. It has been observed that genetic variability is low among the cultivated genotypes and breeding populations of pigeonpea, and this limitation was attributed to domestication and breeding from a relatively lower number of founder genotypes (Bohra et al., 2010; Yohane et al., 2022). High allelic variations and genetic diversity was reported in wild species, with the exception of those from Australia, as compared to cultivated pigeonpea. However, gene flow occurred between the cultivated and wild progenitors in recent times and among the diverse species of a geographical niche in the historical past (Kassa et al., 2012).

Limited genetic diversity in cultivated pigeonpea and narrow genetic base of breeding population, coupled with relatively fewer genomic resources, impeded the progress of genetic gain. It necessitates creating new genetic variations by infusing unexploited alleles from diverse genetic sources through use of contrasting parents and crop wild relatives in particular (Bohra et al., 2022a). Recent advances in genomic resources, including next-generation sequencing technologies and high throughput genotyping assays, have opened new avenues to exploit the hitherto untapped alleles from the germplasm repository. The primary gene pool offers limited remedies to address the issues of new climate change–induced challenges in biotic and abiotic stress tolerance. The secondary and tertiary gene pools provide many opportunities to widen the genetic base and introduce climate-resilient alleles into the breeding population. Interspecific crosses have been particularly successful in extracting cytoplasm-genetic male sterility alleles from the wild species but comparatively underutilized for other traits of economic importance.

Wild relatives are the important sources of genes for economic traits, and these can be brought into the high yielding commercial cultivars for further improvement (Bohra et al., 2022a, 2022b). The wild relatives have high fitness value, but they lack many of the good agronomic attributes that are desirable for economic cultivation, whereas many of the improved cultivars of crops are vulnerable to disease and insect pests, as they lack the defense mechanisms of their wild progenitors. Wild relatives have evolved through continuous natural selection in the unprotected environments, which is why they can survive better in adverse conditions like drought, cold, high temperature, and salinity. They can also fight against natural epiphytotic and insect damage. However, it is difficult to utilize these genes because they are present in the pile of unfavorable genes with regard to agronomy and grain yield. Through pre-breeding interventions, the climate-resilient alleles can be incorporated into the agronomically superior lines in multiple steps and present a way to combat the climate-induced biotic and abiotic stresses.

2.4 ORIGIN, EVOLUTION, AND TAXONOMY

Cytological evidence for 11 homologous pairs of chromosomes in *C. cajan* was documented by Roy(1933). Cytological studies of pollen mother cell established 11 basic chromosomes in pigeonpea (Krishnaswamy & Rangaswami Ayyangar, 1935), and 22 diploid somatic chromosome number (2n = 22) were assigned to *C. cajan* (Naithani, 1941). *Cajanus kertsingii*, the wild relative of African origin, has 2n = 32 chromosomes whereas, all other wild species have 22 somatic chromosomes (Gill & Husaini, 1986). The DNA sequence data of whole genome validated the presence of 11 chromosome pairs and predicted the absence of recent genome-wide duplication events (Varshney et al., 2011). It has been established that the cultivated pigeonpea, *C. cajan*, andits wild species, except *Cajanus kertsingii*,have11 basic chromosome numbers (2x = 2n = 22). Variations in the length and satellite chromosome were reported (Naithani, 1941; Pundir & Singh, 1986). Various studies (Ohri et al., 1994; Greilhuber & Obermayer, 1998; Ohri & Singh, 2002) on karyotype analyses, measurement, and quantification of DNA have shown that *Cajanus* and related genera, viz., *Rhynchosia*, *Dunbaria*, *Flemingia*, and *Paracalyx* contain 4C DNA. The size of somatic chromosomes of cultivated pigeonpea varied from 1.35 to 2.7 μ and chromatin length ranged from 27.6 to 63.5 μ (Naithani, 1941; Mukhopadhyay, 1986). The drafts of the pigeonpea whole genome sequence have given additional information on evolution of pigeonpea (Varshney et al., 2011; Singh et al., 2012). The genome size of pigeonpea was estimated to be 833.07 Mb based on contiguous nucleotide sequences of fixed length *k* statistics (Varshney et al., 2012). Based on studies of interspecific hybrids among different wild species of pigeonpea, it has been proposed that pigeonpea has evolved from a cross between *C. scarabeaoides* and *C. cajanifolius* (Pundir & Singh, 1985a).

Karyotype, number of satellite chromosomes and DNA content of *C. cajan* were strikingly similar to *C. cajanifolius* (Ohri & Singh, 2002). The Polymerase Chain Reaction-Restriction Fragment Length Polymorphism profile of five genera belonging to subtribe cajaninae, viz., *Cajanus*, *Rhynchosia*, *Dunbaria*, *Flemingia*, and *Paracalyx* indicated the involvement of *C. cajanifolius* as the maternal parent in the evolution of pigeonpea (Lakshmi et al., 2000). The esterase enzyme profile of *C. cajan* and six wild species revealed that *C. cajanifolius* is the closest wild relative of pigeonpea (Krishna & Reddy, 1982). The seed protein profile of pigeonpea and its wild relative demonstrated closeness of *C. cajan* with *C. cajanifolius* (Jha & Ohri, 1996). Similar observations were made on the basis of electrophoretic profiles of trypsin and chymo-trypsin inhibitors among 69 genotypes of *C. cajan* and 17 accessions belonging to 7 different wild *Cajanus* species (Kollipara et al., 1994). DNA-based molecular marker studies have also indicated that *C. cajanifolius* is the progenitor species of *C. cajan*. Close association between *C. cajan* and *C. cajanifolius* revealed through RFLP marker studies 24 diverse accessions of pigeonpea wild relatives belonging to four genera (Nadimpalli et al., 1993). The mitochondrial DNA-RFLP, genomic SSR and single-nucleotide polymorphism (SNP) marker studies among diverse accessions of *C. cajan* and its wild relatives have strengthened the theory of *C. cajanifolius* as the closest wild progenitor of cultivated pigeonpea (Sivaramakrishnan et al., 2002; Odeny et al., 2007; Saxena et al., 2010; Kassa et al., 2012; Saxena et al., 2014). India was proposed as the center of origin of pigeonpea (Vavilov, 1951). A wide range of variations has been reported for various traits among the cultivated and crop wild relatives occurring in India. Moreover, various wild species of pigeonpea are distributed throughout the Indian subcontinent (Table 2.1).

TABLE 2.1
Wild Species and Their Distribution in India

S.No.	Name	Distribution	References
1	*C. scarabaeoides*	Widely distributed species across Andaman and Nicobar, Andhra Pradesh, Bihar, Chattisgarh, Himachal Pradesh, Jharkhand, Karnataka, Kerala, Madhya Pradesh, Maharashtra, Orissa, Tamil Nadu, Uttar Pradesh, Uttaranchal, West Bengal	Upadhyaya et al. (2013a)
2	*C. albicans*	Peninsular India	Van der Maesen (1990)
3	*C. cajanifolius*	Andhra Pradesh, Arunachal Pradesh, Assam, Bihar, Chattisgarh, Gujarat, Haryana, Himachal Pradesh, Jammu-Kashmir, Karnataka, Kerala, Maharashtra, Manipur, Meghalaya, Mizoram, Nagaland, Orissa, Pondicherry, Punjab, Rajasthan, Sikkim, Tamil Nadu, Tripura, Uttar Pradesh, West Bengal	http://www.iucnredlist.org/details/full/19891613/0
4	*C. crassus* (King) Maesen var *crassus*	Assam, Central India, and NW Himalaya	Van der Maesen (1990)
5	*C. elonagatus* (Bentham)Maesen	NE India	Van der Maesen (1990)
6	*C. grandiflorus* (Bentham ex Baker) Maesen	Himalayan region	Van der Maesen (1990)
7	*C. mollis*	Himalayan region	Van der Maesen (1990)
8	*C. platycarpus*	Bihar, Gujarat, Haryana, Himachal Pradesh, Jammu-Kashmir, Madhaya Pradesh, Maharashtra, Orissa, Punjab, Rajasthan, and Uttar Pradesh	http://www.legumes-online.net/ildis/aweb/td114/td_24020.htm

Source: Anonymous 2010. WFO Plant List (http://www.wfoplantlist.org).

The proposition of Africa as the center of origin was not favored due to the limited number of wild species and their low frequency and narrow distribution. Only one wild species (C. kerstingi) is endemic to West Africa (Van der Maesen, 1986). Moreover, *C. scarabaeoides* is distrusted predominantly along the coast of Africa. Owing to these observations, Africa is considered the secondary center of origin based on empirical evidences, and India is considered the primary center of origin of pigeonpea (Van der Maesen, 1980, 1986, 1990). Subsequently, pigeonpea spread to the Caribbean islands and then spread all over tropical America. It was postulated that the immigrants dispersed the crop from India to East Africa, then through Egypt via Nile valley to West Africa, and, finally, to America (Odeny, 2007; Kassa et al., 2012). Among 15 wild species found in Australia, 13 are endemic; thereby, Australia is considered the center of diversity. However, the SNP-based analysis revealed low genetic diversity among the wild species of Australia (Kassa et al., 2012), thereby refuting the hypothesis that Australia is the center of genetic diversity.

Cajanus cajanifolius is morphologically similar to *C. cajan* except for the strophiole characteristics (De, 1974). *C. cajanifolius* is regarded as the probable progenitor species of cultivated pigeonpea on the basis of morphological similarities, extent of crossability, chromosome paring and other cytological events, postzygotic development, and fertility of interspecific hybrids (Van der Maesen, 1986; Mallikarjuna et al., 2012; Sinha et al., 2015).

During the process of domestication, there was a huge loss of allelic diversity in the cultivated pigeonpea as compared to the progenitor clade (Saxena et al., 2014) that resulted in narrow diversity of phenotypic traits, including important agronomic traits. Loss of habitat of wild species of pigeonpea is a concern for conserving the diversity of alleles. Therefore, genetic improvement is increasingly dependent on harnessing useful alleles from the wild species (Upadhyaya, 2006; Kassa et al., 2012; Upadhyaya et al., 2013b). The domestication of pigeonpea has been located around Odisha-Andhra-Madhya Pradesh based on wild species distribution (Van der Maesen, 1986) and allelic polymorphism among wild relatives and land races (Saxena et al., 2014). The probable progenitor species of pigeonpea is distributed mainly in the Odisha state of India. Distribution of wild relatives of pigeonpea has latitudinal distribution, whereas domesticated cultivated pigeonpea has broad distribution (Fuller et al., 2019).

The mitotic and meiotic chromosomes of pigeonpea and its wild relatives have demonstrated high homology. Interspecific hybrids have also shown high pairing. The cross-compatible species of Indian origin exhibited more homology among themselves than those of Australian origin. Pollen fertility and seed setting in the compatible interspecific hybrids were reasonably high. Pigeonpea belongs to genus *Cajanus* that include 32–34 species (Khoury et al., 2015), including the cultivated species. The *Cajanus* species occur in Australia, Indian subcontinent including Myanmar and West Africa. Australia is home to 13 endemic species of *C. ajanusviz, C. lanceolatus, C. confertiflorus, C. viscidus, C. acutifolius, C. aromaticus, C. crassicaulis, C. lanuginosus, C. latisepalus, C. reticulates, C. pubescens, C. cinereus, C. marmoratus, and C. mareebensis*,whereas18 species are endemic to Southeast Asia and Myanmar (Mallikarjuna et al., 2011a) and 1species is found exclusively in West Africa (Van der Maesen, 1986). *C. ajanuscajan* and *C. scarabaeoides* are widely distributed throughout Southeast Asia, Pacific Islands, and northern Australia. Some species of *Cajanus*—namely, *C.villosus*, *C. elongatus*, *C. granadiflorus*, and *C. niveus*—reported from northeastern India have now become extinct or rare.

2.5 REPRODUCTIVE BEHAVIOR AND BREEDING APPROACHES

Pigeonpea possesses cleistogamous flowers, which favors self-pollination. Pigeonpea is often cross-pollinated through entomophily. Several factors like flowering habit of a genotype, presence of insect population, temperature, humidity, wind velocity, and wind direction affects natural outcrossing at a given time and space (Bhatia et al. 1981). The most important pollinators are *Apis* spp. (*A. dorsata*, *A. laboriosa*, *A. florae*, and *A. cerana*), *Megachile* spp. (*M. lanata* and *M. flavipes*), and *Xylocopa* spp. (Williams, 1977). Natural outcrossing was observed in pigeonpea (Howard et al., 1919). The rate of outcrossing varies from place to place depending on the extent of pollinator bees and climatic conditions. Outcrossing ranged from 20% to 70% at various locations (Saxena et al., 1990). Even though the seeds of 30-day-old pod attain physiological maturity, but it requires 10–15 additional days for obtaining seed with desirable moisture content. Pod shattering is common in wild and unadapted genotypes, but it may also be noticed in released varieties. There are no reports of inbreeding depression, which is why major breeding methodologies so far adopted are based on exploiting additive genetic variance.

In the National Agricultural Research System, genetic improvement in pigeonpea have been accomplished through pure line breeding. Population improvement programs, tried elsewhere, have not been successful. After long continued efforts of heterosis breeding, hybrids are now a reality in pigeonpea. CMS-based (three line system: "A," "B," and "R") hybrids have been released for cultivation (Saxena & Nadarajan, 2010; Bohra et al., 2025).

Allien gene transfer involves interspecific or intergeneric hybridization followed by backcrossing (Dreissig et al., 2019). Orderly meiotic events and balanced segregation of homologous pairs are necessary to recover the desired recombinants and stable progenies in the succeeding generations for use in breeding programs (Henderson & Bomblies, 2021). The frequency of recombination involving the alien chromosomes may influence the linkage drag, wherein unfavorable alleles are co-transferred in the breeding lines following wide hybridization (Bohra et al., 2022a). Inadequate recombination poses serious bottleneck in recovering the desired recombinants and necessitates large populations and numerous backcross generations. Furthermore, backcross in itself is a limiting factor to fix the favorable alleles; hence, genomic tools are helpful to recover the desired recombinants with negligible linkage drags from the wide crosses (Rabbani & Nayak, 2023).

Crossing among different species may be incompatible to produce hybrids. The compatible crosses show varying degrees of fertility in pollen mother cells and egg cells of the interspecific hybrids. The hybrids may show sterility as a result of imbalanced chromosome segregation and aneuploidy (Bozdag et al., 2021). Incompatibility in distant crosses may arise because of several factors that can be categorized into prezygotic and postzygotic barriers in hybridization. Lack of pollen germination, inadequate pollen tube growth to reach the ovule, and failure of male game to fuse with egg cells are some of the reasons among others for cross-incompatibility. The prefertilization barriers may be overcome with reciprocal crosses, changing the genotypes of the species (Kumar et al., 1990), using pollen mixtures, interventions in pistil and application of growth hormones like GA3 (Kumar et al., 1985a). In some interspecific cross combinations, a zygote is formed but aborted afterward. Embryo rescue can be employed under such situations, as in the case of *C. platycarpu/C. cajan* (Mallikarjuna, 1998). The embryo-rescued hybrids were backcrossed with cultivated parents and BC_1 zygotes were embryo rescued to generate the fertile BC_1 for further genetic manipulations. Tetraploids were also generated for deriving hybrids *C. platycarpu/C. cajan* by treating the apical buds of F_1'swith an aqueous solution of 0.05% colchicine (Mallikarjuna and Moss, 1995; Mallikarjuna et al., 2011b). Successful interspecific hybrids (*C. cajan/C. lineatus*, *C. sericeus*, *C. albicans*, *C. scarabaeoides*, *C. acutifolius*, *C. lanceolatus*, *C. platycarpus*, and *C. volubilis*) and intergeneric hybrids (*C. cajan/Rhynchosia spp.*) of pigeonpea were generated by crossing and various prezygotic and postzygotic interventions, *viz.*, hormone application, genotypic combinations, colchicine treatment, and embryo rescue technique singly or in combinations and often followed by backcrossing to abandon the linkage drag. Subsequently, in many instances, stable wild derivatives were developed for their utilization in breeding programs (Mallikarjuna et al., 2017).

2.6 GENE POOL

The genus *Cajanus* Adans. contains 32–34 taxa, viz., *Cajanus cajan* (L.) Millsp.) *C. cajanifolius* (Haines) Maesen, *C. acutifolius* (F. Muell.) Maesen, *C. albicans* (Wight & Arn.) Maesen, *C. cinereus* (F. Muell.) F. Muell., *C. confertiflorus* F. Muell., *C. lanceolatus* (W. Fitzg.) Maesen, *C. latisepalus* (Reynolds & Pedley) Maesen, *C. lineatus* (Wight & Arn.) Maesen, *C. reticulatus* (Dryand) F. Muell., *C. scarabaeoides* (L.) Thouars, *C. sericeus* (Baker) Maesen, and *C. trinervius* (DC.) Maesen *Cajanus aromaticus* Maesen, *Cajanus crassicaulis* Maesen, *Cajanus crassus* (Prain ex King) Maesen, *Cajanus elongatus* (Benth.) Maesen, *Cajanus geminatus* Pedley ex Maesen, *Cajanus goensis* Dalzell, *Cajanus grandiflorus* (Benth. ex Baker) Maesen, *Cajanus heynei* (Wight & Arn.) Maesen, *Cajanus hirtopilosus* Maesen, *Cajanus kerstingii* Harms, *Cajanus lanuginosus* Maesen, *Cajanus mareebensis* (S. T. Reynolds & Pedley) Maesen, *Cajanus marmoratus* (Banks ex Benth.) F. Muell., *Cajanus mollis* (Benth.) Maesen, *Cajanus niveus* (Benth.) Maesen, *Cajanus platycarpus* (Benth.) Maesen, *Cajanus pubescens* (Ewart & Morrison) Maesen, *Cajanus rugosus* (Wight & Arn.) Maesen, *Cajanus villosus* (Benth. ex Baker) Maesen, *Cajanus viscidus* Maesen, *Cajanus volubilis* (Blanco) Blanco (Khoury et al., 2015; Royal Botanic Gardens, n.d.). Pictorial representation of the taxa are depicted in Figure 2.2. *Cajanus* is divided into three clades: Indian, Australian, and *Scarabaeoides* (Van der Maesen, 1986; Kassa et al., 2012). The taxa of a genus are categorized into primary, secondary, and tertiary gene pool depending on the extent of gene flow between a species and its taxonomic relatives (Harlan & de Wet, 1971). The primary gene pool (GP 1) encompasses all taxa that are easily crossable and produce highly fertile hybrids, thereby allowing free gene flow among the individuals. The secondary gene pool (GP 2) is composed of taxa that are crossable with the cultivated species and

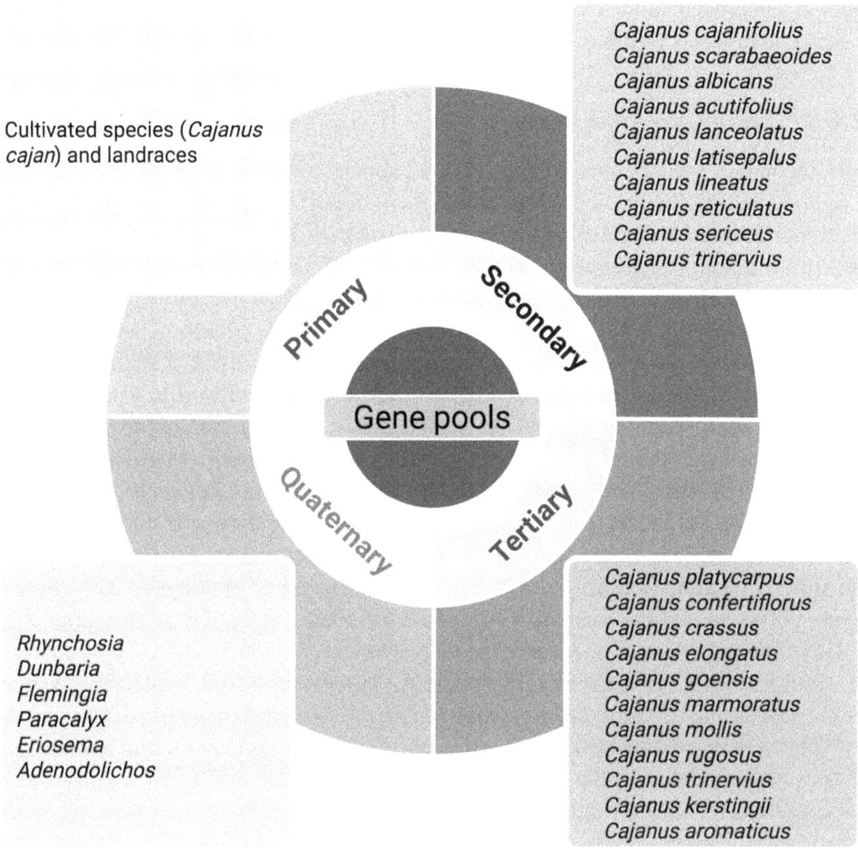

FIGURE 2.2 Gene pool of the genus *Cajanus*.

form fertile or partially fertile hybrids. This led to a variable degree of restriction in gene flow between the cultivated species and other members of the taxa. The gene flow from the members of the tertiary gene pool (GP 3) of a cultivated species is highly restricted, and it is composed of taxa that may be crossable with the aid of embryo rescue, resulting in fertile or partially fertile hybrids.

More recently, Gepts and Papa (2003) offered the concept of quaternary gene pool (GP 4) to include other taxa that are distant relatives of cultivated species and sexually incompatible but may be utilized for gene transfer through advance biotechnological and genomic tools. Thus, GP 4 is described as the noncrossable taxa from the very distant relatives of cultivated species that lie on the outer boundary of the gene pool.

2.6.1 Primary Gene Pool

Cultivated pigeonpea is composed of single species *Cajanus cajan* (L.) Millsp.); hence, no wild taxa belong to the primary gene pool of pigeonpea. The primary gene pool is composed of all cultivated germplasm lines, including local cultivars, land races, and notified varieties of pigeonpea.

2.6.2 Secondary Gene Pool

The secondary gene pool is composed of 12 species (Khoury et al., 2015)—namely, *Cajanus cajanifolius* (Haines) Maesen, in the Indian clade, the putative progenitor of the crop (Kassa et al., 2012); *C. acutifolius* (F. Muell.) Maesen; *C. albicans* (Wight & Arn.) Maesen; *C. cinereus* (F. Muell.) F. Muell.; *C. confertiflorus* F. Muell.; *C. lanceolatus* (W. Fitzg.) Maesen; *C. latisepalus* (Reynolds & Pedley) Maesen; *C. lineatus* (Wight & Arn.) Maesen; *C. reticulatus* (Dryand) F. Muell; *C. scarabaeoides* (L.) Thouars; *C. sericeus* (Baker) Maesen; and *C. trinervius* (DC.) Maesen (Saxena et al., 2005; Mallikarjuna et al., 2011a). These species are crossable with cultivated pigeonpea with relative ease and do not generally need specialized techniques for making crosses in majority of the cases, with a few exceptions (Mallikarjuna et al., 2011a, 2011b). Bivalent formation and chromosome pairing between *C. cajan* and species of the secondary gene pool allows for recombination and normal segregation, thus providing scope for gene introgression. Several useful traits have been identified in various accessions of the secondary gene pool (Upadhyaya 2006; Sujana et al., 2008). In addition, several CMS genes have been derived from this gene pool: A1 cytoplasm from *C. sericeus* (Ariyanayagam et al., 1995), A2 cytoplasm from *C. scarabaeoides* (Tikka et al., 1997), and A3 cytoplasm from *C. volubilis* (Wanjari et al., 1999).The A4 cytoplasm is from *C. cajanifolius* (Saxena et al., 2005) and A5 cytoplasm from crosses between *C. cajan* and *C. acutifolius* (Mallikarjuna & Saxena, 2005). Among the secondary gene pool, *C. scarabaeoides* and *C. cajanifolius* have been used frequently for introgression of specific traits.

2.6.3 Tertiary Gene Pool

Twenty to 21 species of *Cajanus*—namely, *Cajanus aromaticus* Maesen, *Cajanus crassicaulis* Maesen, *Cajanus crassus* (Prain ex King) Maesen, *Cajanus elongatus* (Benth.) Maesen, *Cajanus geminatus* Pedley ex Maesen, *Cajanus goensis* Dalzell, *Cajanus grandiflorus* (Benth. ex Baker) Maesen, *Cajanus heynei* (Wight & Arn.) Maesen, *Cajanus hirtopilosus* Maesen, *Cajanus kerstingii* Harms, *Cajanus lanuginosus* Maesen, *Cajanus mareebensis* (Reynolds & Pedley) Maesen, *Cajanus marmoratus* (Banks ex Benth.) Cajanus mollis (Benth.) Maesen, *Cajanus niveus* (Benth.) Maesen, *Cajanus platycarpus* (Benth.) Maesen, *Cajanus pubescens* (Ewart & Morrison) Maesen, *Cajanus rugosus* (Wight & Arn.) Maesen, *Cajanus villosus* (Benth. ex Baker) Maesen, *Cajanus viscidus* Maesen, *Cajanus volubilis*—were included in the tertiary gene pool (Ramanadan, 1990; Mallikarjuna et al., 2011a; Royal Botanic Gardens, n.d.).Gene introgression lines were derived from two species of the tertiary gene pool. *Cajanus platycarpus* was successfully crossed with cultivated pigeonpea with help of hormone-aided pollinations, embryo rescue, and invitro culture (Mallikarjuna et al., 2011b) to obtain viable hybrids. Extra-early breeding lines were derived from wild derivatives of *C. volubilis* (Mallikarjuna et al., 2014).

2.6.4 Quaternary Gene Pool

The quaternary gene pool comprised ten relatives of *Cajanus* genera in the subtribe cajanae. These genera, viz., *Flemingia*, *Rhynchosia*, *Dunbaria*, *Erisema*, *Paracalyx*, *Adenodolichos*, *Bolusafra*, *Carissoa*, *Chrysoscia*, and *Baukea* (Mallikarjuna et al., 2011a), are rarely used as food legumes. Among the quaternary gene pool, *Rhynchosia* has nutritional and therapeutic properties (Drabu et al., 2011). Several phytochemicals, such as glycosides, carotenoids, coumarins, alkaloids, fatty acids, anthraquinones, flavonoids, dihydrochalcones, steroids and triterpenoids, have been reported in *Rhynchosia* (Bakshu & Raju, 2001).The genera belonging to quaternary gene pool are not crossable to Cajanus species. However, protoplast fusion may offer an avenue to transfer genes from this gene pool into cultivated pigeonpea.

2.7 DIVERSITY IN GERMPLASM

Pigeonpea is grown in different agroclimatic regions of India on a variety of soil and diverse physiographic situations. Variability is observed among the genotypes for various botanical and agronomic traits (Upadhyaya et al., 2007, 2013a). Most of the pigeonpea genotypes are photo-thermo sensitive, and a few behave like photoperiod insensitive. Therefore, days to 50% flowering and physiological maturity do not correspond across the North-South region. Grossly, as mentioned earlier, Indian pigeonpea genotypes have been classified into distinct maturity groups i.e. early, mid-early, medium, and late. Vegetable types (tall, large flower, bigger pods, and seed) are distributed in the North East Hill region, Gujarat, Karnataka, Chhattisgarh, and Andhra Pradesh. Genotypes with non-branching habit (single stem) have also been identified among collections from Madhya Pradesh and other parts of Central and Northern India. A large amount of variability exists for growth habit (determinate and nondeterminate) and branching pattern (erect to spreading). Flower color is predominantly yellow among the accessions of grain type, while purple is predominant among the vegetable types collected from Chhattisgarh, Tripura, and elsewhere. Wide-scale variation exists for seed size and seed color among the germplasm collections.

In order to enhance the efficiency of germplasm utilization for breeding, as well as genetic studies, a large number of genotypes were characterized for morpho-agronomic traits coupled with diversity analysis for making a core collection of genotypes at ICRISAT, and, subsequently, a subset of 146 accessions formed the mini-core collection (Upadhyaya et al., 2006). The mini-core exhibited 80% of diversity among the conserved accessions and is considered representative of existing diversity for genetic studies in association mapping (Gowda et al., 2013). Several genotypes were characterized for disease and insect resistance and tolerance to abiotic stresses—namely, moisture extremities and salinity tolerance, besides morpho-agronomic traits and quality parameters, such as protein and micro-nutrient content. Crop wild species assume importance for harboring useful genes for economic traits, as well as a reservoir of resistance against biotic and abiotic stresses. Many useful traits—namely, resistance to FW, *Phythopthora* stem blight, SMD and pod borer complex, tolerance to drought, high protein content, photo-insensitivity, cleistogamy, and cytoplasmic male sterility, etc.—were documented in wild species (Khoury et al., 2015; Table 2.2). Wild species

TABLE 2.2
Traits Documented in Wild Species

Documented Trait	Wild Species	References
Cytoplasmic male sterility	*C. cajanifolius*, *C. acutifolius*, *C. scarabaeoides*, *C. lineatus*,	Ariyanayagam et al. (1995), Mallikarjuna and Saxena (2005), Saxena and Kumar (2003), Saxena et al. (1998), and Saxena et al. (2005)
Nuclear male sterility	*C. cajanifolius*	Mallikarjuna and Saxena (2005)
Pod borer resistance	*C. acutifolius*, *C. albicans*, *C. platycarpus*, *C. scarabaeoides*, *C. sericeus*	Mallikarjuna et al. (2007), Mallikarjuna et al. (2011a), Mallikarjuna et al. (2007), and Saxena (2005)
Pod fly resistance	*C. acutifolius*, *C. albicans*, *C. scarabaeoides*, *C. sericeus*, *C. lineatus*	Sharma et al. (2003)
Pod wasp resistance	*C. albicans*, *C. scarabaeoides*	Sharma et al. (2003)
Sterility mosaic disease resistance	*C. acutifolius*, *C. albicans*, *C. crassus*, *C. lineatus*, *C. platycarpus*, *C. scarabaeoides*	Mallikarjuna et al. (2011a) and Saxena (2005)
Phytophthora blight resistance	*C. platycarpus*	Mallikarjuna et al. (2005) and Saxena (2005)
Drought tolerance	*C. scarabaeoides*	Upadhyaya (2006)
Salt tolerance	*C. albicans*, *C. platycarpus*, *C. scarabaeoides*, *C. sericeus*	Choudhary et al. (2011)
Aluminum toxicity resistance	*C. platycarpus*, *C. scarabaeoides*	Choudhary et al. (2011)
Cleistogamy	*C. lineatus*	Saxena et al. (1998)
Annuality	*C. platycarpus*	Khoury et al. (2015)
Photoperiod insensitivity	*C. platycarpus*	Mudaraddi et al. (2013)
Dwarfism	*C. scarabaeoides*	Reddy (1990)
Early flowering	*C. platycarpus*, *C. scarabaeoides*	Upadhyaya (2006) and Upadhyaya et al. (2013b)
High seed weight	*C. acutifolius*	Mallikarjuna et al. (2011a)
Broad pods	*C. albicans*	Mallikarjuna et al. (2011a)
High flower and pod setting	*C. platycarpus*	Mudaraddi et al. (2013)
More seeds per pod	*C. albicans*, *C. cinereus*, *C. mollis*	Mallikarjuna et al. (2011a) and Upadhyaya et al. (2013b)
High seed protein/Protein improvement	*C. albicans*, *C. cajanifolius*, *C. crassus*, *C. mollis*, *C. platycarpus*, *C. scarabaeoides*, *C. sericeus*	Mallikarjuna et al. (2011a) and Upadhyaya et al. (2013b)
Good forage source	*C. albicans*, *C. crassus*, *C. mollis*	Mallikarjuna et al. (2011a) and Upadhyaya et al. (2013b)

also exhibit interspecific variation for useful traits among the available accessions; hence, there is prior need to characterize and evaluate them for specific traits and select the best accession for prebreeding (Mallikarjuna et al., 2011a). Cluster analysis of similarity index was performed using the RFLP technique and maize mt-DNA probes on 27 accessions of *Cajanus* species—namely, *C. cajan*, *C. cajanifolius*, *C. sericeus*, *C. acutifolius*, *C. latisepalus*, *C. albicans*, *C. rugosus*, *C. scarabaeoides*, *C. goensis*, *C. mollis*, *C. volubilis*, *C. platycarpus*, *R. minima*, *R. edulis*, *R. sublobata*, and *R. verdcourtii* (Sivaramakrishnan et al., 2002). *C. platycarpus* was grouped separately while all accessions of *C. cajan* fell into the same group with close proximity to *C. cajanifolius*. Variation within the accessions of *C. cajan*, *C. scarabaeoides*, *C. platycarpus*, and *C. acutifolius* were also recorded. Upadhyaya et al. (2011) characterized and evaluated the composite germplasm collection of pigeonpea, and its wild relatives conserved at ICRISAT on the basis of 20 SSR markers and phenotyping for multitude of qualitative and quantitative traits. Two broad clusters were formed using SSR markers and a reference set of 300 of the most diverse genotypes were constellated. Promising accessions were identified for four important agronomic traits.

In a study, SNP assays were performed on 79 accessions of cultivated pigeonpea and 31 accessions pertaining to 13 wild relatives belonging to genus *Cajan*. The sample represented the domesticated accessions (both cultivated and feral) and wild species of Indian and Australian origin collected from widespread geographical regions, viz., Africa, Asia, Latin America, the Caribbean, the Indian subcontinent, and Australia (Khoury et al., 2015). Two accessions of *C. cajanifolius* (C7847 and ICP 15632) exhibited close phylogenetic relationship with domesticated pigeonpea. *C. cajanifolius* accessions have greater affinity toward Indian origin *C. platycarpus* and *C. sericeus* accessions, as well as to individual representatives of *C. albicans* and *C. crassus*. Diversity among the wild species of Indian origins was very high compared to wild species of Australian origin. Furthermore, less differentiation was observed between the domesticated pigeonpea and wild Indian population compared to the wild Australian population.

Fifty four advanced generation (F_8 & F_9) wild derivatives of *Cajanus scarabaeoides*, *Cajanus acutifolius*, and *Cajanus cajanifolius* were evaluated for agronomic traits (Mahadeva Swamy et al., 2024). They observed significant genotypic variation in the advance generation (F_8 & F_9). Heritability estimates ranged from 22.6% to 89.3%. The 54 interspecific wild derivatives were grouped into 4 clusters, indicating significant diversity among the derived lines.

Mudaraddi et al. (2013) studied the diversity among four species of *Cajanus*—namely, *acutifolius*, *scarabaeoides*, *cajanifolius* and *platycarpus* by employing SSR markers. The allelic distribution revealed a close relationship between *C. cajan* and *C. cajanifolius*, whereas *C. scarabaeoides* was the most diverse among these species, contrary to their cross compatibility with cultivated species.

2.8 GERMPLASM REPOSITORY

There are 13,771 accessions of pigeonpea conserved at ICRISAT (Upadhyaya et al., 2007; Gowda et al., 2013). ICRISAT has developed a composite collection of pigeonpea containing 1,000 accessions representative of the diversity of all germplasm

TABLE 2.3
Germplasm Holdings in Different Institutes

Material	Organization	Accessions	References
Cultivated germplasm	ICRISAT	13,771	Upadhyaya et al. (2007) and Gowda et al. (2013)
Wild relatives	ICRISAT	555	Upadhyaya et al. (2007)
Germplasm	ICAR-NBPGR	11,221	Singh et al. (2014)
Germplasm	USDA	4,116	Pazhamala et al. (2015)
Germplasm	KARI-NGBK	1,288	Singh et al. (2013)

collections. Furthermore, a mini-core collection comprising 146 accessions has been grouped from the core collection and other materials. In addition, ICRISAT has 555 accessions of wild relatives belonging to 6 genera and 66 species (Upadhyaya et al., 2007, 2013a). In the national gene bank at the National Bureau of Plant Genetic Resources (NBPGR), 11,221 accessions of Indian collections are conserved (Singh et al., 2014). In addition, NBPGR has also repatriated 5,748 accessions to the national gene bank.U.S. Department of Agriculture (USDA) has maintained 4,116 accessions (Pazhamala et al., 2015) and the Kenya Agricultural Research Institute's National Genebank of Kenya (KARI-NGBK) has 1,288 accessions (Singh et al., 2013). The germplasm holdings in different institutes are provided in (Table 2.3).

The extensive use of few parents in pigeonpea improvement programs has led to the narrowing down of genetic base of the cultivated varieties. Only a small portion of germplasm were utilized for crop breeding; therefore, the majority of the existing germplasm remain unutilized for crop improvement (Majumder and Singh, 2005).

2.9 WIDE HYBRIDIZATION

The cultivated pigeonpea is an often cross-pollinated species and shows variable degrees of natural outcrossing depending on the population of pollinators, genotypes, and environmental conditions. Saxena and Kumar (2010) have reported natural outcrossing and interspecific hybrids in four different *Cajanus* species, viz., *C. scarabaeoides* (ICPW 89), *C. albicans* (ICPW 13), *C. sericeus* (ICPW 162), and *C. lineatus* (ICPW 42), from the secondary gene pool on the basis of occurrence of natural hybrids in the succeeding generations. Maximum outcrossing was observed in *C. lineatus* (17%), followed by *C. albicans* (10%), *C. scarabaeoides* (8.3%), and *C. sericeus* (2.3%) as compared to the cultivated pigeonpea line (ICP 5529), which exhibited 22% outcrossing. The species from the secondary gene pool are crossable with cultivated pigeonpea with varying degrees of ease; some may require external aid like hormone application and embryo rescue to generate hybrids, while others like *C. scarabaeoides* and *C. cajanifolius* do not require any aid. Nonetheless, external interventions increased the frequency pod and seed setting. In many instances of distant hybridization, wild species as female parent deter the hybrid formation (Mallikarjuna & Saxena, 2002). Application of gibberellic acid or kinetin has been used to increase the pod set and seeds per pod in wide hybridization of pigeonpea (Kumar et al., 1985a; Dhanju & Gill, 1985). The species from the tertiary gene pool are not amenable to crossing with cultivated pigeonpea. They pose incompatibility at the prezygotic stage (Jayaprakash & Sarla, 2001) or postzygotic level (Mallikarjuna & Moss, 1995), thereby leading to regular abortion of hybrid embryos (Pundir & Singh, 1987). Application of gibberellic acid to the pollinated pistils was recommended to enhance the embryo formation and recovery of mature embryos that are generally more responsive to the embryo rescue technique (Mallikarjuna et al., 2011b).

2.10 INTERSPECIFIC HYBRIDS

2.10.1 *Cajanus cajanifolius* Hybrids

C. cajanifolius belongs to the secondary gene pool and is considered the probable progenitor of cultivated pigeonpea. The morphology, karyotype, chromosome pairing, and crossability of *C. cajanifolius* are more related to *C. cajan* than other wild relatives (Pundir & Singh, 1985a). A cross between *C. cajan* and *C. cajanifolius* resulted in fertile hybrids with high pollen stainability and ovule fertility (Reddy et al., 1981; Pundir & Singh, 1987; Kumar et al., 1990; Yadav & Padmaja, 2002; Venkatesan & Arunachalam, 2007; Mishra & Panigrahi, 2017; Ravali et al., 2022). *C. cajan/C. cajanifolius* cross was successful when *C. cajan* was taken as the female parent, while reciprocal crosses yielded hybrids that could not set pods (Thiruvengadam & Muthiah, 2007). Male sterility was observed in the F_1's of *C. cajanifolius/C. cajan* (Ariyanayagam et al., 1993).

2.10.2 *Cajanus scarabaeoides* Hybrids

C. scarabaeoides, which belong to the secondary gene pool, is most easily crossable with cultivated pigeonpea (Roy & De, 1967; Pundir & Singh, 1987; Dundas et al., 1989; Kumar et al., 1990; Purohit & Patel, 2025). It harbors useful agronomic traits for drought and heat tolerance to Helicoverpa (Verulkar et al., 1997; Bajpai et al., 2003). Few species carry dense hairs on the pod. Interspecific crosses with cultivated pigeonpea revealed that the erect growth habit of *C. cajan* is dominant over the spreading growth habit of *C. Scarabaeoides* (Pundir & Singh, 1985a, 1985b), 29.2% pod set was noted in *C. cajan* and *C. scarabaeoides*, 8.3% pod set in *C. lineatus/C. scarabaeoides*, and 26.9% pod set in *C. scarabaeoides/C. sericeous* (Pundir & Singh, 1987). Similar results were found in later works (Dundas et al., 1989; Kumar et al., 1990).

2.10.3 *Cajanus sericeus* Hybrids

C. sericeus is another member of the secondary gene pool that carries useful alleles for agronomic importance. Viable hybrids between *C. cajan/C. sericeus* were reported (Reddy 1984, Kumar et al., 1985b; Pundir & Singh, 1987; Ariyanayagam et al., 1993; Yadav & Padmaja, 2002). Successful cross with cultivated pigeonpea may be affected by genotypes (Pundir & Singh, 1987). Simultaneous improvement of seed size and protein content was observed in hybrids of Type 21/*C. sericeus*

(Saxena et al., 1987). Differential effects of *C. cajan* genotypes were noticed for female sterility in backcross generations when hybrids were made between *C. sericeus/C. cajan* (Ariyanayagam et al., 1993). The male sterility mechanism was worked out for *C. sericeus/C. cajan* (Wanjari, 1998).

2.10.4 *Cajanus albicans* Hybrids

C. albicans belongs to the secondary gene pool and offers tolerance to salinity (Subbarao et al., 1990). Fertile hybrids were obtained between *C. cajan/C. albicans*, and regular bivalents were observed (Kumar et al., 1984, 1985b; Yadav, 1986; Pundir & Singh, 1987). In reciprocal cross, the pod set was 7%, and the hybrid retained the twining habit. Male sterility was observed in hybrids when *C. albicans* was taken as the female parent (Ariyanayagam et al., 1993). Kumar et al. (1985b) obtained F_1 hybrids that were intermediate between the parents for morphological traits like leaf shape, growth habit, seed strophiole, and seed mottling. Eleven bivalents exhibited regular disjunction at meiosis.

2.10.5 *Cajanus volubilis* Hybrids

C. volubilis belongs to the tertiary gene pool. Four kinds of male sterile plants were derived from *C. volubilis/*ICPL 83024 cross (Wanjari et al., 1999). Prefertilization barriers were noticed in *C. volubilis/C. Cajan* crosses (Jayaprakash and Sarla, 2001) and application of ε-amino caproic acid (EACA; an amino acid) to the pollen germination medium was suggested, as it resulted in better pollen germination.

2.10.6 *Cajanus acutifolius* Hybrids

C. acutifolius is a member of the secondary gene pool. It is crossable with cultivated pigeonpea, and hybrids were reported (Dundas, 1984; Reddy et al., 2001; Mallikarjuna & Saxena, 2005). Hybrids were obtained between *C. cajan* (Prabhat)/*C. acutifolius* (Dundas et al., 1988; Ariyanayagam et al., 1993) got few fertile seeds from *C. cajan/ C. acutifolius* but did not obtain fertile seed from reciprocal cross. Mallikarjuna and Saxena (2002) reported similar observations.

2.10.7 *Cajanus platycarpus* Hybrids

C. platycarpus belongs to the tertiary gene pool. It has great potential for deriving true photo- insensitive annual and early maturing genotypes of pigeonpea besides resistance to *Phytophthora Drechsleri f* sp *cajani* (Moss et al., 1988) and tolerance to salinity (Subbarao et al., 1990). Reduced homology with *C. cajan* inhibits direct crosses. Small embryos were obtained between *C. platycarpus/C.*acutifolius and *C. platycarpus/C. scarabaeoides*, and bridge cross with other wild species was advocated for transfer to genes from C. *platycarpus* into cultivated pigeonpea (Moss et al., 1988).Successful hybrids were reported between *C. platycarpus C cajan* (Ariyanayagam & Spence, 1978), whereas crosses with cultivated pigeonpea failed (Pundir & Singh, 1987). Special techniques were employed to generate hybrids to extract the segregating generations. Two approaches, namely colchicine treatment and embryo rescue were used to derive hybrids between pigeonpea and *C. platycarpus* (Mallikarjuna & Moss, 1995; Mallikarjuna, 1998; Mallikarjuna et al., 2006). Backcross derivatives of *C. platycarpus* were successfully generated by employing the embryo rescue technique (Mallikarjuna, 2007). Vigorous tetraploids plants were obtained by the treatment of F_1's (*C. platycarpus* × *C. cajan*) with colchicines (Mallikarjuna et al., 2011b). However, the ploidy differences between the derived colchicines-treated hybrid did not allow for further backcrossing with diploid pigeonpea. Open flower segregants were retrieved from the cross between *C. platycarpus* and *C. cajan* (Cherian et al., 2006).

2.10.8 *Cajalnus lineatus* Hybrids

C. lineatus is a member of the secondary gene pool. It resembles cultivated pigeonpea in chromosome morphology and crossable (Reddy & De, 1983; Pundir & Singh, 1987). The cross between *C. cajan* and *C. lineatus* produced a partially fertile hybrid (Kumar et al., 1958).The sterility was ascribed to structural differences in chromosomes of the two species. Homologous pairings were observed with meiotic irregularities. Close lineage was suggested based on homology between the two sets of chromosomes.

2.10.9 *Cajanus trinervius* Hybrids

It belongs to the secondary gene pool. Very low pod set and fertility was observed in cross *C. cajan* (UPAS 120)/*C. trinervius* (Pundir & Singh, 1987).

2.10.10 *Cajanus reticulatus* Hybrids

It belongs to the secondary gene pool. Successful cross was made between *Cajanus cajan*/*C. reticulatus* (Dundas, 1984; Reddy et al., 2001; Mallikarjuna & Saxena, 2005). The hybrid retained the wild traits such as persistent stipules, long pod hairs, pod shattering, brown seeds with gray speckles, and presence of seed strophiole. Meiotic anomalies were observed in the hybrid that restricted the trait transfer into *C. cajan*. During diakinesis and metaphase-I, only 41.9% of the cell had bivalents, whereas several abnormal forms, including quadrivalents, trivalents, and univalents, were detected in high frequency (Reddy et al., 2001). However, these anomalies in the interspecific hybrid are not significant enough to prevent the gene transfer from *C. reticulatus* to the cultivated species through a sexual route.

2.10.11 *Cajanus lanceolatus* Hybrids

C. lanceolatus belongs to the secondary gene pool. Srikanth et al. (2014) obtained hybrids between *C. cajan* and *C. lanceolatus*, which exhibited intermediate morphology and partial to complete male sterility (50%–100%). At meiosis, the fertile hybrids showed homologous paring with high frequency of regular bivalents and normal disjunction between the *C. cajan* and *C. lanceolatus* chromosomes.

2.11 INTERGENERIC HYBRIDS

2.11.1 *Rhynchosia* Hybrid

Rhynchosia bracteata is a crop wild genera of pigeonpea that is consumed by local tribals in the Rollapenta forest, Kurnool district. It is a very hardy species and resistant to Phythopthora and FW diseases. Reports of hybrid between *C. cajan* and *Rhynchosia spp.* is rare. Very low success of obtaining hybrid between *C. cajan* and *Rhynchosia spp.* was reported (Mallikarjuna et al., 2014).

2.12 BREEDING AND GENETIC INTERVENTIONS IN DERIVING ALIEN INTROGRESSION LINES

In a breeding program, fixation of genes and associated traits thereof are dependent on reliable phenotyping for the identification genotypes with desired recombinants. It is all the more important when varieties are designed for high-yield potential, along with good quality traits and multiple resistance for diseases and insect pests at least at par to the best existing cultivars. When unadapted genotypes (germplasm, land races, and crop wild relatives) are crossed with adapted varieties, the resultant initial segregating population and progenies exhibit a lot of linkage drag in the form of undesirable traits in association with the traits that are targeted to transfer into the good agronomic base. Under these circumstances, the new genomic tools aid in the selection process which are otherwise insufficient in intensity and difficult to delineate from the background genetic noise. Elimination of the background genetic noise is one of the main objectives in alien gene transfers from interspecific and intergenetic hybridization.

In this context, the nature of mapping population is also important because each of such populations have their own advantages and disadvantages. For long, biparental F_2's have been the primary mapping population but later were supplemented by RILs and doubled haploids, which are immortal. However, these populations have limitation for deriving QTLs with higher confidence levels (Tanksley, 1993). It is important to locate the QTLs with precision and also to detect the QTLs of smaller effect and pyramid them to generate breeding lines with significant trait performance.

Now, several advanced mapping populations, viz., backcross inbred lines (BILs), near isogenic lines (NILs), advanced intercrossed lines (AILs), chromosome segment substitution lines (CSSLs), multiparent advanced generation inter-cross (MAGIC) populations, association panels and nested association mapping (NAMs), are used for QTL mapping (Brock et al., 2020). BILs, CSSLs, and stepped aligned inbred recombinant strain (STAIRS) were suggested for use in QTL mapping of alien traits (Bohra et al., 2010; Singh & Singh, 2015).CSSLs are derived from intergeneric or interspecific crosses to develop a set of introgressed lines covering the entire genome of wild relative but in the adapted agronomic background.

Koumproglou et al. (2002) designed an advance mapping population wherein single chromosome substitution lines for the haploid set of chromosomes were generated by using telomeric chromosome specific markers for both ends. In the second step, a large number of single recombinant lines (SRLs) were derived by identifying single unique crossover events for each chromosome. Finally, by stacking the SRLs, the chromosomes are stacked sequentially. A set of for particular chromosome makes STAIRS.

Saxena et al. (2020b) have utilized NAM and MAGIC to derive RILs for multilocation evaluation. These populations serve the dual purpose of trait discovery and new cultivar development through genomic interventions.

The limited backcross (BC_2F_1 _ BC_3F_1) populations are more reliable for eliminating the background genetic noise to increase the confidence level of QTL detection without losing the trait expression in case of wild derivates. The BILs are of special mention, as these have been utilized in pigeonpea for developing introgressions lines from *C. acutifolius* and *C. platycarpus* and other species at ICRISAT in recent past. BILs are near homozygous lines and representative of the homologous recombination that carry recurrent genomes randomly. The BILs have the advantage of relative homozygosity for studying environmental interactions more precisely and have commercial value as advance breeding lines for donor traits and potential use as varieties.

2.13 GENOMIC RESOURCES FOR EXPLOITING ALIEN GENES

Several types of molecular markers have been used in pigeonpea for diversity analysis: phylogenetic studies, gene tagging, linkage map, association map, MAS, marker-assisted backcrossing (MABC), and early generation selection (EGS), etC. The first generation molecular markers, viz., RFLP, RAPD, AFLP, were mainly used for diversity analysis besides occasional use for gene tagging.

In the next stage, SSR markers derived from sequence (gSSRs), expressed sequence tags (ESTs-SSRs), and bacterial artificial chromosome (BAC)-end sequences (BES-SSRs) were employed for making linkage maps, gene tagging and QTL analyses (Odeny et al., 2007; Varshney et al., 2012; Bohra et al., 2017). The advent of next-generation sequencing technology provided the third generation DNA markers—viz., diversity arrays technology (DArT) arrays and competitive allele-specific polymerase chain reaction (KASPar) assays were used to generate the SNPs (Yang et al., 2011; Saxena et al., 2014). The transcriptome assembly of pigeonpea had generated intron spanning region (ISR) markers (Kudapa et al., 2012).

With high throughput genome sequencing platforms in place, the concept of genomic-estimated breeding values (GEBVs) has facilitated genomic selection and breeding (Meuwissen et al., 2001; Akdemir et al., 2019; Bohra et al., 2020; Saxena et al., 2021b).

Transfer of favorable alleles from the wild relatives to cultivated crop species are dependent on cross compatibility, normal homologous pairing and bivalent formation, normal segregation of chromosomes during gamete formation, and selection of desirable recombinants in the filial or intermated generations. Molecular markers aid in the identification and fixation of favorable alleles in the breeding population, reduction of linkage drag, and recovery of desirable agronomic base.

Most of the traits of interest in breeding programs are governed by multiple loci and inter alia interaction thereof; hence, they behave as quantitative traits, which is why selection of such traits require skill, technique, procedure, phenotypic evaluation, statistical tools, and desired environment (normal or stress as the case may be). In this context, it is difficult to identify the desirable introgression lines because of several factors like segregation distortion, linkage drag, and trait dilution due to interactions in the new genetic background. Therefore, molecular markers, particularly QTL detecting markers, are very useful for recovery of desirable introgression lines arising out of distant hybridization. Many genomic resources were developed in pigeonpea (Table 2.4).

In the recent past, several SSR markers were developed through mining of bacterial artificial chromosome (BAC)–end sequences and subsequently used in linkage mapping and QTL analysis (Bohra et al., 2010; Bohra et al., 2012). In pigeonpea, the first SSR based genetic map was based on interspecific population, which paved the way for several SSR maps using F_2 population derived from *C. cajan*/*C. cajan* cross.

Marker-assisted backcrossing (MABC) is often employed to transfer resistance or other genes from the donors to the agronomically superior variety with the aim to recover the elite variety along with trait specific donor genes. In MABC, tightly linked or flanking markers are used to target the trait of interest/QTLs for foreground selection and other markers that cover the genome of elite parent are used for background selection.

Wild relatives provide novel genes/QTLs important for breeding in crops, including pigeonpea. Despite harboring potential genetic resources, so far wild

TABLE 2.4
Development of Genomics Resources in Pigeonpea

Genomics Resources in Pigeonpea	References
Biparental mapping populations	Gnanesh et al. (2011), Yang et al. (2011), Bohra et al. 2012), and Saxena et al. (2017, 2018a)
High-density genetic map using various mapping populations	Saxena et al. (2020a)
Whole genome sequence	Varshney et al. (2012)
Whole genome resequencing	Varshney et al. (2017) and Singh et al. (2024)
Pangenome	Zhao et al. (2020)
Chromosome-scale reference genome	Liu et al. (2024)
High-density axiom *Cajanus* SNP array	Saxena et al. (2018b) and Yadav et al. (2019)
62K genic-SNP chip array	Singh et al. (2020b)
Nested association mapping population	Saxena et al. (2020c)
Multiparent advanced generation inter-cross population	Saxena et al. (2020c)
Diagnostic marker kit for FW and sterility mosaic diseases resistance	Saxena et al. (2021)
Marker-assisted backcrossing population	Saxena et al. (2020a)
Genome-wide association studies	Kimaro et al. (2020), Kumar et al. (2022), and Megha et al. (2024)
Transcriptome resource	Rathinam et al. (2019)

Cajanus species were underutilized for improving cultivated *Cajanus* through the introduction of novel alleles/genes/QTLs. By crossing cultivated pigeonpea ICPL87119 with two pigeonpea CWRs (*Cajanus cajanifolius* and *Cajanus acutifolius*), Sharma et al. (2019) developed several introgression lines (ILs). Some evaluated ILs showed improved yield and resistance to FW and SMD across different locations.

Introducing a desired monogenic/oligogenic trait typically takes six to eight years using conventional breeding methods; however, marker-assisted backcrossing (MABC) can circumvent this process, reducing the timeframe to two to three years. Recent advances in pigeonpea genomics research have significantly facilitated plant breeders in employing MABC for introgressing various breeding-important traits, including resistance to FW and SMD, into mega varieties. To introduce combined resistance against SMD and FW, the donor parent ICPL 20096 was used in a backcross breeding program to transfer these resistances into 11 elite pigeonpea varieties, including Bahar, UPAS120, TJT501, and BDN711, via marker-assisted breeding (Saxena et al., 2021). Likewise, advanced backcross QTL approach an alternative approach for introducing complex traits, including drought tolerance from CWRs into the cultivated pigeonpea cultivars. Thus, the MABC tool has the potential to introduce several traits of breeding importance including protruded stigma, earliness, cytoplasmic male sterility and fertility restorer gene(s), and even complex traits, including drought tolerance, thereby enabling the exploitation of hybrid development potential in pigeonpea.

2.14 HARNESSING THE TRAITS OF WILD SPECIES THROUGH BACKCROSS DERIVATIVES AB-QTL

The crop wild species are often associated with high fitness traits, such as pod-shattering behavior for easy dissemination of seed, resistance against biotic, and abiotic stresses, to name a few, but these species are generally inferior in several yield attributing traits. Therefore, the objective is to transfer the specific traits and avoid the unfavorable alleles by making appropriate selection. However, linkage drag is very high, and it is difficult to recover the desirable recombinants from the cultivated x wild species crosses. Few backcrosses to the cultivated parent are useful in alleviating the linkage drag, but there is danger of losing the trait of interest in each backcross generation unless the selection of right plants among backcross F_1 is ensured through an appropriate technique. The genomic and maker-assisted selection is useful for reducing the linkage drag and retrieving the desirable recombinants. Advance backcross QTL hold promise for efficient exploitation of favorable alleles from the secondary and tertiary gene pool by means of detection and incorporation of alien QTL in the gene introgressed lines.

The expression of genes introduced from alien species are affected by the genetic background of the recipient parent, and therefore higher probability of expression can be achieved if QTLs are searched after few backcrosses in the new genetic constellation in the introgressed backcross lines. In wide crosses, two to three back cross with the cultivated species enhances the probability of simultaneously detecting QTLs as well as trait introgression in the new genetic background without losing the power of QTL detection that rely on linkage disequilibrium. Advanced backcross QTL analysis is a molecular marker tool that is generally employed on the advance generation backcross (BC_2 or BC_3) derivatives originating from crosses with unadapted donor germplasm and crop wild relatives for retrieving targeted traits with minimum linkage drag, the undesirable chromosomal fragments linked with target trait DNA and carrying deleterious alleles, in later breeding generations. Based on simulation studies (Tanksley & Nelson, 1996) suggested that AB-QTL will be effective in detecting additive, dominant, partially dominant, or overdominant QTLs while less effective for epistatic QTLs. After identification, the QTLs can be fixed through two to three selling events for generating QTL aided near isogenic lines (QTL-NILs), which when intermated can produce a QTL pyramid to assemble various favorable alleles in an introgressed line. AB-QTL holds a great promise for efficiently introgressing favorable alleles from wild relatives into the cultivated pigeonpea, which has the bottleneck of narrow genetic diversity in the primary gene pool. However, large populations of backcross individuals, a resource intensive and cumbersome activity, need to be analyzed for the detection of QTLs to give room for trapping the favorable alleles without much loss (Varshney et al., 2005). In a genetic experiment, Leprévost et al. (2023) have integrated advanced backcross QTL (AB-QTL) markers with previous information on QTL and GWAS for enhancing and pyramiding of disease resistance genes in field peas. The AB-QTLs are thought to be useful tools in the introgression of economic traits from crop wild species (Mallikarjuna et al., 2014). AB-QTL mapping populations were derived from *C. cajanifolius* and *C. acutifolius*, and advanced breeding lines of pigeonpea were generated for evaluation of traits of interest and agronomic worth (Mallikarjuna et al., 2011a). Another AB-QTL population was made for introgression genes from *C. platycarpus* into cultivated pigeonpea (Mallikarjuna et al., 2011b). In the AB-QTL derivatives of *C. cajanifolius*, DArT analysis was performed on BC_4F_1 progenies wherein introgressed lines showed about 94% genome recovery from the cultivated pigeonpea (Mallikarjuna et al., 2011b). SNP markers were employed in two backcross-derived populations originating from two wild species of pigeonpea. The introgressed lines were evaluated for three years at two locations. Both populations showed strong segregation distortion; hence, a mixed model QTL mapping was used to identify QTLs for yield attributing traits and novel alleles from the related species (Saxena et al., 2020a).

2.15 UTILIZATION OF WILD SPECIES IN PREBREEDING

Wild relatives have evolved through continuous natural selection in the unprotected environment, which is why they can survive better in adverse conditions like drought, cold, high temperature and salinity. They can also fight against natural epiphytotic and insect damage. However, it is difficult to utilize these genes because they are present in the pile of undesirable genes for crop production. Crop wild relatives of pigeonpea have been successfully utilized in the breeding program, particularly prebreeding, to derive lines with enhance traits of economic importance like biotic and abiotic resistance, quality, yield, and yield attributing traits. Climate change is a reality, and it is affecting the crop production through direct means by altering the air and soil temperature and precipitation or by changing the disease and insect-pest scenario. Climate-resilient varieties are the need of the hour. In this context, crop breeding objectives have been devised to utilize resistance sources from wild relatives for resistance to FW, *Phythopthora* blight, SMD, different species of nematodes, tolerance to pod borer complex, pod fly, pod bug, and abiotic stresses like heat, cold, drought, salinity, and waterlogging. Khoury et al. (2015) have reviewed the occurrence and availability of wild germplasm based on publications, documents, herbariums, catalogues of germplasm repositories, and botanical gardens. They have compiled the data in a standardized format and taxonomically verified following GRIN (2012) and Anonymous (2010) as references to identify the gaps in collections and prioritize the explorations with respect to the usefulness of wild taxa in breeding programs.

Prebreeding refers to all activities of genetic improvement that are designed to utilize the unadapted genetic resources, endowed with specific traits of interest from the primary, secondary, tertiary, and quaternary gene pools for incorporating useful alleles into the breeding lines either directly through selection and evaluation or indirectly by means of conventional and/or modern cytogenetical and biotechnological tools. Germplasm encompassing land races, obsolete cultivars, exotic collections, and crop wild relatives provide the basic materials for the prebreeding. Cytogenetic interventions and embryo cultures play an active role in harnessing the favorable alleles from the secondary and tertiary gene pools. Through prebreeding interventions, the climate-resilient alleles can be incorporated into the agronomically superior lines in multiple steps and present a way to combat the climate-induced biotic and abiotic stresses.

2.16 POTENTIAL USE IN BIOTIC STRESS RESISTANCE

Pigeonpea faces several biotic and abiotic stresses that contribute to significant yield losses worldwide (Rana et al., 2016). The major biotic stresses limiting the pigeonpea production worldwide include diseases, viz., FW (caused by *Fusariumudum*), *Phytophthora* stem blight (*Phytophthoracajani*), and SMV, as well as insects, viz., pod borers (*Helicoverpaarmigera* and *Marucavitrata*) and pod flies (*Melanagromyzachalcosoma*). Despite extensive breeding efforts, resistance or tolerance to these stresses remains limited within the cultivated gene pool (Sharma et al., 2020). Wild *Cajanus* species, being genetically distinct from cultivated pigeonpea, serve as valuable reservoirs of useful genes and alleles. These wild relatives harbor a wealth of genetic diversity and genes for biotic and abiotic stress resistance/tolerance (Tables 2.5 and 2.6), including traits that are not present in the cultivated gene pool. This genetic diversity can be used to enrich the primary gene pool of pigeonpea, enabling the development of new, broad-based cultivars with enhanced resistance to biotic stresses. However, cross-incompatibility barriers and the issue of linkage drag impeded wild Cajanus species in crossing programs; hence, they are not utilized adequately in breeding programs. Some of the key wild relatives include *C. platycarpus*, *C. scarabaeoides*, *C. acutifolius*, *C. cajanifolius*, and *C. albicans* (Khoury et al., 2015).

Among different wild relatives, *C. scarabaeoides* (L.) Millspaugh is easily crossable with the cultivated variety and successfully utilized for deriving ILs.

TABLE 2.5
Wild Relatives Tolerant to Biotic Stresses

Resistance/Tolerance to	Source of Resistance	References
Biotic Stress-Diseases		
Phytophthora blight	*C. platycarpus*, *C. scarabaeoides*	Mallikarjuna et al. (2011a) and Singh et al. (2020a)
Sterility mosaic disease	*C. albicans*, *C. lineatus*, *C. sericeus*, *C. volubilis*, *C. crassus*, *C. platycarpus*	Remanandan (1981) and Mallikarjuna et al. (2011a)
Fusarium wilt	*C. scarabaeoides* (ICP12707, ICP15689, ICP15692, ICP15732, ICP15744, ICP15748, and ICP15754)	Nagaraja et al. (2016)
Biotic Resistance-Insect Pests		
Pod borer	*C. acutifolius*, *C. scarabaeoides*, *C. sericeus*, *Rhynchosiaaurea*, *C. sericeus*, and *Flemingia bracteata*	Sharma et al. (2009)
Pod fly and pod wasp	*C. scarabaeoides* (ICPW 141, ICPW 278, and ICPW 280), *R. bracteata* (ICPW 214), *C. albicans* (ICPW 14), *Flemingia stricta* (ICPW 202)	Sharma et al. (2003)
Heterodera cajani	*C. scarabaeoides* (ICPWs 111 and 128), *Flemingia* spp. (ICPWs 194, 202, and 203), and *Rhynchosia rothii* (ICPW 257), *R. densijlora* (ICPW 224), and *R. aurea* (ICPW 210)	Sharma et al. (1993)
Rotylenchulus reniformis	*C. scarabaeoides* (ICPW 38 and 92), *R. aurea* (ICPW 210), *R. minima* (ICPW 237), and *R. rothii* (ICPW 257)	Sharma et al. (1993)

TABLE 2.6
Wild Relatives Tolerant to Abiotic Stresses

	Abiotic Stress	
Drought	*C. albicans*, *C. lineatus*, *C. scarabaeoides*, *C. sericeus*, *R. rothii*	Mallikarjuna, (2003)
Salinity	*C. acutifolius*, *C. scarabaeoides*, *C. sericeus*, *C. platycarpus*	Srivastava et al. (2006)
Photoperiod insensitivity	*C. platycarpus*, *R. rothii*	Mallikarjuna et al. (2006)

C. scarabaeoides serves as a valuable source of high-to-moderate resistance to pod borer, pod fly, pod wasp (Sharma et al., 2003, Pande et al., 2006, Ngugi-Dawit et al., 2020). The resistance mechanism of *C. Scarabaeoides* to pod borer include high levels of antiemesis for oviposition (host nonpreference for oviposition) and antibiosis (negative effect on survival and reproduction of insect; Sujana et al., 2008). In a separate study, Ngugi-Dawit et al. (2020) found that the insecticidal compound in *C. scarabaeoides* is heat-sensitive and identified the presence of insecticidal proteins, including lectin and cysteine proteinase inhibitors. Other wild relatives showing moderate-to-high resistance to pod borer are *C. platycarpus* and *C. acutifolius* (Shanower et al., 1997; Mallikarjuna et al., 2011a; Rathinam et al., 2019). Vanambathina et al. (2021) reported resistance to pod borer in *C. acutifolius*, *C. latisepalus*, *C. lanceolatus*, *C. pubescens*, and *C. reticulatus* var. *reticulatus*. Notably, the resistance to pod borer varies considerably among the different accessions of these wild relatives (Ngugi-Dawit et al., 2020).

At ICRISAT, Sharma et al. (2022) successfully introgressed resistance for pod borer from *C. acutifolius* and *C. scarabaeoides* into two pigeonpea cultivars, viz., ICP 8863 (Asha) and ICPL 87119 (Maruti) through simple and complex cross approaches, which resulted in four backcross prebreeding populations comprising 1,108, 288, 533, and 392 ILs. Out of these, 21 lines showed less than 50% pod borer damage in multilocation trials. Besides this, other wild *Cajanus* species, *C. sericeus* found resistant to pod fly; *C. scaraboides*, *C. acutifolius*, and *C. platycarpus* have resistance to bruchid (Saxena et al., 2018a).

Moderate and high resistance to Phytophthora stem blight pathogen has been reported in *C. platycarpus*, *C. sericeus*, and *C. scarabaeoides* (Mallikarjuna et al., 2005; Pande et al., 2006). (Mallikarjuna et al., 2005), attempted to introgress resistance genes from *C. platycarpus* (ICPW 61) to pigeonpea cultivar ICPL 85030 (a short duration variety) through the embryo rescue technique, resulting in 14 F_2 plants with resistance to Phytophthora pathogen.

FW resistance was observed in *C. scarabaeoides* accessions (ICP12707, ICP15689, ICP15692, ICP15732, ICP15744, ICP15748, and ICP15754; Nagaraja et al., 2016). Broad-based resistance to three isolates of SMV from Patancheru, Bangalore, and Coimbatore was found in accessions of *C. albicans* (ICP 15614, 15615, 15626, 15924, 15925, and 15926) and *C. scarabaeoides* (ICP 15684, 15688, 15700, 15701, 15703, 15725, 15734, 15736, 15737, 15740; Kumar et al., 2005). *C. acutifolius* and *C. cajanifolius* were also reported to possess resistance against SMD (Khoury et al., 2015; Patil & Kumar, 2015). Sharma et al. (2019) generated 75 ILs from crosses between ICPL 87119 (Asha) and wild *C. Cajanifolius* (ICPW 29)/ *C. acutifolius* (ICPW 12), out of these 29 ILs exhibited high resistance for FW and SMD with >10% disease incidence.

2.17 POTENTIAL USE IN ABIOTIC STRESS RESISTANCE

Various abiotic stresses significantly affect pigeonpea production and productivity worldwide, including drought, waterlogging, salinity, temperature extremes (both heat and cold), photoperiod sensitivity, and metal toxicity (Saxena et al., 2019, 2021). These stresses are anticipated to intensify in the future due to the ongoing climate change scenario, posing additional challenges for pigeonpea cultivation (Nandini et al., 2022). The wild relatives of pigeonpea also exhibit tolerance to a few abiotic stresses, offering potential genetic resources for improving the resilience of cultivated pigeonpea to adverse environmental conditions (Khoury et al., 2015). *C. platycarpus, C. scarabaeoides*, and *C. sericea* exhibit tolerance to salinity

(Subbarao et al., 1990; Srivastava et al., 2006). Khoury et al. identified the presence of potential traits for waterlogging tolerance in *C. crassus*, *C. lineatus*, and *C. sericeus*, and drought tolerance in *C. acutifolius*, *C. cinereus*, *C. lanceolatus, C. latisepalus*, *C. lineatus*, and *C. sericeus*.

2.18 EXPLOITATION IN CYTOPLASMIC NUCLEAR MALE STERILITY

Outcrossing of cultivars is generally discouraged, as it results in variation, which may lead to reduced yield or loss of desired traits. Genetic male sterility exists in many plant species. Initially, GMS-based hybrids were developed, which were high yielding, but they were unable to reach farmers' fields due to the inherent challenges associated with maintaining the male sterile line and producing hybrid seeds. Given the challenges in large-scale hybrid seed production with GMS hybrids, the development of CMS became essential. CMS is maternally inherited and linked to specific mitochondrial genes (Budar & Pelletier, 2001).The approach involved inducing CMS by incorporating the pigeonpea genome into wild cytoplasm through hybridization. It was hypothesized that the interaction between the wild cytoplasm and the cultivated nuclear genome would result in male sterility (Saxena & Nadarajan, 2010). This technology has emerged as a promising strategy for significantly boosting the productivity of pigeonpea.

Different pigeonpea wild relatives, viz., *C. sericeus*, *C. scarabaeoides*, *C. volubilis*, *C. cajanifolius*, *C. lineatus*, *C. platycarpus*, *C. reticulatus*, and *C. acutifolius*, have been used as cytoplasm donors for the CMS system (Saxena et al., 2018a). Srivastava et al. (2018) used *C. cajanifolius* (ICPA 2039) as the CMS donor and backcrossed it with ICPL 20176 to develop a stable male sterile line-ICPA 2043. *C. cajan* based male sterile lines have been developed from crosses between cultivated pigeonpea accession as the female parent and wild relatives (*C. acutifolius*, *C. lanceolatus*) as the pollen parent (Mallikarjuna & Saxena 2005; Srikanth et al., 2015).

2.19 SUCCESS OF BACKCROSS IN WILD GENE INTROGRESSION AND TRAIT RECOVERY

Phythopthora blight has emerged as a disease of economic importance in the light of climate change. A good source of resistance is lacking the early and medium maturity group. Hence, *C. cajanifolius* (ICPW 29) and *C. acutifolius* (ICPW 004, and ICPW 12) belonging to the cross-compatible secondary gene pool and *C. platycarpus* (ICPW 68) belonging to the cross-incompatible tertiary gene pool were used as donors to incorporate resistance in the cultivated genotypes—namely, ICPL 87119 and ICPL 85010. The crosses, viz., ICPL 87119/C. *acutifolius* ICPW12 and ICPW 68/*C. platycarpus* ICPL85010, produced 18 ILs (13 from C. *acutifolius* and 5 from *C. platycarpus*) with moderate-to-high levels of resistance. One accession of *C. acutifolius* with antixenosis for oviposition, high expressions of antibiosis tannins and polyphenols in pods, and one accession of *C. scarabaeoides* with Type C and D trichomes (Sharma et al., 2009) were used to generate four-way-wide crosses with the aim to transfer pod borer resistance in the cultivated pigeonpea. Besides generating backcross ILs for resistance against biotic stresses, a considerable variation was created for days to flowering and growth habit, as well as for pod, seed traits, and grain nutrient traits like protein content, grain iron (Fe), zinc (Zn), calcium (Ca), and magnesium (Mg).

The crosses involving Asha and *C. acutifolius* (ICPW 12) and *C. cajanifolius* (ICPW 29) produced breeding lines with a broader genetic base, enhanced resistance, and yield (Mallikarjuna et al., 2017; Sharma et al., 2022). Interspecific crosses are known to produce agronomically undesirable segregants, therefore, for alleviating the linkage drag; the hybrids of wild species were backcrossed to the recipient parent, followed by selfing to fix the introgressed traits in the wild derivatives. The demand for early maturity is rising in light of shift in rainfall pattern. Considerable variation in the interspecific hybrid derived IL were observed for maturity (141–176 days). The short duration photo-insensitive lines can be utilized in program for deriving high yielding short duration varieties suitable for pigeonpea wheat cropping system without penalizing yield of succeeding wheat crop in the central and north India. The ILs (ICPL 15010, ICPL 15019, ICPL 15023, ICPL 15021, ICPL 15077, and ICPIL 17160) of medium-early maturity (150–160 days) can also be good fit in North and Central India in the wheat belt. ILs (ICPL 15085, ICPL 15072, ICPL 15062, ICPL 15067, ICPIL 17164, ICPIL 17165, and ICPIL 17169) showed disease resistance against PSB and good yield (20%–50% superiority over Asha). Based on 3 years of testing, 28 promising ILs in the background of Asha were identified for promotion. Notable among them were ICPP171012, ICPP 171004, ICPP 171102, ICPP 171087, ICPP 171006, and ICPP 171050 (Sharma et al., 2020). These ILs have entered in multilocation trials and have promise as future varieties or at least as donors in breeding programs for developing high-yield varieties for central India, particularly in the cotton and soybean belts, as companion intercrops. The diversity generated in ILs has broadened the genetic base. The results indicate that interspecific cross has generated traits of interest in high frequency and created additional variability in diverse genetic backgrounds to address the issues of climate change (Sharma, 2017). Two *C. acutifolius* derived ILs (ICPL 15028 and ICPL 15036) and one *C. cajanifolius* derived IL (ICPL 15084) derived from ICPW 29 were tested in the AICRP trials.

2.20 CONCLUSION

The productivity of the majority of crops are being adversely affected by global warming, a phenomenon that is a rising trend. Mishra et al. (2017), have studied the impact of temperature and rainfall on pigeonpea productivity over a period of 32 years in Gujrat. The regression analysis indicated that a 1°C rise in mean maximum temperature will reduce the yield by 20.8%. Kwena et al. (2021) have analyzed the maize-pigeonpea cropping system in Kenya, which was projected to witness a rise in 2°C and 11% rainfall by 2100. They predicted that there will be 4%–20% (depending on models used) decline in yield of short duration varieties of pigeonpea. Yadav et al. (2021)modeled the impact of climate change in short and long duration pigeonpea during a period of 85 years from 2010 to 2095. They predicted delayed anthesis and maturity and decreased grain yield (14%–66% for long and short duration genotypes, respectively in reference varieties).

The impact of global warming is multidimensional and leads to climate change in all ecologies. The incidence of *Phythopthora* blight in central India has been on the rise during the past few cropping seasons. Unusual temperature rises during February and April have become a regular feature for major pigeonpea crop ecologies. Late seasonal precipitation also affects the phenology and maturity of pigeonpea in most agro-climatic zones in India. The only way to cope with vagaries of climate and altered crop ecologies (terminal heat and drought, increased intensity, and incidence of pathogens and nematode insect pests, emergence of new pests and diseases, waterlogging, salinity and cold) is to develop climate-resilient biotic and abiotic stress-tolerant varieties to provide food and nutritional security to the burgeoning population. The cultivated germplasm and breeding populations have narrow genetic bases for many economic traits, including disease and insect-pest resistance, nematode resistance, salinity, drought, heat, waterlogging tolerance, and true photo-thermo insensitivity with short annual life cycles. Moreover, pathogens and insect pests evolve over time, and space to render the present-day, disease-resistant cultivars is susceptible.

Under the prevailing circumstances, there is a need to search resistance to biotic and abiotic stresses in crop wild relatives. Climate change warrants developing cultivars with increased plasticity and ameliorating climate resilience by infusing genes for biotic and abiotic stress tolerance from the germplasm in general and wild relatives in particular for combating various climate change–induced environmental stresses (Khoury et al., 2015; Sharma, 2017). Though prebreeding provides opportunities to introgress the genes of interest in the cultivated pigeonpea for deriving wild derivatives, but crop wild relatives were not utilized to the fullest potential in spite of the enormous potential to faceoff the challenges posed by climate change–induced biotic and abiotic stresses. Cross-incompatibility, very poor successes in crossing, low recovery of traits of interest, high linkage drag, and agronomically inferior progenies with poor adaptability are some of the reasons for under exploitation of wild relatives in regular breeding programs (Sharma et al., 2013). Difficulties associated with possibility and pod set can be resolved through appropriate interventions (Mallikarjuna & Moss, 1995; Mallikarjuna 1998; Mallikarjuna et al., 2011a, 2011b), whereas issues of linkage drag and adaptability can be addressed through modern molecular tools. In fact, wild relatives from the secondary, tertiary, and quaternary gene pools have been utilized and documented by Mallikarjuna et al. (2017), a remarkable achievement for a legume crop pigeonpea. Several economic traits have been introgressed successfully into the cultivated pigeonpea from many secondary and few tertiary gene pool species.

Advance breeding lines were generated with useful alien genes broadening the genetic base. Availability of molecular tools, advances in next-generation sequencing technology, and efficient and precise marker systems and genomic tools have paved the way for overcoming linkage drag effectively to generate superior wild derivatives with a desirable agronomic base. Markers including QTLs for biotic and abiotic stress tolerance and plant architectural and maturity traits (CMS systems, multiple disease resistance, and pest resistance) were developed. The success story of introgression of useful genes (agronomic traits and disease and insect-pest resistance) from wild species into agronomically superior genotypes have demonstrated the utility of crop wild relatives in breeding program in the development of climate-resilient cultivars (Reddy et al., 1997; Mallikarjuna et al., 2017; Sharma, 2017; Saxena et al., 2018a; Sharma et al., 2020).

Concerted efforts might result in significant success in generating intergeneric fertile hybrids for utilizing useful genes from *Rhynchosia* and other genera. Coordinated and focused multidisciplinary efforts are required to harness the full potential of wild relatives of pigeonpea, particularly for deriving truly photo-thermo insensitive short life span annual pigeonpea ILs from *C. platycarpus* and extremely hardy biotic and abiotic tolerant climate-resilient pigeonpea plants from *Rhynchosia bracteata*.

REFERENCES

Akdemir D, Beavis W, Fritsche-Neto R, Singh AK, Isidro-Sánchez J (2019) Multi-objective optimized genomic breeding strategies for sustainable food improvement. *Hered* 122(5): 672–683.

Anonymous (2010) WFO Plant List (wfoplantlist.org).

Anonymous (2022) *AICRP on Pigeonpea annual report 2021–2022*. ICAR-IIPR, Kanpur.

Anonymous (2024) Directorate of Economics and Statistics (DES), Ministry of DAC&FW, GOI. website, https://desagri.gov.in/about-us/profile-of-directorate-of-economics-statistics

Ariyanayagam RP, Spence JA (1978) A possible gene source for early, day-length neutral pigeon peas, Cajanus cajan (L.) M illsp. *Euphytica* 27: 505–509.

Ariyanayagam RP, Rao AN, Zaveri PP (1993) Gene-cytoplasmic male sterility in pigeonpea. *International Pigeonpea Newsletter* 18:7–11.

Ariyanayagam RP, Rao AN, Zaveri PP (1995) Cytoplasmic-Genic Male-Sterility in interspecific matings of Cajanus. *Crop Sci* 35(4): 981–985.

Bajpai GC, Singh J, Tewari SK (2003) Wide hybridization and its genetic and cytogenetic consequences in pigeonpea–A review. *Agricultural Reviews* 24(4): 265–274.

Bakshu LM, Raju RRV (2001) Antimicrobial activity of *Rhynchosia beddomei*. *Fitoterapia* 72(5): 579–582.

Bhatia GK, Gupta SC, Green JM, Sharma D (1981) Estimates of natural cross pollination in Cajanus cajan (L.) Millsp. Several experimental approaches. *ProceIntWorkshop Pigeonpeas* 2: 129–136.

Bohra A, Mallikarjuna N, Saxena K, Upadhyaya HD, Vales MI, Varshney RK (2010) Harnessing the potential of crop wild relatives through genomics tools for pigeonpea improvement. *J Plant Biol* 37(1): 83–98.

Bohra A, Saxena RK, Gnanesh BN, Saxena KB, Byregowda M, Rathore A, KaviKishor PB, Cook DR, Varshney RK (2012) An intra-specific consensus genetic map of pigeonpea [Cajanus cajan (L.) Millspaugh] derived from six mapping populations. *Theor Appl Genet* 125: 1325.

Bohra A, Jha R, Pandey G, Patil PG, Saxena RK, Singh IP, Mishra RK, Mishra A, Singh F, Varshney RK & Singh NP (2017) New hypervariable SSR markers for diversity analysis, hybrid purity testing and trait mapping in pigeonpea [Cajanus cajan (L.) Millspaugh]. *Front Plant Sci* 8: 377.

Bohra A, Saxena KB, Varshney RK, Saxena RK (2020) Genomics-assisted breeding for pigeonpea improvement. *Theor Appl Genet* 133(5): 1721–1737.

Bohra A, Kilian B, Sivasankar S, Caccamo M, Mba C, McCouch SR, Varshney RK (2022a) Reap the crop wild relatives for breeding future crops. *Trends Biotechnol* 40(4): 412–431.

Bohra A, Tiwari A, Kaur P, Ganie S, Raza A, Roorkiwal M, Mir R, Fernie A, Smykal P, Varshney R (2022b) The key to the future lies in the past: Insights from grain legume domestication and improvement should inform future breeding strategies. *Plant Cell Physiol* 63(11): 1554–1572.

Bohra A, Tiwari A, Pareek S, Joshi R, Satheesh Naik SJ, Kumari K, Verma RL, Parihar AK, Patil PG, Dixit GP (2025) Past and future of cytoplasmic male sterility and heterosis breeding in crop plants. *Plant Cell Rep* 22: 33.

Bozdag GO, Ono J, Denton JA, Karakoc E, Hunter N, Leu JY, Greig D (2021). Breaking a species barrier by enabling hybrid recombination. *Curr Biol* 31(4): R180–R181.

Brock MT, Rubin MJ, Penna DD, Weinig C (2020) A nested association mapping panel in *Arabidopsis thaliana* for mapping and characterizing genetic architecture. *G3* 10 (10): 3701–3708.

Budar F, Pelletier G (2001) Male sterility in plants: occurrence, determinism, significance and use. *Life Sci.* 324: 543–550.

Cherian CA, Mallikarjuna N, Jadhav DR, Saxena KB (2006) Open flower segregants selected from Cajanus platycarpus crosses. *J SAT Agric Res* 2: 1–2.

Choudhary A, Sultana R,Pratap A, Nadarajan N, Jha U (2011). Breeding for abiotic stresses in pigeonpea. *J. Food Legum* 24: 165–174.

De DN (1974) Pigeonpea. In: J. Hutchinson (Ed.) *Evolutionary Studies in World Crops, Diversity and Change in the Indian Subcontinent* (pp. 79–87). Cambridge University Press, London.

Dhanju MS, Gill BS (1985) Intergeneric hybridization between Cajanus cajan and Atylosia platycarpa.

Drabu S, Chaturvedi S, Sharma M (2011) Analgesic activity of methanolic extract from aerial parts of Rhynchosia capitata DC.

Dreissig S, Mascher M, Heckmann S (2019) Variation in recombination rate is shaped by domestication and environmental conditions in barley. *Mol Biol Evol* 36(9): 2029–2039.

Dundas IS (1984) Cytogenetic investigations involving pigeonpea (Cajanus cajan (L.) Millsp.) and some native Australian Atylosia species.

Dundas IS, Britten EJ, Byth DE, Gordon GH (1988) The use of tangled pachytene chromosomes for karyotype analysis in Atylosia. *J Hered* 79(3): 175–178.

Dundas IS, Saxena KB, Singh L, Faris DG (1989) Crossability relationships and possible bridging routes involving Asian and Australian Atylosia species.

Fuller DQ, Murphy C, Kingwell-Banham E, Castillo CC,Naik S (2019) Cajanus cajan (L.) Millsp.origins and domestication: the South and Southeast Asian archaeobotanical evidence. *Genet Resour Crop Evol* 66(6): 1175–1188.

Ganiga J, Sharma M, Reddy N, Telangre R (2020) Temporal progress and spatial distribution of phytophthora blight of pigeonpea in Deccan plateau of India. *J Agrometeorol* 22: 63–66.

Gepts P, Papa R (2003) Possible effects of (trans) gene flow from crops on the genetic diversity from landraces and wild relatives. *Environ Biosafety Res 2*(2): 89–103.

Ghritlahre SK, Singh A, Singh IP, Singh F, Katiyar PK, Singh NP (2024) Current Status and Way Forward for Scaling up Productivity of Pigeonpea in Madhya Pradesh: A Review. *Agric Rev* 45(2).

Gill LS, Husaini SWH (1986) Cytological observations in Leguminosae from southern Nigeria. *Willdenowia* 15: 521–527.

Gnanesh BN, Bohra A, Sharma M, Byregowda M, Pande S, Wesley V, Varshney RK (2011) Genetic mapping and quantitative trait locus analysis of resistance to sterility mosaic disease in pigeonpea [Cajanus cajan (L.) Millsp.]. *Field Crops Res* 123(2): 53–61.

Gowda CL, Upadhyaya HD, Sharma S, Varshney RK, Dwivedi SL (2013) Exploiting genomic resources for efficient conservation and utilization of chickpea, groundnut, and pigeonpea collections for crop improvement. *The Plant Genome*, 6(3): 1–28.

Greilhuber J, Obermayer R (1998) Genome size variation in *Cajanus cajan* (Fabaceae): a reconsideration. *Plant Syst Evol* 212: 135–141.
GRIN (Germplasm Resources Information Network), USDA, ARS, National GeneticResources Program, 2012. National Germplasm Resources Laboratory, Beltsville, Maryland. Online resource.http://www.ars-grin.gov/~sbmljw/cgi-bin/index.pl?language=en (accessed March 2013).
Harlan JR, de Wet JM (1971) Toward a rational classification of cultivated plants. *Taxon*, 20(4): 509–517.
Henderson IR, Bomblies K (2021) Evolution and plasticity of genome-wide meiotic recombination rates. *Annual Regenet* 55(1): 23–43.
Howard A, Howard GLC, Khan AR (1919) Studying the pollination of Indian crops I. *Memoirs Dept Agri* 10: 195–200.
Jayaprakash P, Sarla N (2001) Development of an improved medium for germination of Cajanus cajan (L.) Millsp.pollen in vitro. *J Exp Bot* 52(357): 851–855.
Jha SS, Ohri D (1996) Phylogenetic relationships of Cajanus cajan (L.) Millsp.(pigeonpea) and its wild relatives based on seed protein profiles. *Genet Resour Crop Evol* 43: 275–281.
Kassa MT, Penmetsa RV, Carrasquilla-Garcia N, Sarma BK, Datta S, Upadhyaya HD, Cook DR (2012) Genetic patterns of domestication in pigeonpea (Cajanus cajan (L.) Millsp.) and wild Cajanus relatives. *PloS One* 7(6): e39563.
Khoury CK, Castañeda-Alvarez NP, Achicanoy HA, Sosa CC, Bernau V, Kassa MT, Struik PC (2015) Crop wild relatives of pigeonpea [Cajanus cajan (L.) Millsp.]: Distributions, ex situ conservation status, and potential genetic resources for abiotic stress tolerance. *Biol Conserv* 184: 259–270.
Kimaro D, Melis R, Sibiya J, Shimelis H, Shayanowako A (2020) Analysis of genetic diversity and population structure of pigeonpea [Cajanus cajan (L.) Millsp] accessions using SSR markers. *Plants* 9(12): 1643.
Kollipara KP, Singh L, Hymowitz T (1994) Genetic variation of trypsin and chymotrypsin inhibitors in pigeonpea [Cajanus cajan (L.) Millsp.]and its wild relatives. *Theor Appl Genet* 88: 986–993.
Koumproglou R, Wilkes TM, Townson P, Wang XY, Beynon J, Pooni HS, Kearsey MJ (2002) STAIRS: a new genetic resource for functional genomic studies of Arabidopsis. *Plant J* 31(3): 355–364.
Krishna TG, Reddy LJ (1982) Species affinities between Cajanus cajan and some Atylosia species based on esterase isoenzymes. *Euphytica* 31: 709–713.
Krishnaswamy N, Rangaswami Ayyangar GN (1935) Chromosome numbers in Cajanus indicus, Spreng. *Curr Sci* 3: 614–615.
Kudapa H, Bharti AK, Cannon SB, Farmer AD, Mulaosmanovic B, Kramer R, Varshney RK (2012) A comprehensive transcriptome assembly of pigeonpea (Cajanus cajan L.) using Sanger and second-generation sequencing platforms. *Mol Plant* 5(5): 1020–1028.
Kumar LSS, Thombre MV, D'Cruz R (1958) Cytological studies of an intergeneric hybrid of Cajanus cajan (Linn.) Millsp. and Atylosia lineata, W. & A. *Proceedings/Indian Academy of Sciences*, 47(4): 252–262.
Kumar PS, Subrahmanyam NC, Faris DG (1984) Nucleolar variation in a pigeonpea intergeneric hybrid: evidence for allosyndetic recombination. *Can J Genet Cytol* 26(5): 499–505.
Kumar PS, Subrahmanyam NC, Faris DG (1985a) Intergeneric hybridization in pigeonpea. I. Effect of hormone treatments. *Field Crops Res* 10: 315–322.
Kumar PS, Subrahmanyam NC, Faris DG (1985b) Morphological variation and inheritance in a pigeonpea intergeneric hybrid. *Curr Sci* 54(7).
Kumar PS, Subrahmanyam NC,Faris DG (1990) Intergeneric hybridization in pigeonpea. II. Effect of cultivar on crossability and hybrid fertility. *Field Crops Res* 24: 189–193.

Kumar PL, Latha TKS, Kulkarni NK, Raghavendra N, Saxena KB, Waliyar F, Rangaswamy KT, Muniyappa V, Doriswamy S, Jones AT (2005) Broad-based resistance to pigeonpea sterility mosaic disease in wild relatives of pigeonpea (Cajanus: Phaseoleae). *Ann Appl Biol* 146(3): 371–379.

Kumar K, Anjoy P, Sahu S, Durgesh K, Das A, Tribhuvan KU, Sevanthi AM, Joshi R, Jain PK, Singh NK, Rao AR, Gaikwad K (2022) Single trait versus principal component based association analysis for flowering related traits in pigeonpea. *Sci Rep* 2(1): 10453.

Kwena KM, Karuku GN, Ayuke FO, Esilaba AO (2021) Impact of climate change on maize and pigeonpea yields in semi-arid Kenya. In *Environ Issues and Sust Dev*. IntechOpen.

Lakshmi M, Senthilkumar P, Parani M, Jithesh MN, Parida AK (2000) PCR-RFLP analysis of chloroplast gene regions in Cajanus (Leguminosae) and allied genera.*Euphytica* 116: 243–250.

Leprévost T, Boutet G, Lesné A, Rivière JP, Vetel P, Glory I, Miteul H, Le Rat A, Dufour P, Regnault-Kraut C, Sugio A, Lavaud C, Pilet-Nayel ML (2023) Advanced backcross QTL analysis and comparative mapping with RIL QTL studies and GWAS provide an overview of QTL and marker haplotype diversity for resistance to Aphanomyces root rot in pea (Pisum sativum). *Front Plant Sci* 14: 1189289.

Liu C, Ding X, Wu Y, Zhang J, Huang R, Li X, Liu G, Liu P (2024) Chromosome-scale reference genome of an ancient landrace: unveiling the genetic basis of seed weight in the food legume crop pigeonpea (Cajanus cajan). *Hortic Res* 11(9): 201.

Mahadeva Swamy HK, Amaresh Nunavath A, Bajpai GC, Verma SK, Gopalareddy K, Maruthi RT, Sreenivas V, Behera C (2024) Genetic diversity and association analysis for agronomic and yield related traits in interspecific progenies of pigeonpea (Cajanus cajan). *J Sci Res Rep* 30(6): 679–689.

Majumder ND, Singh F (2005) Pigeonpea improvement in India. In *Souvenir 4th International Food Legume Research Conference*. (18–22).

Mallikarjuna N (1998) Ovule culture to rescue aborting embryos from pigeonpea (Cajanus cajan (L) Millspaugh) wide-crosses. *Indian J Exp Biol* 36(2): 225–228.

Mallikarjuna N (2003) Wide hybridization in important food legumes. In: P.K. Jaiwal, R.P. Singh (Eds.) *Improvement Strategies of Leguminosae Biotechnology*, pages 155–170. Kluwer Acad Publishers.

Mallikarjuna N (2007) Production of fertile progeny from interspecific incompatible cross Cajanus platycarpus × C. cajan. *J SAT Agric Res* 3(1): 1–3.

Mallikarjuna N, Moss JP (1995) Production of hybrids between Cajanus platycarpus and Cajanus cajan. *Euphytica* 83: 43–46.

Mallikarjuna N, Saxena KB (2002) Production of hybrids between *Cajanus acutifolius* and C. cajan. *Euphytica* 124(1): 107–110.

Mallikarjuna N, Saxena KB (2005) A new cytoplasmic nuclear male-sterility system derived from cultivated pigeonpea cytoplasm. *Euphytica* 142: 143–148.

Mallikarjuna N, Jadhav D, Reddy MV, Dutta-Tawar U (2005) Introgression of Phytophthora blight disease resistance from Cajanus platycarpus into short duration pigeonpea [Cajanus cajan (L.) Millsp.]. *Indian J Genet Plant Breed* 65(04): 261–263.

Mallikarjuna N, Jadhav D, Reddy P (2006) Introgression of Cajanus platycarpus genome into cultivated pigeonpea. C. cajan. *Euphytica* 149: 161–167.

Mallikarjuna N, Sharma HC, Upadhyaya HD (2007) Exploitation of wild relatives of pigeonpea and chickpea for resistance to Helicoverpa armigera. *J SAT Agric Res* 3(1): 4.

Mallikarjuna N, Saxena KB, Jadhav DR (2011a) Cajanus. In: C Kole (Ed.) *Wild Crop Relatives: Genomic and Breeding Resources*. Springer, Heidelberg, Berlin pp 21–33.

Mallikarjuna N, Senapathy S, Jadhav DR, Saxena K, Sharma HC, Upadhyaya HD, Rathore A, Varshney R (2011b). Progress in the utilization of Cajanus platycarpus (Benth.) Maesen in pigeonpea improvement. *Plant Breed* 130(5): 507–514.

Mallikarjuna N, Saxena K, Lakshmi J, Varshney R, Srikanth S, Jadhav D (2012) Differences between Cajanus cajan (L.) Millspaugh and C. cajanifolius (Haines) van der Maesen, the progenitor species of pigeonpea. *Genet Resour Crop Evol* 59: 411–417.

Mallikarjuna N, Srikanth S, Kumar CS, Srivastava RK, Saxena RK, Varshney RK (2014) Pigeonpea. In: M Singh et al. (Eds.) *Broadening the Genetic Base of Grain Legumes.* Springer, India Edition 1: 149–159.

Mallikarjuna N, Saxena RK, Byre Gowda M, Varshney RK (2017) Wide crossing technology for pigeonpea improvement. In: Varshney RK, Saxena RK, Jackson S (Ed.) *The Pigeonpea Genome, Compendium of Plant Genomes.* Springer, Heidelberg, Berlin pp 31–39.

Megha SN, Sharma M, Langyan S, Singh NK (2024) Genome wide association study of antioxidant activity in pigeonpea germplasm. *Discover Food* 4(1): 82.

Meuwissen TH, Hayes BJ, Goddard M (2001) Prediction of total genetic value using genome-wide dense marker maps. *Genetics* 157(4): 1819–1829.

Mishra SK, Panigrahi J (2017) Identification of inter-specific hybrid between Cajanus cajan (L.) and C. cajanifolius (H.) using cyto-morphological and DNA markers. *Electron J Plant Breed* 8(1): 63–71.

Mishra S, Singh R, Kumar R, Kalia A, Panigrahy SR (2017) Impact of climate change on pigeon pea. *Econ Aff* 62(3): 455–457.

Moss JP, Singh AK, Sastri DC, Dundas IS (1988) Wide hybridization in legumes at ICRISAT.

Mudaraddi B, Saxena KB, Saxena RK, Varshney RK (2013) Molecular diversity among wild relatives of Cajanus cajan (L.) Millsp. *Afr J Biotechnol* 12(24): 3797–3801.

Mukhopadhyay S (1986) Intergeneric relationship between three genera of Leguminosae. *J Indian Bot Soc* 65(1): 124–129.

Nadimpalli RG, Jarret RL, Phatak SC, Kochert G (1993) Phylogenetic relationships of the pigeonpea (Cajanus cajan) based on nuclear restriction fragment length polymorphisms. *Genome* 36(2): 216–223.

Nagaraja NR, Kumar MA, Prasad CM, Sinha P, Jain J (2016) Genetic diversity studies and screening for Fusarium wilt (Fusarium udum Butler) resistance in wild pigeonpea accessions, Cajanus scarabaeoides (L.) Thouars. *Ind J Plant Genet Resour* 29(02): 121–129.

Naithani SP (1941) Cytological studies on Indian pulses. Part I-the somatic chromosomes and thé prochromosomes of Cajanus. *Proc Natl Acad Sci India* 11: 67–73.

Nandini B, Venkatesh Reddy UG, Mallikarjuna BP, Manu B, Vaijayanthi PV, Ashwini M, Surendra P, Vijayakumar AG, Kumar CJ, Manjunath L, Ghosh S, Ganguly S, Kundu Chaudhuri R, Chakraborti D (2022) Genomic design for abiotic stress resistance in pigeonpea. In:C Kole (Eds.) *Genomic Designing for Abiotic Stress Resistant Pulse Crops.* Cham: Springer International Publishing. 169–248.

Ngugi-Dawit A, Hoang TML, Williams B, Higgins TJ, Mundree SG (2020) A wild Cajanus scarabaeoides (L.) Thouars, IBS 3471, for improved insect-resistance in cultivated pigeonpea. *Agron* 10(4): 517.

Odeny DA (2007) The potential of pigeon pea (Cajanus cajan (L.) Millsp.) in Africa. *Nat Resour Forum* 31(4): 297–305.

Odeny DA, Jayashree B, Ferguson M, Hoisington D, Crouch J, Gebhardt C (2007) Development, characterization and utilization of microsatellite markers in pigeonpea. *Plant Breed* 126(2): 130–136.

Ohri D, Singh SP (2002) Karyotypic and genome size variation in Cajanus cajan (L.) Millsp. (pigeonpea) and some wild relatives. *Genet Resour Crop Evol* 49: 1–10.

Ohri D, Jha SS, Kumar S (1994) Variability in nuclear DNA content within pigeonpea. *Pl Syst Evol* 189: 211–216.

Pande S, Sharma M, Pathak M, Rao JN, Tomar O (2006) Resistance to Phytophthora blight in the improved pigeonpea lines at ICRISAT, Patancheru, India. *J SAT Agri Res* 2(1): 1–3.

Patil BL, Kumar PL (2015) Pigeonpea sterility mosaic virus: a legume-infecting Emaravirus from South Asia. *Mol Plant Pathol* 16(8): 775–786.

Pazhamala L, Saxena RK, Singh VK, Sameerkumar CV, Kumar V, Sinha P, Varshney RK (2015) Genomics-assisted breeding for boosting crop improvement in pigeonpea (Cajanus cajan). *Front Plant Sci* 6: 50.

Pundir RPS, Singh RB (1985a) Cytogenetics of F1 hybrids between Cajanus and Atylosia species and its phylogenetic implications. *Theor Appl Genet* 71: 216–220.

Pundir RPS, Singh RB (1985b) Crossability relationships among Cajanus, Atylosia and Rhynchosia species and detection of crossing barriers. *Euphytica* 34: 303–308.

Pundir RPS, Singh RB (1986) Karyotype analysis of Cajanus, Atylosia and Rhynchosia species. *Theor Appl Genet* 72: 307–313.

Pundir RPS, Singh RB (1987) Possibility of genetic improvement of pigeonpea (Cajanus cajan (L.) Millsp.) utilizing wild gene sources. *Euphytica* 36: 33–37.

Purohit Y, Patel PR (2025) Wide hybridization between *Cajanus cajan* and *Cajanus scarabaeoides*. *Int J Adv Biochem Res* 9: 811–815.

Rabbani MT, Nayak G (2023) Recruiting distant hybridization for reshaping meiotic recombination. *Plant Breed Biotech* 11: 168–184.

Ramanadan P (1990) Pigeonpea: Genetic resources. In: YL Nene, SD Hall, VK Shelia (Ed.) *The Pigeonpea*, pp 349–374. CAB International, Wallingford, Oxon, UK.

Rana DS, Dass A, Rajanna GA, Kaur R (2016) Biotic and abiotic stress management in pulses. *Indian J Agron* 61(4): S238–S248.

Rathinam M, Mishra P, Vasudevan M, Budhwar R, Mahato A, Prabha AL, Sreevathsa R (2019) Comparative transcriptome analysis of pigeonpea, Cajanus cajan (L.) and one of its wild relatives Cajanus platycarpus (Benth.) Maesen. *PLoS One* 14(7): e0218731.

Ravali P, Kalaiyarasi R, Hemavathy AT, Sudha M (2022) Cytological investigation on pre and post fertilization barriers in interspecific cross of pigeonpea. *Electron J Plant Breed* 13(2): 724–730.

Ravikumara BM, Ramanagouda G, Naik MK, Telangre R, Sharma M (2022) Distribution and pathogenic diversity in *Fusarium udum* Butler isolates: the causal agent of pigeonpea *Fusarium* wilt. *BMC Plant Biol* 22: 147.

Reddy LJ (1984) Fusion of centromeres and star formatons at pachytene of Cajanus× Atylosia hybrids. *Heredity* 53(2): 435–439.

Reddy LJ (1990) Pigeonpea: Morphology. In: YL Nene, DH Susan, VK Sheila (Eds.) *The Pigeonpea*, pp 47–87. CAB International, UK.

Reddy LJ, De DN (1983) Cytomorphological studies in Cajanus cajan x Atylosia lineata. *Indian J Genet Plant Breed* 43(1): 96–103.

Reddy LJ, Green JM, Sharma D (1981) Genetics of Cajanus cajan (L.) Mills sp. × Atylosia spp. In: *Proceedings of the International workshop on Pigeonpeas, 15–19 Dec 1980. ICRISAT Center*, Vol 2, pp. 39–50. Patancheru, AP, India.

Reddy LJ, Saxena KB, Jain KC, Umaid Singh US, Green JM, Sharma D, Nene YL(1997) Registration of high-protein pigeonpea elite germplasm ICPL 87162.

Reddy LJ, Kameswara Rao N, Saxena KB (2001) Production and characterization of hybrids between Cajanus cajan× C. reticulatus var. grandifolius. *Euphytica* 121(1): 93–98.

Remanandan P (1981) The wild genepool of Cajanus at ICRISAT, present and future. In: *Proceedings of International Workshop on Pigeonpeas*. Volume 2, 15–19 December 1980, ICRISAT Center, ICRISAT, Patancheru, AP, India. pp 29–38.

Roy B (1933) Studies in the development of the female gametophyte in some leguminous crop plants of India.

Roy A, De DN (1967) Intergeneric hybridization of Cajanus and Alylosio. *Sci Cult* 31: 93–95.

Royal Botanic Gardens, Kew https://powo.science.kew.org/taxon/urn:lsid:ipni.org:names:331442–2

Saxena KB (2005) Pigeonpea. In: RJ Singh, PP Jauhar (Ed.) *Genetic Resources, Chromosome Engineering, and Crop Improvement* (pp 86–115). CRC Press, Taylor and Francis, New York.
Saxena KB, Kumar RV (2003) Development of a cytoplasmic nuclear male-sterility system in pigeonpea using C. scarabaeoides (L.) Thouars. *Ind J Genet Plant Breed 63*(03): 225–229.
Saxena KB, Kumar RV (2010) Insect-aided natural out-crossing in four wild relatives of pigeonpea. *Euphytica* 173: 329–335.
Saxena KB, Nadarajan N (2010) Prospects of pigeonpea hybrids in Indian agriculture. *Electron J Plant Breed* 1(4): 1107–1117.
Saxena KB, Faris DG, Singh U, Kumar RV (1987) Relationship between seed size and protein content in newly developed high protein lines of pigeonpea. *Plant Foods Hum Nutr* 36: 335–340.
Saxena KB, Singh L, Gupta MD (1990) Variation for natural out-crossing in pigeonpea. *Euphytica* 46: 143–148.
Saxena KB, Srivastava DP, Tikka SBS (1998) Breaking yield barrier in pigeonpea through hybrid breeding. In: *Proceedings of National Symposium on Biotic and Abiotic Stress of Pulse Crops*. 26–28 June 1998, Indian Inst Pulses Res, Kanpur: 55–64.
Saxena KB, Kumar RV, Srivastava N, Shiying B (2005) A cytoplasmic-nuclear male-sterility system derived from a cross between Cajanus cajanifolius and Cajanus cajan. *Euphytica* 145: 289–294.
Saxena RK, Prathima C, Saxena KB, Hoisington DA, Singh NK, Varshney RK (2010) Novel SSR markers for polymorphism detection in pigeonpea (Cajanus spp.). *Plant Breed* 129(2): 142–148.
Saxena RK, Von Wettberg E, Upadhyaya HD, Sanchez V, Songok S, Saxena KB, Paul K, Varshney RK (2014) Genetic diversity and demographic history of Cajanus spp. illustrated from genome-wide SNPs. *PLoS One* 9(2): e88568.
Saxena RK, Kale SM, Kumar V, Parupali S, Joshi S, Singh V, Garg V, Das RR, Sharma M, Yamini KN, Ghanta A, Rathore A, Kumar CVS, Saxena KB, Varshney RK (2017) Genotyping-by-sequencing of three mapping populations for identification of candidate genomic regions for resistance to sterility mosaic disease in pigeonpea. *Sci Rep* 7(1): 1813.
Saxena KB, Saxena RK, Sharma S, Sameer Kumar CV, Sultana R, Von Wettberg EB, Varshney RK (2018a) The drivers and methodologies for exploiting wild Cajanus genome in pigeonpea breeding. *Euphytica* 214: 1–20.
Saxena RK, Rathore A, Bohra A, Yadav P, Das RR, Khan AW, Varshney RK (2018b) Development and application of high-density Axiom Cajanus SNP array with 56K SNPs to understand the genome architecture of released cultivars and founder genotypes. *The Plant Genome* 11(3): 180005.
Saxena KB, Choudhary A, Srivastava R, Bohra A, Saxena R, Varshney RK (2019) Origin of early maturing pigeonpea germplasm and its impact on adaptation and cropping systems. *Plant Breed* 138: 243–251.
Saxena RK, Hake A, Hingane AJ, Kumar CS, Bohra A, Sonnappa M, Varshney RK (2020a) Translational pigeonpea genomics consortium for accelerating genetic gains in pigeonpea (Cajanus cajan L.). *Agron* 10(9): 1289.
Saxena RK, Kale S, Mir RR, Mallikarjuna N, Yadav P, Das RR, Molla J, Sonnappa M, Ghanta A, Narasimhan Y, Rathore A, Kumar CVS, Varshney RK (2020b) Genotyping-by-sequencing and multilocation evaluation of two interspecific backcross populations identify QTLs for yield-related traits in pigeonpea. *Theor Appl Genet* 133: 737–749.
Saxena RK, Molla J, Yadav P, Varshney RK (2020c) High resolution mapping of restoration of fertility (Rf) by combining large population and high density genetic map in pigeonpea [Cajanus cajan (L.) Millsp]. *BMC Genomics* 21: 1–8.

Saxena RK, Hake A, Bohra A, Khan AW, Hingane A, Sultana R,Singh IP, Naik SJS, Varshney RK (2021) A diagnostic marker kit for Fusarium wilt and sterility mosaic diseases resistance in pigeonpea. *Theor Appl Genet* 134: 367–379.

Saxena KB, Bohra A, Choudhary AK, Sultana R, Pazhamala L, Saxena RK (2021a) The alternative breeding approaches for improving yield gains and stress response in pigeonpea [*Cajanus cajan* (L.) Millsp.]. *Plant Breed* 140: 76–86.

Saxena RK, Jiang Y, Khan A, et al. (2021b) Characterization of heterosis and genomic prediction based establishment of heterotic pattern for developing better hybrids in pigeonpea. *Plant Genome* 14: e20125.

Shanower TG, Yoshida M, Peter JA (1997) Survival, growth, fecundity, and behavior of Helicoverpa armigera (Lepidoptera: Noctuidae) on pigeonpea and two wild Cajanus species. *J EconEntomol* 90(3): 837–841.

Sharma HC (2016) Host plant resistance to insect pests in pigeonpea: potential and limitations. *Legume Perspectives* (11): 24–29.

Sharma S (2017) Prebreeding using wild species for genetic enhancement of grain legumes at ICRISAT. *Crop Sci* 57(3): 1132–1144.

Sharma SB, Remanandan P, Jain KC (1993) Resistance to cyst nematode (Heterodera cajani) in pigeonpea cultivars and in wild relatives of Cajanus. *Ann Appl Biol* 123(1): 75–81.

Sharma HC, Pampapathy G, Reddy LJ (2003) Wild relatives of pigeonpea as a source of resistance to the pod fly (Melanagromyza obtusa Malloch) and pod wasp (Tanaostigmodes cajaninae La Salle). *Genet Resour Crop Evol* 50: 817–824.

Sharma HC, Sujana G Manohar RaoD (2009) Morphological and chemical components of resistance to pod borer, Helicoverpa armigera in wild relatives of pigeonpea. *Arthropod-Plant Interact* 3: 151–161.

Sharma S, Upadhyaya HD, Varshney RK, Gowda CLL (2013) Pre-breeding for diversification of primary gene pool and genetic enhancement of grain legumes. *Front Plant Sci 4*: 309.

Sharma M, Ghosh R, Telangre R, Rathore A, Saifulla M, Mahalinga DM, Saxena DR, Jain YK (2016) Environmental influences on pigeonpea-Fusarium udum interactions and stability of genotypes to Fusarium wilt. *Front Plant Sci* 7: 253.

Sharma S, Paul PJ, Kumar CS, Rao PJ, Prashanti L, Muniswamy S, Sharma M (2019) Evaluation and identification of promising introgression lines derived from wild Cajanus species for broadening the genetic base of cultivated pigeonpea [Cajanus cajan (L.) Millsp.]. *Front Plant Sci* 10: 1269.

Sharma S, Paul PJ, Sameer Kumar CV, Nimje C (2020) Utilizing wild Cajanus platycarpus, a tertiary genepool species for enriching variability in the primary genepool for pigeonpea improvement. *Front Plant Sci* 11: 1055.

Sharma S, Jaba J, Rao PJ, Prasad S, Gopal NTVV, Sharma HC, Kilian B (2022) Reaping the potential of wild cajanus species through pre-breeding for improving resistance to pod borer, Helicoverpa armigera, in cultivated pigeonpea (Cajanus cajan (l.) millsp.). *Biology* 11(4): 485.

Singh BD, Singh AK (2015) Mapping populations. In: BD Singh & AK Singh (Eds.) *Marker-Assisted Plant Breeding: Principles and Practices*. Springer, New Delhi.

Singh NK, Gupta DK, Jayaswal PK, Mahato AK, Dutta S, Singh S, Sharma TR (2012) The first draft of the pigeonpea genome sequence. *J Plant Biochem Biotechnol* 21: 98–112.

Singh N, Tyagi RK, Pandey C (2013) Genetic resources of pigeonpea: conservation for use. *National Bureau of Plant Genetic Resources (NBPGR), New Delhi*, 1–49.

Singh M, Gautam NK, Rana MK, Dahiya OP, Dutta M, Bansal KC (2014) Pigeon pea genetic resources and its utilization in India, current status and future prospects. *J Plant Sci Res* 1: 107.

Singh G, Singh I, Taggar GK, Rani U, Sharma P, Gupta M, Singh S (2020a) Introgression of productivity enhancing traits, resistance to pod borer and Phytopthora stem blight from Cajanus scarabaeoides to cultivated pigeonpea. *Physiol Mol Biol Plants* 26: 1399–1410.

Singh S, Mahato AK, Jayaswal PK, Singh N, Dheer M, Goel P, Raje RS, Yasin JK, Sreevathsa R, Rai V, Gaikwad K, Singh NK (2020b) A 62K genic-SNP chip array for genetic studies and breeding applications in pigeonpea (Cajanus cajan L. Millsp.). *Scientific Reports* 10(1), 4960.

Singh A, Ramakrishna G, Singh NK, Abdin MZ, Gaikwad K (2024) Genomic insight into variations associated with flowering-time and early-maturity in pigeonpea mutant TAT-10 and its wild type parent T21. *Int J Biol Macromol* 257: 128559.

Sinha P, Singh VK, Suryanarayana V, Krishnamurthy L, Saxena RK, Varshney RK (2015) Evaluation and validation of housekeeping genes as reference for gene expression studies in pigeonpea (*Cajanus cajan*) under drought stress conditions. *PloS One* 10(4): e0122847.

Sivaramakrishnan S, Kannan S, Reddy LJ (2002) Diversity in selected wild and cultivated species of pigeonpea using RFLP of mtDNA. *Euphytica* 125: 21–28.

Srikanth S, Rao MV, Mallikarjuna N (2014) Interspecific hybridization between Cajanus cajan (L.) Millsp. and C. lanceolatus (WV Fitgz) van der Maesen. *Plant Genet Res* 12(2): 255–258.

Srikanth S, Saxena RK, Rao MV, Varshney RK, Mallikarjuna N (2015) Development of a new CMS system in pigeonpea utilizing crosses with Cajanus lanceolatus (WV Fitgz) van der Maesen. *Euphytica* 204: 289–302.

Srivastava N, Vadez V, Upadhyaya HD, Saxena KB (2006) Screening for intra and inter specific variability for salinity tolerance in pigeonpea (Cajanus cajan) and its related wild species. *J SAT Agric Res* 2(1): 1–12.

Srivastava RK, Saxena KB, Kumar RV (2018) Development of a Cajanus cajanifolius-based CMS Hybrid Technology in Pigeonpea. *J Crop Sci Biotechnol* 21: 443–449.

Subbarao GV, Johansen C, Jana MK, Kumar Rao, JVDK (1990) Effects of the sodium/calcium ratio in modifying salinity response of pigeonpea (Cajanus cajan), *J Plant Physiol* 136(4): 439–443.

Sujana G, Sharma HC, Rao DM (2008) Antixenosis and antibiosis components of resistance to pod borer Helicoverpa armigera in wild relatives of pigeonpea.*Int J Trop Insect Sci* 28: 191–200.

Tanksley SD (1993) Mapping polygenes. *Annual Review of Genetics* 27(1): 205–233.

Tanksley SD, Nelson JC (1996) Advanced backcross QTL analysis: a method for the simultaneous discovery and transfer of valuable QTLs from unadapted germplasm into elite breeding lines. *Theor Appl Genet* 92: 191–203.

Thiruvengadam V, Muthiah A (2007). Interspecific hybridization between Cajanus cajan and Cajanus cajanifolius. *Crop Breed Appl Biotechnol* 7(2).

Tikka SBS, Parmar LD, Chauhan RM (1997) First record of cytoplasmic-genic male sterility system in pigeonpea (Cajanus cajan (L.) Millsp.) through wide hybridization.

Upadhyaya HD (2006) Improving pigeonpea with the wild. *SA Trends ICRISAT's Monthly Newsletter* 62: 1–1.

Upadhyaya HD, Reddy LJ, Gowda CLL, Reddy KN, Singh S (2006) Development of a mini core subset for enhanced and diversified utilization of pigeonpea germplasm resources. *Crop Sci* 46(5): 2127–2132.

Upadhyaya HD, Reddy KN, Gowda CLL, Sube S (2007) Phenotypic diversity in pigeonpea core collection. *Genet Resour Crop Evol* 54: 1167–1184.

Upadhyaya HD, Reddy KN, Sharma S, Varshney RK, Bhattacharjee R, Singh S, Gowda CLL (2011) Pigeonpea composite collection and identification of germplasm for use in crop improvement programmes. *Plant Genet Res* 9(1): 97–108.

Upadhyaya HD, Reddy KN, Sastry DVSSR, Gowda CLL (2013a) The wild genepool of pigeonpea at ICRISAT genebank-status and distribution. *Ind J Plant Genet Resour 26*(03): 193–201.

Upadhyaya HD, Reddy KN, Singh S, Gowda CLL (2013b) Phenotypic diversity in Cajanus species and identification of promising sources for agronomic traits and seed protein content. *Genet Resour Crop Evol* 60: 639–659.

Van der Maesen LJG (1980) India is the native home of the pigeonpea. In: JC Arends, G Boelama, CT de Grant, AJM Leeuwenberg (Ed.) *Libergratulatorius in Honerem HCD de Wit*, pp 257–262, Agricultural University Miscellaneous Paper, vol 19, Wageningen, The Netherlands.

Van der Maesen LJG (1986) Cajanus DC.and Atylosia W. & A. (Leguminosae), Agricultural University Miscellaneous Papers 85–4 (1985). Agricultural University, Wageningen, The Netherlands.

Van der Maesen LJG (1990) Pigeonpea: origin, history, evolution and taxonomy. In: YL Nene, SD Hall, VK Sheila (Ed.) *The Pigeonpea*, pp 15–46, CAB International, Wallingford, UK.

Vanambathina P, Rachaputi RC, Sultanbawa Y, Phan ADT, Henry RJ, Brier H (2021) Biochemical basis of resistance to pod borer (*Helicoverpa armigera*) in Australian wild relatives of pigeonpea. *Legum Sci* 3(4): e101.

Varshney RK, Graner A, Sorrells ME (2005) Genomics-assisted breeding for crop improvement. *Trends Plant Sci* 10(12): 621–630.

Varshney RK, Chen W, Li Y, Bharti AK, Saxena RK, Schlueter JA, Donoghue MT, Azam S, Fan G, Whaley AM, Farmer AD, Sheridan J, Iwata A, Tuteja R, Penmetsa RV, Wu W, Upadhyaya HD, Yang SP, Shah T, Saxena KB, Jackson SA (2011) Draft genome sequence of pigeonpea (Cajanus cajan), an orphan legume crop of resource-poor farmers. *Nat Biotechnol* 30(1): 83–89.

Varshney RK, Kudapa H, Roorkiwal M, Thudi M, Pandey MK, Saxena RK, Pande S (2012) Advances in genetics and molecular breeding of three legume crops of semi-arid tropics using next-generation sequencing and high-throughput genotyping technologies. *J Biosci* 37: 811–820.

Varshney RK, Saxena RK, Upadhyaya HD, Khan AW, Yu Y, Kim C, Datta SK (2017) Whole-genome resequencing of 292 pigeonpea accessions identifies genomic regions associated with domestication and agronomic traits. *Nat Genet* 49(7): 1082–1088.

Vavilov NI (1951) The origin, variation, immunity and breeding of cultivated plants. *Soil Sci.* 72(6): 482.

Venkatesan T, Arunachalam M (2007) Interspecific hybridization between *Cajanus cajan* and *Cajanus cajanifolius*. *Crop Breed Appl Biotechnol* 7: 204–211.

Verulkar SB, Singh DP, Bhattacharya AK (1997) Inheritance of resistance to podfly and podborer in the interspecific cross of pigeonpea. *Theor Appl Genet* 95: 506–508.

Wanjari KB (1998) *Utilization of alien species for development of cytoplasmic male sterility in pigeonpea*. In *Proceeding of Late Vasantrao Naik Memorial Seminar, College of Agriculture, Nagpur (India)*(10–11).

Wanjari KB, Patil AN, Prema Manapure PM, Manjayya JG, Manish Patel MP (1999) Cytoplasmic male sterility in pigeon pea with cytoplasm from Cajanus volubilis. *Ann Plant Physiol* 13(2): 170–174.

Williams IH (1977) Behavior of insects foraging on pigeon pea (cajanus-cajan-(l) millsp) in India. *Trop Agric* 54(4): 353–363.

Yadav TS (1986) Cytomorphological studies of Fl hybrids of Cajanus cajan x Atylosia albicans. *J Indian Bot Soc* 65: 77–78.

Yadav PBS, Padmaja V (2002) Interspecific hybridization and evolutionary relationships of Cajanus cajan (L.) Millspaugh and four of the wild species. *Cytologia* 67: 67–73.

Yadav P, Saxena KB, Hingane A, Kumar CS, Kandalkar VS, Varshney RK, Saxena RK (2019) An "Axiom Cajanus SNP Array" based high density genetic map and QTL mapping for high-selfing flower and seed quality traits in pigeonpea. *BMC Genomics* 20: 1–10.

Yadav MK, Patel C, Singh RS, Singh KK, Balasubramanian R, Mall RK, Yadav SK (2021) Assessment of climate change impact on different pigeonpea maturity groups in north Indian condition. *J Agrometeorol* 23(1): 82–92.

Yang SY, Saxena RK, Kulwal PL, Ash GJ, Dubey A, Harper JD, Upadhyaya HD, Gothalwal R, Kilian A, Varshney RK (2011) The first genetic map of pigeon pea based on diversity arrays technology (DArT) markers. *J Genet* 90: 103–109.
Yohane EN, Shimelis H, Laing M, Shayanowako A (2022) Genetic diversity and grouping of pigeonpea [Cajanus cajan Millspaugh] Germplasm using SNP markers and agronomic traits. *PLoS One* 17(11): e0275060.
Zhao J, Bayer PE, Ruperao P, Saxena RK, Khan AW, Golicz AA, Nguyen HT, Batley J, Edwards D, Varshney RK (2020) Trait associations in the pangenome of pigeon pea (Cajanus cajan). *Plant Biotechnol J* 18(9): 1946–1954.

3 Applications and Challenges of Mutation Breeding for Creating Climate-Resilient Pigeonpea

Dhanasekar Punniyamoorthy and Souframanien Jegadeesan

3.1 INTRODUCTION

Pigeonpea (*Cajanus cajan* (L.) Millspaugh) is one of the most significant annual legumes widely cultivated in the rainfed tropical and subtropical regions of Asia and Africa, known for its resilience to drought and adaptability to diverse soil types (Bohra et al., 2020; Choudhary and Nadarajan, 2011). Pigeonpeas are known globally by a variety of vernacular names like *red gram*, *tur*, *arhar*, *Angola pea*, *Congo pea*, *yellow dhal*, *oil dhal*, *mbaazi*, *gandul*, *no-eye pea*, etc. They are vital in promoting food and ecological security through their protein- and nutrient-rich seeds and their ability to fix atmospheric nitrogen. Its versatility (providing food, fodder, and firewood) and indispensable contribution to sustainable agriculture make it a crop of choice among impoverished and marginal farmers (Gaur et al., 2018b). Also, pigeonpea has a great nutritive value (Table 3.1).

Despite its importance, the breeding tactics are inefficient in breaking the productivity plateau owing to several biotic and abiotic constraints. This is further exacerbated and aggravated by the serious and challenging consequences emanating from climate change. Climate change substantially influences agriculture, affecting crop productivity, soil health, water availability, and farming practices (Prabhu et al., 2023). Therefore, it becomes imperative to develop resilient and adaptive farming systems to tide over climate change. Breeding strategies relying solely on conventional breeding are incompetent in mitigating rapidly the adversities of climate change. Any genetic improvement program ultimately relies on the existing genetic variability to address a given challenge. Recombination and mutation are fundamental events leading to the generation of genetic variability. The latter is potentially endowed with the ability to create new variation, unlike the former, which depends on the existing variability (Begna, 2021). Spontaneous mutations occurring naturally

DOI: 10.1201/9781003603092-3

TABLE 3.1
Nutritional Value of Pigeonpea per 100 g of Seeds (USDA Database, 2018)

Composition	Amount per 100 g
Water	10.6 g
Energy	343 kcal
Protein	21.7 g
Total lipid (fat)	1.49 g
Ash	3.45 g
Carbohydrate	62.8 g
Dietary fiber	15 g
Calcium	130 mg
Magnesium	183 mg
Phosphorus	367 mg
Iron	5.23 mg
Potassium	1392 mg
Sodium	17 mg
Copper	1.06 mg
Manganese	1.79 mg
Zinc	2.76 mg
Thiamin	0.643 ng
Riboflavin	0.187 mg
Niacin	2.96 mg
Pantothenic acid	1.27 mg
Vitamin B6	0.283 mg
Folate(B9)	456 µg

are very rare and are of little use to expeditiously redress the changing agricultural and ecological demands. Induced mutagenesis through various physical and chemical mutagens comes in handy in accelerating the rate of mutations and keeping pace with the dynamic needs. Thus, this chapter attempts to elaborate on the applications and challenges of mutation breeding in evolving climate-resilient pigeonpeas.

3.2 TAXONOMY, ORIGIN, AND GENOME OF PIGEONPEA

The dicotyledonous cultivated pigeonpeas belong to the order Fabales of Clade Rosids, family Fabaceae (Syn. Leguminosae), subfamily Faboideae (Syn. Papillionoideae), tribe Phaseoleae, subtribe Cajaninae, genus *Cajanus*, subgenus *Cajanus*, section *Cajanus*, and species *cajan* (van der Maesen, 1991). Pigeonpea (*Cajanus cajan*) is the most widely cultivated food legume species within this subtribe, apart from the lesser cultivated ones, *C. indicus* and *C. scarabaeoides*. Though there are several theories on the origin of pigeonpea, India is unambiguously considered the primary center of origin, with Africa as the secondary center, based on the vast diversity, existence of wild species, including the progenitor, linguistic evidence, and widespread use in daily cuisines (van der Maesen, 1980). The most probable progenitor of *C. cajan* is postulated to be *C. cajanifolius*, which had given rise to the former through a single gene mutation event. In the nineteenth century, pigeonpeas got introduced into East Africa through the emigrants, from where they got into the Nile Valley and West Africa en route to the Americas (Odeny, 2007). As of today, pigeonpea is widely cultivated in South Asia, Southeast Asia, Africa, Latin America, Australia, and Oceania regions. The diploid genome of pigeonpea comprises 11 pairs of chromosomes ($2n = 2X = 22$), with an estimated genome size of 833.07 Mb and 48,680 genes. The average coding sequence length of the genes is slightly less than 1 kbp with 3.6 exons per gene (Varshney et al., 2012). Genome-wide duplication events detected in soybean were not observed in pigeonpea.

3.3 IMPORTANCE AND NUTRITIVE VALUE OF PIGEONPEA

Worldwide, pigeonpea is grown on 5.4 million hectares with annual production of 4.6 million tons and productivity of 851 kg ha−1 (FAOSTAT 2023). India is a major producer of pigeonpea, accounting for 72.2% of the global pigeonpea production. In India, pigeonpea is the second most important pulse crop, next only to chickpea, reckoning 16% of the total pulse production (~30 Mt) of the country, with productivity matching almost the global value. As of 2023, the major pigeonpea-producing states are Karnataka, Maharashtra, Uttar Pradesh, Gujarat, and Jharkhand, contributing 5 to 30% of All-India production (MoA&FW 2024). Pigeonpea is nutritionally, agriculturally, economically, and ecologically important. Rich in proteins and other vital nutrients like iron, calcium, and vitamins, pigeonpea contributes to food security in vegetarian diet-based developing regions and helps combat malnutrition. It contains all the amino acids, more than the minimum required (mg/g of protein), except for methionine and cystine (Dhanasekar et al., 2021). With its resilience to water stress, nitrogen-fixing capability, and suitability for crop rotation and intercropping systems, pigeonpea enhances overall farm productivity and sustainability (Saxena et al., 2019, 2021). Pigeonpeas augment the income of the farmers through the sale of seeds, fodder, and firewood. The deep roots not only prevent soil erosion but also improve soil structure. Thus, the multifunctional benefits make it a cornerstone of sustainable agriculture and rural livelihoods.

3.4 FLORAL BIOLOGY AND BREEDING BEHAVIOR

Pigeonpea possesses a unique floral morphology among the legumes that permits both self- and insect-aided cross-pollination to varying extents and is thus classified as an often cross-pollinated crop. The extent of cross-pollination rates varies significantly (0%–60%) across different countries and environments, influenced by a complex interplay of biological and ecological factors. The typical papilionaceous flower consists of five sepals, five petals (one outer *standard*, two *wings*, and two innermost petals united at the base to form a structure called a *keel*), diadelphous stamens (9+1), and a pistil consisting of a stigma, style, and ovary. The anthesis usually starts from 06.00 h and continues even up to 17.00 h, with peak anthesis occurring between

9.00 and 10.00 h (Sharma and Green, 1980). The flowers remain open for 6 to 36 h, and the stigma remains receptive between −68 h and + 20 h after anthesis, typically exhibiting protogyny. Fertilization occurs on the day of pollination, and the multiple ovules (up to 9) eventually develop into seeds in the ovary. For artificial hybridization, depending on the prevailing temperature, buds likely to open the next day (winter days) or two days later (higher temperatures) need to be selected for emasculation, as there is an increased chance of pollen bursting at higher temperatures.

3.5 IMPACT OF CLIMATE CHANGE

Pigeonpea productivity is extremely impacted by climate change, affecting its growth, yield, and adaptability in cultivated areas (Penna et al., 2023). The climatic vagaries affect biotic, abiotic, and edaphic factors and thereby the overall crop physiology and productivity as detailed in the following sections.

3.5.1 Temperature Extremes

Pulse crops in general and pigeonpea in particular are reported to be sensitive to high-temperature stress at the flowering stage, and a few days of exposure to temperatures exceeding 30°C–35°C causes heavy yield losses through flower drop or pod damage (Siddique et al., 1999). High temperatures prevailing during critical stages of pigeonpea growth, like flowering and pod formation, lead to flower drop, poor pollination, and pod abortion, resulting in a drastic reduction in yields. Accelerated growth and maturity attributing to elevated temperatures also provide less of a window for biomass and yield accumulation. The rise in maximum temperatures had a significant and negative impact on pigeonpea yield (Mishra et al., 2017). Studies reveal that a 1°C increase in maximum temperature could lead to a 20.8% decrease in pigeonpea yield. Similarly, low temperatures of <7°C also affect productivity in similar lines. Low temperatures cause sterility, poor dehiscence of pollen, rupturing of pollen membranes, reduced photosynthesis, and also pod abortion.

3.5.2 Erratic Rainfall Patterns

Erratic patterns in rainfall result in drought, flooding, and delayed or early onset of monsoons. Drought stress resulting from inadequate rainfall during the growing season causes water stress, resulting in impaired germination, root development, and pod filling. Excess rainfall and poor drainage result in flooding and waterlogging, leading to poor plant growth, poor aeration, nutrient leaching, and the spread of diseases like *Phytophthora* blight and root rot. Shifts in monsoon patterns, like early or delayed onset, result in disrupted sowing, harvesting, and other intercultural schedules, leading to poor productivity.

3.5.3 Emergence of New Pests and Diseases

Climate change has led to a shift in the type and range of pests and diseases. Hitherto minor pests and diseases have become major owing to the suitability of the changing climate for their proliferation and adaptation. Warmer temperatures and higher humidity favor the multiplication of pests like *Helicoverpa armigera*, *Maruca* pod borers, aphids, and pod flies. The prevalence of diseases is also dramatically affected by climate change. The pathogens causing *Fusarium* wilt and sterility mosaic are becoming more virulent. Diseases like dry root rot (root-rot complex) are increasingly becoming a menace in recent times due to climate change, especially under prolonged dry spells. *Phytophthora* blight has gained prominence primarily under flooding and waterlogging conditions.

3.5.4 Salinity and Soil Degradation

The rising sea entailing global warming and poor irrigation practices degrades the soil by increasing the soil salinity, which negatively affects pigeonpea root growth, nutrient uptake, and overall productivity. Erratic changes in the weather patterns also lead to loss of soil fertility, soil erosion, and nutrient depletion, thereby influencing the productive potential of pigeonpea.

3.5.5 Impact on Pollination

Changing weather patterns also influence the activity of insect pollinators like bees, which considerably affect the pollination of pigeonpea, as it is an often-cross-pollinated crop, thereby disturbing the pod set.

3.5.6 Geographic Shifts in Suitability

Because of changes in the climate, regions becoming hotter and more prone to water stress become unfit for pigeonpea cultivation in due course, leading to a decrease in traditional areas under pigeonpea cultivation, whereas erstwhile cooler or temperate regions may turn suitable for pigeonpea cultivation, resulting in expansion in nonconventional areas.

3.5.7 Impact on Quality

Heat and water stress can affect the seed size, seed weight, and viability, leading to reduced market value. The stresses can reduce biomass accumulation due to reduced leaf area and photosynthetic activity.

The compounded effects of climate change on pigeonpea productivity threaten food security, livelihoods, and economic stability in regions reliant on this crop. Comprehensive adaptation measures are essential to safeguard its production and role in sustainable agriculture.

3.6 TARGET BREEDING TRAITS TO ADDRESS CLIMATE CHANGE IN PIGEONPEA

Climate change is a global concern posing significant challenges to food production and food security. The shift in weather patterns over a long period of time results in changes in climate, leading to temperature extremes, erratic rainfalls, floods, rise in

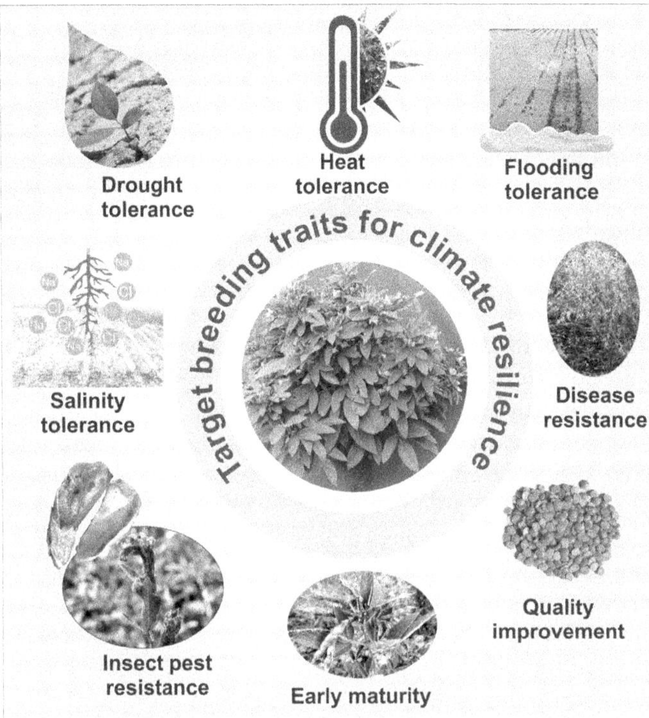

FIGURE 3.1 Target breeding traits for addressing climate resilience in pigeonpea.

sea level, droughts, new pests, and diseases. To mitigate these impacts, pigeonpea breeding focuses on traits that enhance resilience, productivity, and sustainability (Figure 3.1).

The following are the key target traits.

3.6.1 Early Maturity

The longer the maturity period of pigeonpeas, the higher the risk exposure to climate vagaries. So, the foremost trait that can be easily taken up for breeding climate resilience is to look for early or extra early pigeonpea varieties that can easily escape climate variability, especially during the fag end of the crop. Early varieties can complete their life cycle before the onset of severe drought conditions, they can avoid exposure to late-season high temperatures, and can be harvested well in time prior to heavy monsoon rains or flooding. Reduced crop duration also enables pest and disease avoidance during late seasons. Shorter growing periods also optimize water and nutrient usage, especially in rainfed systems. Early maturing varieties like TJT-501 became very popular in Central India owing to their ability to escape frost damages during crop maturity (Souframanien and Ganapathi, 2021). Therefore, the maturity period is one of the key targets in pigeonpea breeding programs for climate resilience.

3.6.2 Drought Tolerance

Pigeonpea genotypes with high yield stability under water-limited conditions need to be selected. Traits such as deeper and denser roots for improved water uptake under receding water conditions and lower transpiration for efficient water use through stomatal regulation and osmotic adjustment can help identify pigeonpea genotypes with water stress tolerance, especially during continued dry spells due to climate change (Bakala et al., 2024).

3.6.3 High Temperature and Cold Tolerance

Tolerances to high temperature and cold remain critical traits for enhancing pigeonpea's resilience to climate change. Genotypes with high photosynthetic efficiency, stable reproductive structures (reduced flower and pod abortion), the ability to regulate canopy temperature, the capability of producing heat-stable enzymes, increased pollen viability, synchronized flowering, and a high seed set under elevated temperatures can help tide over increased frequencies of heatwaves, a common phenomenon of climate change. High temperatures often coincide with water scarcity, and therefore, breeding genotypes with tolerance to both heat and water stress could ensure sustained productivity under climate change (Dhanasekar and Souframanien, 2023). Likewise, climate change can also lead to extremes of cold, and therefore, pigeonpeas must be bred for both high temperatures capable of enduring between 32 and 45°C and also withstanding cold temperatures between 2 and 10°C.

3.6.4 Flood and Waterlogging Tolerance

Tolerance against water logging is another important trait that enables pigeonpea to endure climate change impacts, particularly in regions prone to erratic rainfall, heavy monsoons, and poor drainage. Waterlogging tolerance allows pigeonpea to withstand anaerobic soil conditions caused by excess water, which otherwise impair root function and nutrient uptake. Tolerant varieties are characterized by traits such as reduced flower and pod drop during extended wet conditions, tolerance to diseases like *Phytophthora* root rot and stem decay that thrive in waterlogged conditions, formation of aerenchymatic cells and lenticels (ICP5028, ICPH2431, ICPL87119, ICPH2740, MAL15) in roots that help in oxygen transport under hypoxic conditions, and production of secondary adventitious roots (ICPL149 and ICPL20241) near the soil surface, improving water and nutrient absorption during floods (Hingane et al., 2015). Swelling of stems near root-shoot junctions under waterlogging conditions was often correlated with lenticel formation that enhances the endurance. The tolerant genotypes should also exhibit rapid recovery once submergence conditions recede.

3.6.5 Salinity Tolerance

Salinity tolerance is especially useful in regions affected by soil salinity due to rising sea levels because of global warming and in regions with poor drainage and erratic rainfall. Usually, saline-tolerant genotypes show little reduction in transpiration rate with increasing salinity, although inconsistently. An increase in leaf osmolytes like proline, glycine betaine, and soluble sugars under salinity could also contribute to subdued adverse effects of salinity. Much of salinity tolerance is attributed to efficient regulation of Na and Cl. The saline-tolerant genotypes usually exhibit five to ten times increased concentrations of shoot sodium with two to three times higher root sodium concentrations than susceptible species. Unlike the sensitive genotypes, the tolerant lines show an increase in potassium concentrations and maintain leaf magnesium concentrations with increasing salinity. Therefore, plants that regulate Na^+ and K^+ ions effectively prevent saline toxicity and ensure unhindered metabolic functions (Subbarao et al., 1990). The seedling stage of pigeonpea and 80 mM NaCl concentration have been found to be optimum for identifying pigeonpea genotypes tolerant to salt stress under controlled conditions (Joshi et al., 2022). Moreover, genotypes with deep and efficient roots can also minimize salt uptake while accessing nutrients and water from lower soil layers.

3.6.6 Horizontal Resistance to Pests and Diseases

Rising temperatures and erratic rainfall can increase the frequency and severity of pest and disease outbreaks. Resistant varieties with horizontal resistance against several pests and diseases can endure such pressures, maintaining productivity. Climate change also leads to the emergence of new pests and diseases that were hitherto of minor importance. For instance, the *Maruca* pod borer until a decade ago was not a major pest of pigeonpea. But with climate change, it is assuming the status of the most dangerous pest in recent years, especially in Central India. Therefore, stacking of genes through introgressive breeding, gene pyramiding, and enhancing the broad spectrum of resistance against several pests and diseases could help pigeonpea genotypes from succumbing to untimely pest and disease outbreaks. Moreover, genotypes with enhanced secondary metabolites could also help pest deterrence to a certain extent.

3.6.7 Other Traits

Some of the other traits that can also contribute to climate resilience include enhanced nutrient use efficiency, improved carbon sequestration, and better-quality traits under stress conditions.

By targeting these traits, pigeonpea breeding can help mitigate the adverse effects of climate change, ensuring food security and agricultural sustainability in the face of evolving challenges.

3.7 LIMITATIONS OF CONVENTIONAL BREEDING

Conventional breeding has tremendously and reliably contributed to the increase in the productivity of pigeonpea over the years. But there are certain limitations of conventional breeding that forbids the utility of traditional breeding in meeting the dynamic challenges of climate change. Introspection of genetic relatedness among the pigeonpea varieties released in India shows that 60% and 38% of the pigeonpea varieties have been developed through pedigree and pure line selections, respectively. A limited number of genotypes have been repeatedly used in the development of varieties. For instance, the genotype T-190 has been directly or indirectly involved in the development of 21 varieties. Likewise, the genotypes UPAS-120, ICP8863, T-21, and C-11, were generously used in varietal development (Naik et al., 2020). Such insensible use of limited genotypes has apparently resulted in a narrow genetic base of released varieties, rendering them vulnerable to climatic uncertainties and outbreaks of new insect pests and diseases (Boukar et al., 2020). The available genetic variability within the exploitable gene pool has been exhausted, and further genetic improvement through conventional breeding is very limited unless novel germplasm from unexploitable gene pools and new genetic variability generated through mutation breeding or biotechnological interventions are employed (Souframanien and Dhanasekar, 2023). The often-cross-pollinated nature of pigeonpea also makes it more resource-demanding to offset the ill effects in conventional breeding and is also mainly responsible for the loss of genetic purity, progeny unpredictability, slow progress in trait stabilization, risk of undesirable trait introgression, loss of local adaptation, and genetic erosion. Moreover, the long gestational period of pigeonpea development (>12 years) through conventional breeding slows down the breeding process and makes it inefficient to keep pace with the rapid dynamic challenges of climate change. Breeding for climate resilience often involves the development of resistance against various abiotic stressors that are very complex and concern the simultaneous introgression of a number of genes, and conventional breeding is incompetent for handling such a magnitude and multitude of tasks in a limited time. Even though various genetic treasures are available in different gene pools, they are often unexploitable due to cross-incompatibility with wild relatives, and conventional breeding alone is inadequate to tackle such issues (Sharma et al., 2006; Bohra et al. 2022). The linkage drag associated with many of the beneficial genes from uncultivable sources also poses a daunting challenge crippling conventional breeding. Traditional breeding also doesn't leverage predictive modeling or big data analysis tools to anticipate the effect of climate change on future crop performance. To overcome these limitations, it is imperative to integrate non-traditional breeding tools like mutation breeding, genomic selection, marker-assisted breeding, genetic engineering, and speed breeding to keep pace with the challenges of rapidly changing climate.

3.8 MUTATION BREEDING TO THE RESCUE OF CONVENTIONAL BREEDING

Mutation breeding is generating allelic variants of various genes in an organism's genome through induced mutagenesis and identifying stable variants meeting a selection criterion or a trait of interest, mimicking the accelerated process of spontaneous mutation and natural selection occurring in nature. The superior mutants with good agronomical values can be released directly as a variety. Alternatively, others could be used to hybridize with cultivars or mutants to develop mutant derivatives with a prospective varietal release (Table 3.2).

Mutation breeding has the potential to overcome some of the shortcomings of conventional breeding. It could be beneficial in situations where there is low genetic

TABLE 3.2
List of Pigeonpea Mutant Varieties Including Direct Mutants, Mutant x Mutant, Mutant x Cultivar, or Cultivar x Mutant Derivatives

S no.	Variety	Year of Release	Mutagen/Pedigree	Salient Features	Reference
1	T-1	1953	Spontaneous, Landrace	Early	Saxena et al. (2019)
2	T-21	1961	T-1[a] x T-190	150–170 days maturity, semi-spreading, loose plant type, indeterminate, small seeded	Pathak and Singh (1961)
3	CO-1	1970	Spontaneous, Perambalur local	Early (135–140 days), photo-insensitive, brown seeded	Saxena et al. (2019)
4	Prabhat	1972	T-1[a] x T-190	110–115 days, determinate, dwarf, clustered fruiting	Lal and Sinha (1972)
5	3D 8111	1973	Spontaneous, UC 5543-1	Semi-dwarf, early maturing (106–140 days), high yielding	Saxena et al. (2019)
6	3D 8127	1973	Spontaneous, UC 1381	Semi-dwarf, early maturing (106–140 days), high yielding	Saxena et al. (2019)
7	3D 8104	1973	Spontaneous, UC 5103	Semi-dwarf, early maturing (106–140 days), high yielding	Saxena et al. (2019)
8	UPAS 120	1974	Spontaneous, P 4758	120–140 days, indeterminate, open-tall, semi-spreading, small seeded, synchronous maturing,	Singh et al. (1974)
9	Pant A-1; -2; -3	1976	Spontaneous, UPAS 120	120–130 days, semi-erect, determinate, dwarf	Saxena et al. (2019)
10	CO-2	1977	Spontaneous, PB- 4728	115–120 days, dwarf, terminal cluster, tolerant to *Fusarium* and *Rhizoctonia* root-rot	Saxena et al. (2019)
11	CO-3	1977	0.6% EMS, CO-1	130 days, large seed, high yield, early, resistant to root rot, tolerance to wilt and pod borers	Sheriff et al. (1977)
12	Hy-2	1978	Spontaneous, PI4628	Early (140–150 days), large white seeds	Saxena et al. (2019)
13	Hy-4	1978	Spontaneous, PI4839	Early (140–150 days), large brown seeds	Saxena et al. 2019
14	CO-4	1981	Spontaneous, S-80	130–134 days, indeterminate, axillary cluster, resistant to wilt, root rot and YMV	Saxena et al. (2019)
15	AL 15	1982	Spontaneous, P 8-9	130–140 days, small seeded, determinate, terminal cluster, tolerant to pod fly	Saxena et al. (2019)
16	Hy-5	1982	Spontaneous, PI3701	Early (130–135 days)	Saxena et al. (2019)

(*Continued*)

TABLE 3.2 (Continued)
List of Pigeonpea Mutant Varieties Including Direct Mutants, Mutant x Mutant, Mutant x Cultivar, or Cultivar x Mutant Derivatives

S no.	Variety	Year of Release	Mutagen/Pedigree	Salient Features	Reference
17	CO-5	1984	160 Gy gamma rays, CO-1	100–115 days, determinate, early maturing, photoperiod insensitivity, tolerance to water stress, moderately resistant to pod fly, root rot, and SMD	Souframanien and Dhanasekar (2023)
18	TAT- 5	1984	15 Gy fast neutrons, T-21	Early maturing (140 days), large seeds (>17 g/100 seeds)	Souframanien and Dhanasekar (2023)
19	WB20	1984	Prabhat[a] x B517	Resistant to *Alternaria*, suitable for pre-rabi cultivation	Kumar et al. (2022)
20	Trombay Vishakha 1 (TT 6)	1985	Fast neutrons, T-21	35% increase in seed size compared to parent, compact and erect	Souframanien and Dhanasekar (2023)
21	TAT-10	1985	25 Gy fast neutrons, TT2[a] x TT-9[a]	Extra early (115–120 days), medium-large seeds	Souframanien and Dhanasekar (2023)
22	Pusa 84	1985	Pusa Ageti x T-21[a]	140–150 days, medium tall indeterminate, semi-spreading	Kumar et al. (2022)
23	Manak	1985	T-21[b] x UPAS120[a]	Tall, maturity (150–152 days), seed yield 20–22 q/ha	MoA&FW, GoI (2024)
24	ICPL87 (Pragati)	1986	T-21[b] x JA277	120–130 days, determinate, spreading, brown seeds	MoA&FW, GoI (2024)
25	Pusa 85	1987	No. 148 x UPAS120[a]	Tall, maturity (150 days), bold seeded, seed yield 20–25 q/ha	MoA&FW, GoI (2024)
26	Pusa 33	1989	C11 x UPAS120[a]	Tall, erect, early maturity, high yield	Kumar et al. (2022)
27	ICPL 151 (Jagriti)	1989	ICP6997 x Prabhat[a]	120–130 days, determinate, semi-spreading, cream-colored seeds, medium-large, tolerance to salinity	MoA&FW, GoI (2024)
28	Pusa 855	1993	Spontaneous, T-21	150–180 days, indeterminate, large seeds, tolerant to wilt & SMD	Saxena et al. (2019)
29	Pusa 855	1993	T-21	140–160 days, early, indeterminate medium-large seeds	MoA&FW, GoI (2024)
30	CO 6	1993	250 Gy gamma rays, SA-1	Indeterminate, tolerant to pod borer, dwarf type, medium seed size, photo-insensitive	Souframanien and Dhanasekar (2023)
31	Vamban 1	1993	(Prabhat[a] × Hy 3A) × (T- 21[b] × 102)	Determinate, suitable for intercropping with peanut	MoA&FW, GoI (2024)
32	Pusa 9	1993	UPAS 120[a] × 3673	Indeterminate, resistant to SMD and *Alternaria* blight, suitable for pre-rabi cultivation, sensitive to Al toxicity	MoA&FW, GoI (2024)

33	Durga (ICPL84031)	1995	Prabhat[a] × UPAS120[a]	Determinate	MoA&FW, GoI (2024)
34	COPH-2	1997	Gamma rays, Derived (male sterile line of CO-5[a])	Hybrid suitable for both irrigated and rainfed conditions	Souframanien and Dhanasekar (2023)
35	Paras (H82-1)	1998	EE 76 × UPAS120[a]	Indeterminate	MoA&FW, GoI (2024)
36	APK 1	1999	Spontaneous, ICPL87101	Nonshattering, nonlodging, resistant to SMD, large seeds	Saxena et al. (2019)
37	Laxmi (ICPL85063)	2001	BDN 1 × (T-21[b] × JA 275)	120–130 days, semi-spreading, indeterminate, suitable also for rabi planting (80–90 days), resistant to SMD	Kumar et al. (2022)
38	Pusa 992	2004	Spontaneous, ICPL90306	120–160 days, suitable for early sowing, medium-large, round brown seeds, tolerant to moisture stress	Saxena et al. (2019)
39	CO(RG)-7	2004	Fast neutrons, (ICP 8863 x AL 101) x (PA 1218 x TT6[a])	Early, tolerant to wilt, sterility mosaic, and *Phytophthora* blight	Souframanien and Dhanasekar (2023)
40	VBN (Rg)3	2005	Vamban 1[b] x Gulburga	Very early, resistant to SMD, tolerant to pod borer and high protein content	MoA&FW, GoI (2024)
41	Pusa 2001	2006	GP No-148 x UPAS 120[a]	Medium tall, maturity (140–150 days), seed yield 18–20 q /ha	MoA&FW, GoI (2024)
42	TT-401	2007	Fast neutrons, ICPL84008 x TT-6[a]	High yield, tolerance to *Fusarium* wilt, pod borer, and pod fly damage	Souframanien and Dhanasekar (2023)
43	VL Arhar 1 (ICPL88039)	2007	ICP 6 × Pant A-2[a]	Indeterminate, semi-erect, resistant to wilt and *Alternaria* blight	MoA&FW, GoI (2024)
44	TJT-501	2009	Fast neutrons, ICPL84008 x TT-6[a]	Resistant to *Fusarium* wilt and *Phytophthora* blight	Souframanien and Dhanasekar (2023)
45	PKV-TARA	2013	Fast neutrons, ICPL84008 x TT-6[a]	tolerant to *Fusarium* wilt and SMDs	Souframanien and Dhanasekar (2023)
46	CO8	2017	APK1[a] x LRG41	Early, high-yielding resistant to SMD and tolerant to root rot, moderately resistant to pod fly and pod borer, drought tolerance	MoA&FW, GoI (2024)
47	CO-9	2019	CO6[a] x IC52547	Large seeds (9.9g/100 seed), high protein, moderately resistant to wilt, SMD, *Maruca* and pod fly	Souframanien and Dhanasekar (2023)

[a] Mutant.
[b] Mutant derivative.

FIGURE 3.2 Gamma rays induced mutation in pigeonpea cultivar TT-201: (1) and (2). Dwarf mutants (3). Basal branching mutant (4). Unifoliate leaf mutant (5). Multifoliate leaf mutants with 3, 4, 5, 6, 7, 8, and 9 leaflets (6). White testa color mutant (R) and parent with brown testa (L).

diversity. Induced mutation can help broaden genetic diversity by creating novel alleles (Figure 3.2).

Thus, it caters to newer breeding challenges, especially when desirable traits are absent in the natural population or have been lost during evolution (Souframanien and Dhanasekar, 2023). It can also help break linkages between beneficial and unwanted and unfavorable genes. Mostly, pleiotropy is undesirable, but it could prove rewarding, especially in mutation breeding when mutations affecting a single gene could bring out simultaneous changes in several traits (Lobo, 2008). Mutations can aid studies through forward and reverse genetics to unravel the underlying molecular mechanisms leading to mutant phenotypes that could be suitably exploited by genome editing tools to swiftly bring out targeted and desired changes in different genetic backgrounds efficiently (Jankowicz-Cieslak and Till, 2015). Conventional breeding can be time-consuming, often requiring multiple generations to achieve desired traits. Mutation breeding can produce desired variations in a shorter time frame, accelerating the development of new cultivars. Unlike conventional breeding, where multiple generations may be needed to fix a trait, mutation breeding can directly induce stable mutations, leading to rapid gain in homozygosity within a shorter time frame.

Initial studies in induced mutagenesis in pigeonpea were limited to identifying effective and efficient doses of different physical and chemical mutagens and the effect of these mutagens on inducing genetic variation for agronomically important traits. The origin of the cultivated form of pigeonpea from a wild species, *Cajanus cajanifolius*, has been attributed by different taxonomical, cytological, and genomic studies to various natural mutational and selection events (van der Maesen 1980, Varshney et al., 2017). Early maturing pigeonpeas probably originated through mutation (Saxena et al., 2019). A spontaneous mutant "T-1" was identified in 1953 in Gorakhpur District of Uttar Pradesh and was used in hybridization to develop the first early maturing pigeonpea cultivar "T-21" in 1961 (Pathak & Singh, 1961). Since then, several early maturing spontaneous mutants were identified and released as varieties (Saxena et al., 2019) (Table 3.2). The success of any mutation breeding program depends on selecting the right parental material, appropriate type (physical or chemical), and dose of mutagen. The mutagen dose should be optimized to induce maximum variability while avoiding unintended damages that could be achieved by conducting a "lethal dose (LD)" study. Appropriate doses suboptimal to LD50 (lethal dose for 50% of the population), usually GR50 (50% growth reduction), are chosen to maximize the chances of recovering useful mutations. The LD50 values for both physical and chemical mutagens in pigeonpea are given in Table 3.3.

Among the physical mutagens, gamma radiation has been widely used, while in the case of chemical mutagens, EMS finds prominence. Comparative studies involving physical and chemical mutagens invariably showed higher effectiveness of gamma rays over EMS or DES (Nadarajan and Ramalingam, 1982; Ariraman et al., 2015; Etther et al., 2019). The frequency of viable mutations was found to be higher at 100 Gy–200 Gy gamma rays (Khan and Veeraswamy, 1974; Ariraman et al., 2015; Sangle, 2016), 25–60 mM or 0.05% EMS (Ariraman et al., 2015; Sangle, 2016; Khan and Veeraswamy, 1974), or 0.02% SA (Sangle, 2016). Mutations affecting plant stature (tall, dwarf), leaves (large and small compact leaves, simple or multifoliate compound (Figure 3.2), pod size (small, large, two-seeded to five seeded pod types), branching (little to profuse), early or late flowering, pod color (light green or dark

TABLE 3.3
LD50 Values of Important Physical and Chemical Mutagens in Pigeonpea

S no.	Mutagen	LD_{50} Dose	Reference
1	Gamma rays	300 Gy	Palshetkar et al. (2022)
		200 Gy	Ariraman et al. (2014)
		300–400 Gy	Giri (2014), Gaur et al. (2018a)
		400 Gy	Bhatnagar and Gupta (2014)
2	Ethylmethanesulfonate (EMS)	0.2%	Palshetkar et al. (2022)
		0.04%–0.05% for 6 h	Omosun et al.(2021)
		25 mM	Ariraman et al. (2014)
		30–40 mM for 8 h	Giri (2014)
3	Methylmethanesulfonate (MMS)	0.01%–0.05% (v/v) for 4–6 h	Jadhav and Hegde (2016)
4	Sodium azide (NAN_3)	1.60 mM–2.69 mM	Chaudhary et al. (2021)

FIGURE 3.3 TJT-501, a very popular early maturing and high-yielding pigeonpea mutant variety in full bloom at a farmer's field in Jabalpur, Madhya Pradesh.

black), biotic and abiotic stress tolerance, and flooding tolerance have been widely obtained (Sangle, 2016, Gaur et al., 2024).

The success of mutation breeding in pigeonpea improvement is evident from the list of varieties developed (Table 3.2). It has been a valuable tool in developing varieties with traits such as early maturity, determinate and prostrate growth habits, disease resistance especially to *Fusarium* wilt, sterility mosaic disease (SMD) and root-rot diseases, drought tolerance, and improved yield. The utility of mutagens in creating a variety of morphological, biochemical, agronomical, biotic, and abiotic stress-tolerant mutants has been reported in the literature. For instance, the dearth of early maturing and high-yielding varieties in Madhya Pradesh and adjoining areas of Uttar Pradesh resulted in heavy yield losses due to the coincidence of frost damage during the flowering time. Shown in Figure 3.3, the release of an early maturing mutant variety "TJT-501" developed jointly by Bhabha Atomic Research Centre, Trombay, and Zonal Agricultural Research, Khargone (earlier under the jurisdiction of JNKVV, Jabalpur, and presently under RVSKVV, Gwalior) resulted in a significant shift in pigeonpea cultivation in Madhya Pradesh (Chauhan et al., 2016).

The area and productivity of pigeonpea almost doubled in this state, and the seed replacement rate increased nearly five times from 10.48% to 48.11% in the last decade (Souframanien and Ganapathi, 2021). Mutagenesis has the potential to induce variability in both positive and negative directions and therefore provides opportunities for contrasting selections (such as early and late flowering, tall and dwarf stature, fertile and sterile types, and determinate and indeterminate plant types) within the same mutation breeding program (Mahala et al., 1990).

By broadening the genetic base and overcoming the limitations of traditional breeding methods, mutation breeding continues to play a pivotal role in addressing the challenges of pigeonpea cultivation under changing climatic and agronomic conditions. Explorations for using novel mutagens like electron beams, ion beams, proton beams, and space mutagenesis must be encouraged to broaden the spectrum and type of mutations.

3.9 IMPORTANT CONSIDERATIONS IN MUTATION BREEDING

3.9.1 Selection and Purity of Parental Material

Selection and purity of parental material are of foremost importance underlying the success of mutation breeding. Depending on the objective, the parental genotypes should be chosen in such a manner that, excepting for the trait of interest, the genotype should have all elite agronomical and adaptation values. For instance, if the objective is to have a large seed size with a high-yielding background, the parental genotype can be decided in two ways. First, a genotype with large seed size but lacking in other yield-attributing characters can be used to induce mutations for all other traits except for seed size. Secondly, a genotype that is high-yielding but with a small seed size could be used to induce mutations only for large seed size. Among these, the second option is expected to be more successful, as only a single trait needs to be improved through mutation breeding. With pigeonpea often being cross-pollinated, the purity of the parental genotype cannot be ensured unless the parental material is grown under protected conditions like nets (to keep off the insect pollinators) for three or four generations prior to taking up irradiation. Also, selecting parental lines well-suited to the target environment ensures that beneficial mutations enhance local adaptability and productivity.

3.9.2 Selection of Appropriate Type and Dose of Mutagen

Physical mutagens like gamma rays, X-rays, and fast or thermal neutrons, and chemical mutagens like sodium azide, ethylmethanesulphonate, and methylmethanesulphonate have been used successfully to induce mutations in pigeonpea. But keeping in mind the infrastructural requirements, ease of handling of mutagens, penetration of mutagens through the irradiated biological material, and mutation efficiency, appropriate mutagens need to be selected. Gamma rays have been widely used in pigeonpea mutation breeding with great success (Souframanien and Dhanasekar, 2023). The use of an appropriate dose is also equally important, as it determines the rate of recovery of useful mutations and avoids lethality. Sublethal doses below LD50 values are known to avoid excessive damage or sterility (Souframanien et al., 2020). Table 3.3 shows the LD50 values reported by various workers in pigeonpea. Moreover, the dose is genotype specific, and hence, dose-response studies need to be undertaken to identify the most effective dose range for inducing mutations without affecting plant viability.

3.9.3 When and Where to Look for Mutation

Identifying and isolating beneficial mutations in mutation breeding requires a well-planned approach regarding timing (when) and locations (where) for observation and selection. The M_1 generation (first generation immediately after irradiation) should be observed for immediate mutagenic effects (e.g., chlorophyll mutations, sterility, growth abnormalities) to get a fair idea about the deposition of mutagenic energy within the biological system. This generation does not show the expression of mutation, as most of them are in a heterozygous condition (unless dominant mutations).

The second generation (M_2) is the key generation for mutation detection as mutations become homozygous, resulting in phenotypic expression. Selection for desirable traits or mutations should be undertaken in this generation or delayed to later generations (M_4 or M_5) for quantitative traits. The stability of the mutations should be studied in the M_3 generation and beyond. The lookout for mutations should be undertaken in each of the stages, seedling, vegetative, flowering, and harvest, depending on the trait of interest. Any of the organs (seeds, roots, leaves, stems, flowers, pods, branches) can carry mutations. For identifying biotic or abiotic stress resistance, pigeonpeas have to be evaluated in hot spots for diseases and pests or in stress-prone environments like drought, salinity, flooding, or high temperatures. Controls and replications are required to ensure accurate detection and validation. Since mutation is probability and chance based, screening of sufficiently large populations becomes necessary to identify rare beneficial mutations.

3.9.4 Pigeonpea Biology

Though pigeonpea is self-pollinated, its often cross-pollinated status involving insect pollination (Sultana and Saxena, 2021) demands due care in handling mutant populations. The mutant populations across different generations need to be grown under protected conditions from insects to avoid cross-pollination and cross-contamination, along with maintaining sufficient isolation distance between different populations. The mutants in M_2 and further generations should be bagged to ensure complete self-pollination. Long life cycle and photoperiod sensitivity also demand careful planning of mutation experiments.

3.9.5 Resource Intensive

Mutation breeding of pigeonpea is resource intensive, as the crop agronomy and phenology require a lot of land area for screening a sufficiently large population, taking into consideration the spacing, often cross-pollination behavior, isolation distances, long duration, and number of generations. The mutant population also needs to be protected from biotic and environmental stresses, making it a cost-, labor-, and time-intensive process.

3.10 CHALLENGES OF MUTATION BREEDING IN CLIMATE CHANGE SCENARIO

Mutation breeding holds potential for developing climate-resilient crops, but it faces several constraints that limit its effectiveness in addressing complex climate challenges. Most mutations induced are random and often deleterious. The frequency of beneficial mutations is extremely low, requiring large populations and extensive screening. Induced mutagenesis often produces pleiotropic effects, where mutations beneficial for one trait negatively impact others (crucial genes, regulatory regions, or metabolic processes). This can lead to trade-offs, such as reduced yield or poor adaptability, which are detrimental to climate-resilient breeding objectives (Loewe, 2008). Since mutations are mostly recessive, traits governed by dominant genes are extremely difficult to improve through mutation breeding, though not impossible. Most of the climate-smart traits (drought tolerance, heat tolerance, flooding tolerance, etc.) are governed by highly intricate polygenic genes (Ravikiran et al., 2025), making them less responsive to mutation breeding, which is less suited for addressing complex quantitative traits. The success of mutation breeding also depends on rapid and simple screening methodologies, and the lack of infrastructure simulating real life climate change conditions forbids the utility of mutation breeding alone in climate resilience genetic improvement programs. Mutation breeding is largely a trial-and-error, limited-target approach, and therefore, it cannot directly target genes conferring resilience to climate change. Climate change demands rapid development of resilient crop varieties, while mutation breeding of pigeonpea involves long cycles of generation advancement (long-duration crop), screening, and field trials for mutant stabilization and confirmation, making it slower compared to modern molecular breeding techniques. The baseline genetic variability in pigeonpea for resilience traits is limited, which reduces the potential for meaningful improvements through mutation breeding. While mutation breeding can contribute to climate-resilient agriculture, it is not a standalone solution. Its effectiveness depends on careful integration with modern breeding tools (genomics, genome editing, speed breeding, and biotechnological), efficient resource utilization, and a strategic focus on traits with high impact and feasibility.

3.11 MUTATION BREEDING CAN AID MODERN TOOLS TO ADDRESS CLIMATE CHANGE

Induced mutagenesis is not a standalone solution but rather a part of a comprehensive approach to climate-smart agriculture. Modern tools can significantly enhance the efficiency and precision of mutation breeding in pigeonpea for climate resilience by addressing its limitations. These tools allow better targeting, screening, and incorporation of desirable traits into breeding programs. Multi-omics tools (genomics, transcriptomics, proteomics, and metabolomics) can help identify genes and regulatory regions associated with climate-resilience traits and help prioritize genes or loci for targeted mutagenesis (Liu et al., 2022). Highly reliable molecular markers (SSRs, SNPs, etc.) can help track and validate mutations linked to climate-resilient traits and also enable marker-assisted selection (Bohra et al. 2017, Saxena et al. 2018). Genome-wide association studies (GWAS) and whole genome sequencing (WGS; Varshney et al., 2012) can help identify genetic variations associated with key traits and provide insights into the genetic basis of complex traits like drought resilience (Varshney et al., 2017). Gene expressions studies under specific stress conditions through transcriptomics and metabolomics can identify candidate proteins and metabolites for mutagenesis in stress-response pathways (Pahal et al., 2024; Karmakar et al., 2024). Gene editing tools like CRISPR (Prasad et al., 2024), can make use of the knowledge derived through reverse genetic approaches like TILLING (allelic variants producing desired phenotypes) for rapid and precise editing of those genes to produce suitable climate-smart phenotypes (Varshney et al., 2010). Double haploid technology produces homozygous lines in a single generation and can fast-track the development of stable mutant lines with desired traits (Srividya et al., 2023). Tissue culture technologies can aid in vitro mutagenesis for mutation induction at the cellular level and help in rapid propagation of promising mutant lines (Krishna et al., 2010). High-throughput phenotyping and stress screening platforms can automate and accelerate screening of large mutant populations using sensors, imaging, and data analytics under simulated climate stress conditions. Artificial intelligence (AI) and machine learning (ML)-based tools can analyze large datasets from genomic, transcriptomic, and phenotypic studies to predict promising mutants or genes for specific traits and optimize mutation breeding strategies and improve decision-making in mutant selection (Pawar et al., 2023; Gaithond and Murthy 2024). Climate modeling tools like Agricultural Production Systems Simulator (APSIM), InfoCrop, and Climate Models can predict the performance of mutant varieties under different climatic scenarios. A pigeonpea simulation model has been created that simulates the way weather, soil conditions, and agronomic management affect the development, growth, nitrogen accumulation, and yield of a variety of maturity types, ranging from extra-short to medium-duration (Robertson et al., 2001). A model for prediction of *Helicoverpa armigera* incidence during future climate change periods using MarkSim multimodel data anticipated a higher number of generations with shortened generation time in future climate periods due to an increase in projected temperature (Mathukumalli et al., 2016). In recent years, speed breeding has opened new opportunities to hasten growth and development of plants, thus reducing the breeding cycle length and accelerating selection decisions in plant breeding programs (Aggarwal et al. 2025). Integrating mutation with speed breeding can accelerate generation turnover under controlled conditions for rapid advancement of mutant lines and reduce breeding time for keeping pace with the dynamic challenges of climate change. A speed breeding protocol with a precise seed/pod chip-based genotyping technique in pigeonpea has been optimized by ICRISAT, catering to different maturity groups and facilitating the rapid development of climate-resilient varieties and hybrids (Gangashetty et al., 2024).

Hence, modern tools like genomics, high-throughput phenomics, tissue culture, and bioinformatics can significantly improve the efficiency of mutation breeding in pigeonpea. These tools not only enhance the precision of mutation induction and selection but also accelerate the development of climate-resilient pigeonpea varieties.

3.12 CONCLUSION

Breeding for climate-smart agriculture necessitates a paradigm shift in the outlook of the breeders. Breeding strategies need to be revisited considering the adversities of climate change and to offset the limitations of conventional breeding by adopting multipronged approaches. Mutation breeding is an effective tool, particularly for inducing novel traits when natural variability is insufficient. However, it requires patience, precision, and integration with other breeding methods to keep pace with the dynamic challenges of climate change. Mutation breeding is indispensable for creating variability in quantitative traits, but by leveraging climate-resilient technologies and exploring biotechnological interventions (like CRISPR/Cas9) that largely influence qualitative traits, it can provide a holistic approach and complement conventional breeding efforts to accelerate the development of climate-smart and combat-ready pigeonpea varieties capable of enduring highly unprecedented, unproductive, and unpredictable climate change.

REFERENCES

Aggarwal A, Jeena AS, Mehra K, Kumar B, Kashyap S, Yadav DK, Maurya AK, Venkatesh SC, Singla P, Bohra A (2025) Speed-bred crops for food security and sustainable agriculture. *Planta* 262: 34.

Ariraman M, Gnanamurthy S, Dhanavel D, Bharathi T, Murugan S (2014) Mutagenic effect on seed germination, seedling growth and seedling survival of pigeon pea (*Cajanus cajan* (L.) Millsp). *Int Lett Nat Sci* 21: 41–49.

Ariraman M, Bharathi T, Dhanavel D (2015) Induced mutagenic effectiveness and efficiency studies on pigeonpea (*Cajanus cajan* L. Millsp.). *Food Biol* 4: 1–4.

Bakala HS, Devi J, Singh G, Singh I (2024) Drought and heat stress: Insights into tolerance mechanisms and breeding strategies for pigeonpea improvement. *Planta* 259: 123.

Begna T (2021) Application of mutation in crop improvement. *Int J Res Agron* 4: 1–8.

Bhatnagar SK, Gupta S (2014) Mutagenesis in pigeonpea (*Cajanus cajan*) with gamma radiation. *Radiat Res Environ Stud* 5: 12–15.

Bohra A, Jha R, Pandey G, Patil PG, Saxena RK, Singh IP, Singh D, Mishra RK, Mishra A, Singh F, Varshney RK, Singh NP (2017) New hypervariable SSR markers for diversity analysis, hybrid purity testing and trait mapping in pigeonpea [*Cajanus cajan* (L.) Millspaugh]. *Frontiers in Plant Science* 8: 377.

Bohra A, Saxena KB, Varshney RK, Saxena RK (2020) Genomics assisted breeding for pigeonpea improvement. *Theor Appl Genet* 133: 1721–1737.

Bohra A, Tiwari A, Kaur P, Ganie S, Raza A, Roorkiwal M, Mir R, Fernie A, Smykal P, Varshney R (2022) The key to the future lies in the past: Insights from grain legume domestication and improvement should inform future breeding strategies. *Plant and Cell Physiology* 63(11): 1554–1572

Boukar O, Abberton M, Oyatomi O (2020) Introgression breeding in cowpea [*Vigna unguiculata* (L.) Walp.]. *Front Plant Sci* 11: 567425.

Chaudhary L, Sharma R, Kumar M (2021) Estimation of LD50 and effect of sodium azide on germination and seedling parameters of different cultivars of *Cajanus cajan* (L.) Millspaugh. *Toxicol Environ Health Sci* 13: 279–285.

Chauhan JS, Singh BB, Gupta S (2016) Enhancing pulses production in India through improving seed and variety replacement rates. *Indian J Genet Plant Breed* 76: 410–419.

Choudhary AK, Nadarajan N (2011) *Breeding improved cultivars of pigeonpea in India.* Kanpur: Indian Institute of Pulses Research (IIPR).

Dhanasekar P, Souframanien J (2023) Genetics and genomics of drought and heat tolerance in cowpea, mung bean and black gram. In *Legumes: Physiol Mol Biol Abiotic Stress*, ed. P.M.A. Samy, A. Ramasamy, V. Chinnusamy, and B. Sunil Kumar, 202–233. Singapore: Springer.

Dhanasekar P, Souframanien J, Suprasanna P (2021) Breeding cowpea for quality traits: A genetic biofortification perspective. In *Breed Enh Nutr Bioact Comp Food Legumes*, ed. Gupta DS, Gupta S, Kumar J 157–179. Cham: Springer.

Etther Y, Gajukar SJ, Akhare A, Patil, AN, Jambhulkar, SJ, Gawande M (2019) Mutagenic effectiveness and efficiency of gamma rays, ethyl methyl sulfonate and their synergistic effects in pigeonpea (*Cajanus Cajan* L.). *J Pharmacogn Phytochem* 8: 489–493.

FAO. (2023) http://faostat.fao.org

Gaithond R, Murthy V (2024) Exploring deep learning technique for detection of sterility mosaic disease in pigeonpea. *J Electr Syst* 20–9s: 1996–2001.

Gangashetty PI, Belliappa SH, Bomma N, et al. (2024) Optimizing speed breeding and seed/pod chip based genotyping techniques in pigeonpea: A way forward for high throughput line development. *Plant Methods* 20: 27.

Gaur AK, Singh I, Singh S, Reddy KS (2018a) Studies on effects of gamma ray doses on germination in pigeonpea [*Cajanus cajan* (L.) Millspaugh] under laboratory and field conditions. *Int J Chem Stud* 6: 1975–1977.

Gaur AK, Singh I, Singh S, Sharma S (2018b) Genetic analysis of M_4 mutant lines of pigeonpea (*Cajanus cajan* (L.) Millspaugh) developed through gamma irradiation. *J Food Legumes* 31: 71–74.

Gaur AK, Singh I, Kaur M, Singh S (2024) Induction of genetic variability through gamma rays in pigeonpea (*Cajanus cajan* (L.) Millspaugh). *J Food Legumes* 29: 212–215.

Giri SP (2014) Studies of mutagenic sensitivity in pigeonpea (*Cajanus cajan* (L.) Millsp.). *Biosci discov* 5: 227–229.

Hingane AJ, Saxena KB, Patil SB, et al. (2015) Mechanism of water-logging tolerance in pigeonpea. *Indian J Genet Plant Breed* 75(2): 208–214.

Jadhav MS, Hegde SN (2016) Mutagenic response of pigeonpea (*Cajanus cajan* L.) to chemical mutagenesis. *J Plant Mut* 8: 15–22.

Jankowicz-Cieslak J, Till BJ (2015) Forward and reverse genetics in crop breeding. In *Advances in Plant Breeding Strategies: Breeding, Biotechnology and Molecular Tools*, ed. Al-Khayri J, Jain S, Johnson D 215–240. Cham: *Springer*.

Joshi R, Ramawat N, Sah RP, et al. (2022). Assessment of salt tolerance potential at the germination and seedling stages in pigeonpea (*Cajanus cajan* L.). *J Indian J Genet Plant Breed* 82: 311–323.

Karmakar S, Selvaraj S, Gayen D, Baig MJ (2024) A detailed proteomics and metabolomics landscape sheds light on the mechanistic insights into the resistance response of transgenic pigeon pea against wilt stress. *Plant Cell Environ* doi:10.1111/pce.15261

Khan WMA, Veeraswamy R (1974) Mutations induced in red gram (*Cajanus cajan* (L) Millsp.) by gamma radiation and EMS. *Radiat Bot* 14: 237–242.

Krishna G, Reddy PS, Ramteke PW, et al. (2010) Progress of tissue culture and genetic transformation research in pigeon pea [*Cajanus cajan* (L.) Millsp.]. *Plant Cell Rep* 29: 1079–1095.

Kumar A, Kumar S, Kumar RR, et al. (2022) Pigeon pea genetic resources and it's utilization: Current status and future prospects. In *Exploration in Genomics a Contemporary Approach*, ed. Katkani D, Tiwari S, S Sharma 157–180. Agra: SR Scientific Publication.

Lal S, Sinha SC (1972) Prabhat- an extra early variety of red gram. *Farm Par J* 7: 18–24.

Liu C, Tai Y, Luo J, et al. (2022) Integrated multi-omics analysis provides insights into genome evolution and phosphorus deficiency adaptation in pigeonpea (*Cajanus cajan*). *Hortic Res.*

Lobo I (2008) Pleiotropy: One gene can affect multiple traits. *Nat Educ* 1: 10.

Loewe L (2008) Genetic mutation. *Nat Educ* 1: 113.

van der Maesen LJG (1980) India is the native home of the pigeonpea. In *Liber Gratulatorius in nonerem*, ed. Arends JC, G Boelema, CT Groot, and AJM Leeuwenberg, 257–262, Wa- geningen: H.C.D. de Wit, Agricultural University.

van der Maesen LJG (1991) Pigeonpea: origin, history, evolution and taxonomy. In *The Pigeonpea*, ed. Nene YL, 15-–46. Oxon, UK: CABI.

Mahala SVS, Mor BR, Yadav JS (1990) Induced genetic variability for oil content in mustard. (*Brassica juncea* L. Czern and Coss). *Oil Crops Newsletter* 7: 13–15.

Mathukumalli SR, Dammu M, Sengottaiyan V, et al. (2016) Prediction of *Helicoverpa armigera* Hubner on pigeonpea during future climate change periods using MarkSim multimodel data. *Agric For Meteorol* 228–229: 130–138.

Mishra S, Singh R, Kumar R, Kalia A, Panigrahy SR (2017) Impact of climate change on pigeon pea. *Economic Affairs* 62: 455–457.

MoA&FW. (2024) Agricultural Statistics at a Glance 2023. Ministry of Agriculture and Farmers welfare, Government of India, 40–41. (Accessed https://desagri.gov.in/wp-content/uploads/2024/09/Agricultural-Statistics-at-a-Glance-2023.pdf).

MoA&FW, GoI (2024) Ministry of Agriculture and Farmers Welfare, Government of India. https://seednet.gov.in/SeedVarieties/CentralVariety.aspx (accessed 31.12.2024).

Nadarajan, N., and R.S. Ramalingam. 1982. Mutagenic effectiveness and efficiency in *Cajanus cajan* (L) Millsp. *Madras Agric J* 69: 71–75.

Naik SJS, Singh IP, Bohra A, et al. (2020) Analyzing the genetic relatedness of pigeonpea varieties released over last 58 years in India. *Indian J Genet Plant Breed* 80: 70–76. doi:10.31742/IJGPB.80.1.9

Odeny DA (2007) The potential of pigeonpea (*Cajanus cajan* (L.) Millsp.) in Africa. *Nat Resour Forum* 31: 297–305.

Omosun G, Ekundayo EO, Okoro IA, Ojimelukwe PC, Egbucha KC, Akanwa FE (2021) Preliminary study on the effect of different concentrations of EMS on two pigeon pea (*Cajanus cajan* L. Millsp.) Accession. *Sci Afr* 22: 58–66.

Pahal S, Srivastava H, Saxena S, et al. (2024) Comparative transcriptome analysis of two contrasting genotypes provides new insights into the drought response mechanism in pigeon pea (*Cajanus cajan* L. Millsp.). *Genes Genom* 46: 65–94.

Palshetkar MG, Sawardekar SV, Dalvi VV, Narangalkar AL, Dhekale JS (2022) Effect of mutagenic treatments on seed germination, seedling growth and survival of pigeon pea [*Cajanus cajan* (L.) Millsp.]. *Pharma Innov* 11: 1160–1164.

Pathak GN, Singh KP (1961) Type 21 arhar (pigeonpea) out does Type 1. *Indian Farming* 11: 15.

Pawar SY, Ghante PH, Hingole DG, Patil LP, Thomse SR (2023) Exploring artificial intelligence technique for detection of pigeonpea sterility mosaic disease. *Pharma Innov* SP-12: 482–489.

Penna S, Shirani Bidabadi S, Jain SM (2023) Mutation breeding to promote sustainable agriculture and food security in the era of climate change. In *Mutation Breeding for Sustainable Food Production and Climate Resilience*, ed. Penna S, and Jain SM, 1–23, Singapore: Springer.

Prabhu KR, Kumar A, Yumkhaibam RS, Janeja HS, Krishna B, Talekar N (2023) A review on conventional and modern breeding approaches for developing climate resilient crop varieties. *J Appl Nat Sci* 15: 987–997.

Prasad K, Gadeela H, Bommineni PR, et al. (2024) CRISPR/Cas9-mediated mutagenesis of phytoene desaturase in pigeonpea and groundnut. *Funct Integr Genomics* 24: 57.

Ravikiran KT, Thribhuvan R, Anilkumar C (2025) Harnessing the power of genomics to develop climate-smart crop varieties: A comprehensive review. *J Environ Manage* 373: 123461.

Robertson MJ, Carberry PS, Chauhan YS, Ranganathan R, O'Leary GJ (2001) Predicting growth and development of pigeonpea: a simulation model. *Field Crop Res* 71: 195–210.

Sangle SM (2016) Induced mutagenesis in pigeonpea (*Cajanus cajan* (L.) Millsp.). *Int J Curr Res* 8: 33953–33957.

Saxena RK, Rathore A, Bohra A, Yadav P, Das RR, Khan AW, Singh VK, Chitikineni A, Singh IP, Sameer Kumar CV, Saxena KB, Varshney RK (2018) Development and application of high density Axiom®CajanusSNP Array with 56K SNPs for understanding genome architecture of released varieties and founder genotypes for redefining future pigeonpea breeding programmes. *The Plant Genome* 11: 180005.

Saxena K, Choudhary AK, Srivastava RK, Bohra A, Saxena RK, Varshney RK (2019) Origin of early maturing pigeonpea germplasm and its impact on adaptation and cropping systems. *Plant Breed* 138: 243–251.

Saxena KB, Bohra A, Choudhary AK, Sultana R, Pazhamala L, Saxena RK (2021) The alternative breeding approaches for improving yield gains and stress response in pigeonpea [*Cajanus cajan* (L.) Millsp.]. *Plant Breed* 140: 76–86.

Sharma D, Green JM (1980) Pigeonpea, In *Hybridization of crop plants*, ed. Fehr, W.R., and H.H. Hadley, 471–481. WI, USA: American Society of Agronomy and Crop Science Society of America.

Sharma KK, Sreelatha G, Dayal S (2006) Pigeonpea (*Cajanus cajan (L.) Millsp.*). In *Agrobacterium Protocols. Methods Mol Biol* 343(4): 359.

Sheriff NM, Khan WMA, Annappan RS (1977) Red gram Co.3 - an economic mutant strain for Tamil Nadu. *Madras Agric J* 64: 561–564.

Siddique KHM, Loss SP, Regan KL, Jettner RL (1999) Adaptation and seed yield of cool season grain legumes in Mediterranean continents of South Western Australia. *Aust J Agric Res* 50: 75–387.

Singh BB, Gupta SC, Singh BD (1974) Note on 'UPAS-120' an early-maturing mutant of pigeonpea. *Indian J Agric Sci* 54: 233–234.

Souframanien J, Dhanasekar P (2023) Potential of mutation breeding in genetic improvement of pulse crops. In *Mutation Breeding for Sustainable Food Production and Climate Resilience*, ed. Penna, S., and S.M. Jain, 445–485. Singapore: Springer. doi:10.1007/978-981-16-9720-3_15

Souframanien J, Ganapathi TR (2021) Mutation breeding in India: Accomplishments and socio-economic impact. *Plant Breeding and Genetics Newsletter* 46: 23–25.

Souframanien J, Saha AJ, Dhole VJ, Dhanasekar P, Misra G (2020) Genetic improvement of pulse crops through induced mutation and biotechnological approaches. *IANCAS Bulletin* XV: 71–80.

Srividya A, Sandal SS, Walia P (2023) Doubled haploids in crop improvement: Unraveling strategies, advancements and prospects for enhanced genetics. *International Journal of Plant & Soil Science* 35: 348–358.

Subbarao, GV, Johansen C, Jana MK, Kumar Rao JVDK (1990) Physiological basis of differences in salinity tolerance of pigeonpea and its related wild species. *J Plant Physiol* 137: 64–71.

Sultana R Saxena KB (2021) Natural cross-pollination is both a boon and bane for pigeonpea breeders. *Legume Res* 44: 493–500.

Varshney RK, Penmetsa RV, Dutta S, et al. (2010) Pigeonpea genomics initiative (PGI): an international effort to improve crop productivity of pigeonpea (*Cajanus cajan* L.). *Mol Breed* 26: 393–408.

Varshney R, Chen W, Li Y, et al. (2012) Draft genome sequence of pigeonpea (*Cajanus cajan*), an orphan legume crop of resource-poor farmers. *Nat Biotechnol* 30: 83–89.

Varshney R, Saxena R, Upadhyaya H, et al. (2017) Whole-genome resequencing of 292 pigeonpea accessions identifies genomic regions associated with domestication and agronomic traits. *Nat Genet* 49: 1082–1088.

4 Pigeonpea Hybrids for Enhanced Resilience and Productivity

K B Saxena, R Sultana, Pallavi Sinha, Vijay A. Dalvi, Rachit K. Saxena, and Prakash Gangashetty

4.1 INTRODUCTION

The semiarid tropical regions of south Asia and Africa, where the population explosion is dangerously high, suffer from malnutrition that is linked to protein-energy starvation. According to FAO (2021) there is no respite for this global health-related issue, and it was revealed in 2022 that over 700 million people faced hunger, an increase of 122 million cases of starvation in 2019. This institution also projected that in 2030 almost 600 m people will be chronically under-nourished, and perhaps half of them may fail to survive due to severe poverty-driven malnutrition. This scenario assumes greater significance in the backdrop of under-par (24 g/head/day) protein availability among the rural population, and it is approximately half of the normal requirement of an adult (NIN, 2010). It is believed that this situation has emerged due to reasons such as uncontrolled population growth, productivity stagnation of cereals and legumes, and expansive animal protein resources. Shalendra et al. (2013) revealed that over the decade (2000 to 2010), there has been about 10%–11% decline in the consumption of cereals, pulses, and sugar in both rural and urban India. The aforementioned factors pose serious challenges with respect to global food and nutritional security. In this context pulses, known for their high carbohydrate and protein, offer potential scope.

The first pioneering research on nutritional qualities of pulses was carried out by Pal (1939), and his results showed that pigeonpea (*Cajanus cajan* (L.) Millsp.) was the best source of biological protein for humans and further suggested that it should be eaten with rice to make a balanced diet. Some researchers, including Daniel et al. (1970), Hulse (1977), and Bidinger and Nag (1981), opined that increasing the pulse supplement in cereal-based diets can markedly improve their nutritional quality. Among the pulses, pigeonpea ranks high with respect to both the production and nutritional sustainability owing to its nutritive grains, drought tolerance, important soil amelioration properties, and multiple usage (Saxena et al., 2021). Globally, pigeonpea is cultivated at 6.36 m ha in 22 countries with production of 5.48 m tons and average productivity of 860 kg/ha (FAOSTAT, 2021). Besides India, Malawi, Myanmar, Tanzania, and Kenya are the other major pigeonpea growing areas.

DOI: 10.1201/9781003603092-4

In India, pigeonpea is extensively (5.6 m ha) cultivated and consumed in a variety of regional recipes. In fact, the annual production of 3.2 m t is not enough to meet the consumption demands of over one billion Indians. Hence, the country annually imports over 500,000 tons of pigeonpea from Myanmar and Africa (http://www.ipga). In spite of its high importance in the national economy and nutritional security, pigeonpea remains a second-grade crop as far as the national research and development priorities are concerned. Due to nonavailability of high-yielding cultivars, only landraces were cultivated for decades. However, in the early part of the twentieth century, pure line breeding activities were launched at some Indian research centers. Soon significant accomplishments were recorded in breeding for earliness, plant type, seed traits, and disease resistance (Ramanujam and Singh, 1981). But unfortunately, the productivity of pigeonpea remained low and unchanged, and the issue of yield plateau in pigeonpea is a major concern, and it still haunts the scientists.

4.2 WHY IS PIGEONPEA CONSIDERED A SUSTAINABLE LEGUME?

- The continuous exposure of undulating farming lands to natural vagaries such as fluctuation in radiations, soil erosion-inducing winds, spells of excessive rain or drought, and faulty crop husbandry often leads to loss of valuable topsoil, organic matter, and nutrients. Over a period of time, such agricultural territories have become unfit for the economic cultivation of food crops. These farming lands are either abandoned, converted into grazing lands, or used for cultivating some drought-tolerant, low-value kharif crop. Also, some early maturing pigeonpea cultivars have been successfully introduced in these areas with the twin objectives of rejuvenating eroded soils and, in addition, harvesting some quantity of grains for domestic consumption. This aspect of dryland agriculture has been discussed by a number of reviewers and lately by Saxena et al. (2021). The key points revealed that the pigeonpea crop contains some special traits, listed next, which help in enhancing sustainability of the crop.
- Deep roots of pigeonpea can extract soil moisture from deep layers.
- Root nodules fix atmospheric nitrogen at 40–50 kg/ha.
- Its roots also release soil-bound phosphorus to make it available to the standing crop.
- Deep and extensive lateral root systems of pigeonpea facilitate the arrest of loose topsoil to reduce the soil erosion.
- In the rainy season, the canopy coverage is rapid and helps in reducing erosion.
- Long-term cultivation of pigeonpea improves soil structure and water infiltration.
- The pigeonpea plants help in recycling the nutrients and adding organic matter through huge leaf fall during reproductive period.
- The hard soil pans, formed at the plow depth in heavy soils, can be penetrated easily by the deep root system of pigeonpea. This improves the drainage and root growth.

Growing a successful pigeonpea crop with zero tillage has been well demonstrated in the barren and eroded land areas of Uttarakhand and Chinese hills. These crops can potentially yield around 600–700 kg/ha grains. But this productivity level is considered not good enough for large-scale adoption, and, therefore, breeding of new cultivars with traits that can ably encounter various abiotic stress factors is required.

4.3 MAJOR ABIOTIC STRESSES IN BREEDING HYBRID PIGEONPEA

There are three major abiotic constraints that adversely affect productivity and resilience in pigeonpea under different ecosystems. In general, the losses from each of them may be inconsistent but significant. These constraints are drought, waterlogging, and soil salinity. The information on these critical aspects is briefly presented here.

4.3.1 DROUGHT

Drought is a universal crop production constraint, particularly under subsistence agriculture where irrigation sources are limited or not available. It is a complex scientific phenomenon, and, therefore, genetic enhancement for drought tolerance is a challenge, particularly under rainfed environments due to inconsistency in the precipitation over the seasons and locations. Besides these, the large genotype x environmental interactions also affect the expression of yield and its components.

In the case of pigeonpea, a major rainy-season legume of rainfed agriculture, often drought spells of different durations are experienced either at the seedling, mid-crop growth, or reproductive stage. In addition, changes in climate parameters also add to this issue by influencing the time, duration, and intensity of drought events. Pigeonpea is cultivated both on Vertisols and Alfisols. The Vertisols are deep and hold <200 mm of plant-available water to a depth of 1.5 m at the end of the rainy season. In contrast, the Alfisols are usually <1 m deep and hold about 90 mm plant-available water up to a depth of 1 m (Reddy and Virmani, 1981). For commercial cultivation in a particular ecosystem, pigeonpea cultivars are recommended in such a way that the available moisture is fully utilized. Hence, the longer duration pigeonpeas are grown in Vertisols and conversely in Alfisols, the earlier maturing types are cultivated. But still the drought incidences may harm the crops in these two ecosystems.

The phenotype-based screening for drought tolerance under field conditions is very difficult and results in poor genetic advance due to season-to-season variability in the timing and intensity of drought being unpredictable. In contrast, the screening for terminal drought could be relatively easier. Loomis and Connor (1992) established three groups on the basis of mechanisms involved in encountering drought. These are (i) escaping the drought, where the crop tries to complete its reproductive growth before the soil water deficit becomes too severe; (ii) drought avoidance, where the crop either minimize the water loss from their tissues or enhance water absorption even under drought conditions; and (iii) drought tolerance, where the crop enhances the physical and/or physiological capability of their cells to continue metabolism even at low plant water status.

Due to the unpredictable nature of the time, duration, and frequency of droughts, the genetic enhancement for drought tolerance is still an unachievable goal in most field crops. In comparison to cereals, the quantum of drought research in pigeonpea (both laboratory and field) is far less, and even basic knowledge on the critical issues is lacking. However, limited studies conducted with pigeonpea hybrids have demonstrated that the hybrids have clear advantage over inbred cultivars in tolerating drought (Saxena et al., 1992; Lopez et al., 1996; Yu and Tuinstra 2001, Saeki et al., 2016). Therefore, at present and in the near future, an obvious drought-tolerant variety development approach should hover around drought avoidance and hybrid breeding.

4.3.2 Waterlogging

In pigeonpea, both short- and long-term waterlogging are serious production constraints, particularly in high rainfall areas and poorly drained soils. The losses are irreversible, as the root growth of the plants is seriously hampered. This situation can inflict huge losses on yields, especially in poorly drained habitats. The screening for waterlogging resistance has helped in identifying a few tolerant germplasm and hybrids (Sultana et al., 2013). It seems that this production constraint has not been given its due research attention and allocations. Waterlogging tolerance in pigeonpea involves various morphological, physiological, and metabolic adaptations parameters, and due research is needed to decipher the key genetic and physiological aspects.

4.3.3 Soil Salinity

Soil salinity is a growing challenge for pigeonpea production, particularly in India where saline-alkaline soils cover approximately 7.0 M ha. The crop has a salinity threshold level of <1.3 dS/m. In fact, at present, it is not a priority research area due to its complex nature and various interactions involving sodium and chloride exclusion from shoots and retention in roots, etc. Notably, an inter-specific hybrid between *Cajanus albicans* and pigeonpea genotype ICP 3783 was found resistant to salinity, and the resistance was found to be controlled by a dominant gene trait (Subbarao et al., 1991).

4.4 DEVELOPMENT OF HYBRIDS: A GIANT STEP IN BREEDING PIGEONPEA

Historically, the value of hybrid vigor in enhancing yield was globally recognized when the commercial maize hybrids saved the world from hunger and malnutrition in the early half of the twentieth century. According to Troyer (1991), after the introduction of maize hybrids, the on-farm yields of maize increased by more than fivefold. The phenomenal success of hybrid maize in the USA led the plant breeders to apply this technology in other crops, such as cotton, sorghum, pear millet, sunflower, brassica, and various horticultural crops. In the legumes, however, only limited efforts were made in the past to develop breeding technology. These crops include faba bean (Bond, 1987), soybean (Palmer et al., 2001), and pigeonpea (Saxena, 2015).

In order to smash the low yield plateau in pigeonpea, plant breeders at ICRISAT planned to exploit hybrid vigor in the crop by using its natural cross-pollination (Saxena et al., 1990). It was, indeed, a Herculean task since no hybrid was ever bred in any food legume. The pigeonpea hybrid development programs were launched with the breeding of stable cytoplasmic nuclear male sterility (CMS) systems. From these efforts, two CMS systems (A_2 and A_4) were successfully bred, and, as a follow-up, the process of breeding hybrids began in medium and early maturity groups. Since it was a new proposition, some odd issues persisted that would need amicable solutions. To overcome these, there is a need to address these critical issues. We hope that these efforts will lead to further enhancement of productivity, resilience, and adaptation of new pigeonpea hybrids.

4.4.1 Breakthrough in Breeding Cytoplasmic Nuclear Male Sterility Systems

In a given plant system, the CMS arises due to certain deleterious interactions between its cytoplasmic and nuclear genomes (Bohra et al., 2016, 2025). The hybrid breeding programs based on the CMS system revolve around three distinct genotypes—namely, male sterile (A-), its maintainer (B-), and fertility restorer (R-) lines. To develop such a system, the breeding programs were launched both at ICRISAT and Indian Council of Agricultural Research (ICAR) by placing the nuclear genome of a cultivated genotype into the cytoplasm of wild *Cajanus* species of the secondary gene pool. Tikka et al. (1991) reported the selection of male-sterile plants within an open-pollinated F_2 population involving *C. scarabaeoides* as female and pigeonpea as male parent. This male sterility, designated as A_2, was maintained by an inbred line ICPL 288. This male-sterile source exhibited high stability across diverse environments, and, subsequently, it was used in developing hybrids in the Gujarat state of India. Using the A_2 CMS system, a pigeonpea hybrid GTH 1, with >30% standard heterosis, was released for cultivation in Gujrat state (Chauhan et al., 2004). Subsequently, (Saxena et al., 2005), bred another CMS system from a cross between a wild species (*C. cajanifolius*) and pigeonpea inbred line ICP 28. The F_1 hybrid plants of this cross were completely male sterile with and perfect female fertility. This CMS system, designated as A_4, was highly stable across environments. Four pigeonpea hybrids IPH 15-03 (2019), IPH 09-5 (2020), and Pusa Arhar Hybrid 5 (Pusa Arhar Yamuna) (2022), and BDNPH 2018-05 (2024) were developed and released for commercial cultivation. The pigeonpea hybrids developed recently at ICAR-Indian Institute of Pulses Research (IIPR), Kanpur are shown in Figures 4.1 and 4.2. The pigeonpea hybrid IPH 15-03 showed 28.3% yield superiority over the best performing genotypes over four years of multilocation field evaluations. Similarly, another hybrid IPH 9-05 registered 30.5% yield advantage over the best performing pigeonpea genotypes over multiple years of field-testing.

FIGURE 4.1 The pigeonpea hybrid IPH 9-05.

FIGURE 4.2 The pigeonpea hybrid IPH 15-03.

4.4.2 Heterosis in Early, Medium, and Late-Maturing Hybrids

As observed in most crops, the elite pigeonpea hybrids have also demonstrated greater productivity and adaptation than inbred cultivars in most growing environments (Saxena et al., 2018). Considering the requirements of crop diversification, pigeonpea breeders developed a number of hybrids in the three major maturity groups for multilocation testing (Table 4.1). The sets of early maturing hybrids were tested at 25 locations across India. The top three early maturity hybrids were ICPH 2433 (2306 kg/ha), ICPH 2438 (2127 kg/ha), and ICPH 2363 (2048 kg/ha). These hybrids recorded, respectively 54%, 42%, and 36% standard heterosis. In the medium duration group, hybrids ICPH 3491, ICPH 3497 and ICPH 3481 were found promising with, respectively, 57%, 44%, and 41% standard heterosis in 18 testing locations. In comparison to early and medium groups, the number of hybrids tested in the long-maturity group were fewer. Their testing was done only at five locations, and hybrids ICPH 2307 (53% heterosis), ICPH 2306 (39% heterosis), and 2896 (38% heterosis) exhibited significant hybrid vigor.

4.4.3 On-Farm Superiority of Hybrids under Diverse Ecosystems

ICRISAT, in collaboration with various ICAR centers and nongovernmental organizations, organized extensive on-farm testing of the three high-yielding medium maturing pigeonpea hybrids ICPH 2671, ICPH 2740, and ICPH 3762. These trials were conducted by rainfed cultivators using the best local cultivar as a control. Each trial was sown in one acre plot using farmers' own cultural practices. All three hybrids expressed considerable heterosis for grain yield over the local control cultivar, suggesting high adaptation of hybrids under diverse ecosystems (Saxena and Tikle, 2015). Hybrid ICPH 2671, released in the state of Madhya Pradesh, was evaluated in 1541 on-farm trials in three states (Table 4.2). The mean superiority of the

TABLE 4.1
Hybrid Advantage for Yield Recorded in Early, Medium and Late-Maturing Hybrids in Multilocation Trials

Maturity Group	Hybrid Number	No. of Locations	Mean Yield (kg/ha)	Stand. Hete (%)
Early	ICPH 2433	25	2306**	54
	ICPH 2438	25	2127**	42
	ICPH 2363	25	2048**	36
Medium	ICPH 3491	18	2919**	57
	ICPH 3497	18	2686**	44
	ICPH 3481	18	2637**	41
Late	ICPH 2307	05	2855**	53
	ICPH 2306	05	2600**	39
	ICPH 2896	05	2579**	38

Source: Various ICRISAT Pigeonpea Breeding Reports.

** Significantly superior over the check

TABLE 4.2
Demonstration of Yield (kg/ha) Advantages Recorded in Three Released Pigeonpea Hybrids over Inbred Cultivar in the 1772 On-Farm Trials Conducted under Rainfed Conditions in Four States of India

Hybrid	State	Farmers (No.)	Hybrid Yield	Control Yield	Hybrid Adv (%)
ICPH 2671	Maharashtra	782	969	717	35
	Andhra Pradesh	399	1411	907	55
	Madhya Pradesh	360	1940	1326	46
	Total/mean	1541	1437	983.3	46.1
ICPH 2740	Andhra Pradesh	47	1999	1439	39
	Gujarat	40	1633	1209	35
	Total/mean	87	1816	1324	37.2
ICPH 3762	Odisha	144	1726	813	112

hybrid over the control was 46.1%. Another hybrid ICPH 2740, released in Andhra Pradesh, was tested in 87 trials in two states. On average, this hybrid recorded 37.2% superiority over the inbred control. For hybrid ICPH 3762, the third hybrid released in the state of Odisha, a total of 144 on-farm trials were organized in five districts and the hybrid (1726 kg/ha), on average, expressed 112% yield advantage over the best adapted cultivars. The on-farm data obtained from these trials showed that the pigeonpea hybrids can be a good option for enhancing the production and productivity of pigeonpea. From various published research, it is concluded that "*[i]n pigeonpea significant hybrid vigour for grain yield is available and it should be exploited for the benefit of farming communities.*"

4.5 SOME KNOWN PARAMETERS ASSOCIATED WITH PRODUCTIVITY AND ADAPTATION IN PIGEONPEA HYBRIDS

The phenomenon of hybrid vigor is quite complex, and in spite of extensive research in the areas of quantitative genetics, physiology, and biochemistry research, the mystery about the expression of hybrid vigor for yield is still behind the curtains. However, a number of critical observations were made in the past toward understanding the phenomenon of hybrid vigor. These, of course, may vary across the genetic materials and environments used in generating empirical data sets and some of them related to hybrid productivity, adaptation, and resilience are briefly discussed in the following text.

4.5.1 Seedling Parameters Related to Resilience

While studying the effects of seedling hybridity on the expression of some key adaptative traits in pigeonpea, (Saxena et al., 2020), revealed that the genetic hybridity in pigeonpea seedlings not only enhanced their germination but also rooting parameters and overall growth and development.

4.5.1.1 Rapid Germination

Pigeonpea has a hypogeal type of germination where the cotyledons remain in the soil, and the shoot is carried above the soil level by elongation of epicotyl. Studies conducted by Mudaraddi and Saxena (2012) showed that the hybrid seeds germinated three to six days prior to inbred seeds. In addition, it was also observed that white-colored seeds germinated earlier than those having dark seeds. The white seeded genotypes have thin seed coats, which facilitate rapid water absorption and germination. Rajjou et al. (2012), while studying the seedling traits, revealed that the main contributor to seed germination success was the quality of m-RNA that was stored during embryo maturation on the mother plant.

4.5.1.2 Radicle and Plumule Growth

In an experiment conducted under controlled conditions, Mudaraddi and Saxena (2012) revealed the effects of hybridity on growth parameters of radicle and plumule. They observed that in ten-day-old seedlings, the radicles in the hybrids were 17.8% longer than that of inbred seeds, but in contrast, the mean plumule length in hybrids was 6.1% less than inbreds. The total dry weight accumulated by the seedlings were also 23.69% greater than inbred cultivars. Besides, the hybrids recorded 14.79% greater seedling vigor index than the control cultivar. Almost similar results were also reported by Thakre et al. (1991). Studies conducted by Romagnoti et al. (1990) showed the presence of different proteins and mRNAs in the primary root tips of hybrid plants but not in the inbred parents. Abreu et al. (2019), also reported heterosis in maize seedlings more expressively for root length and number of seminal roots. The selection for these traits in breeding populations can help in imparting resilience to pigeonpea inbred and hybrid cultivars. The selection for these traits in breeding populations can help in imparting resilience to pigeonpea inbred and hybrid cultivars.

4.5.1.3 Emergence of Aerenchyma, Lenticels, and Adventitious Roots

Waterlogging is a widespread production constraint for pigeonpea. Most of the genotypes are susceptible to this constraint, and potentially, it can inflict huge losses to yields, especially in high rainfall and poorly drained habitats. The significant biological consequence of waterlogging is the deficiency (hypoxia) or complete absence (anoxia) of oxygen in the soil; this sets anerobic conditions and restricts the plant growth and development, and consequently the seed yield. The tolerant pigeonpea genotypes can adapt to transient waterlogging by inducing certain built-in mechanisms. The common adaptation features of plants to waterlogging condition are the development of adventitious roots and lenticels with the formation of aerenchyma cells (Shimamura et al., 2010).

4.5.1.4 Shoot and Root Growth

In a field study (Saxena et al., 1992), reported that between four and seven weeks, the hybrid plants exhibited significant levels of hybrid vigor over the pure line control both for roots and shoots (Table 4.3). The roots of hybrid plants recorded 24.21 to 32.33% heterotic advantage over inbred controls, while the corresponding values for shoot varied from 28.49% to 52.94%. These traits of the hybrid seedling not only

TABLE 4.3
Mean Root and Shoot Dry Weights (g/plant) of Hybrid and Inbred Cultivars at Two Seedling Stages

DAS/ Plant Portion	Hybrid	Cultivar	% Hybrid Advantage
30 Days			
Shoot	1.597[a]	1.195	33.64
Root	0.407[a]	0.315	29.20
50 Days			
Shoot	7.690[a]	5.985	28.49
Root	2.055[a]	1.553	32.33

[a] Significantly superior to inbred line seedling.

help in rapid seedling establishment but also make them more resilient than inbred seedlings. Yu and Tuinstra (2001) observed that in comparison to inbred lines, the hybrid sorghums expressed greater tolerance to low temperature during their germination and early seedling stages.

According to Saeki et al. (2016), the presence of differentially expressed genes between hybrid and its parents leads to a rapid growth in different hybrid plant parts. They further opined that once the shoot becomes photosynthetically active, the superiority of hybrid was linked with the increase in assimilation rate or leaf area ratio. They also reported that after the first week of germination, an increase in the number and size of photosynthetic palisade mesophyll cells was responsible for the expression of hybrid vigor in Chinese cabbage.

4.5.2 Adult Plant Traits Linked to Resilience and Adaptation

4.5.2.1 Greater Drought Tolerance

The hybrid plants are known to have an extra degree of resilience to encounter various stresses much better than those of pure line cultivars and produce high and stable yields. Saxena et al. (1992), using a line-source irrigation system, reported that the hybrid pigeonpea plants tolerated both extremely dry and wet conditions much better than those of inbred cultivars. Lopez et al. (1996) also reported that under adverse conditions, besides improved root traits, the pigeonpea hybrids also maintained relatively high-water content during their growth and development periods, and these contributed to their enhanced capacity of drought tolerance. They further opined that as compared to inbred genotypes, genetic hybridity provided more resilience and better environmental buffering to hybrid plants to compete well with the fast-growing rainy-season weeds or intercropped cereals.

4.5.2.2 Enhanced Expression of Disease Resistance

There are two major pigeonpea diseases. To assess the comparative resilience of resistant hybrids with respect to their productivity, both the resistant hybrids and inbred cultivars were evaluated simultaneously in disease-sick and disease-free fields

TABLE 4.4
Grain Yield (t/ha) of Resistant Hybrids and Inbred Cultivars in Disease-Free and Diseased-Sick Fields at ICRISAT

	Disease-Free Field			Disease-Sick Field		
Genotype	**1993**	**1994**	**Mean**	**1993**	**1994**	**Mean**
Hybrids						
IPH 1326	2.6	2.5	2.53	2.1	1.2	1.64
IPH 1395	2.2	2.3	2.26	2.0	1.5	1.72
IPH 1327	2.4	1.8	2.13	1.9	1.5	1.67
Mean	2.4	2.2	2.31	2.0	1.4	1.68
Inbreds						
87119	2.6	1.9	2.25	1.1	0.7	0.88
87051	1.5	1.6	1.60	1.3	0.9	1.11
Mean	2.05	1.75	1.93	1.2	0.8	1.00
Hyb adv (%)	17.1	25.7	19.7	66.7	75.0	68.0

Source: ICRISAT Pigeonpea Breeding Report 1996–1997.

at ICRISAT. The plant mortality in the susceptible controls in the sick nursery ranged between 95% and 100%, suggesting very high levels of virulent inoculums. Under disease-free fields, both the resistant and susceptible test entries recorded <1% disease score with respect to Fusarium wilt and sterility mosaic virus.

The data recorded at the two test environments showed large differences in productivity and the extent of heterosis. In disease-free field, the hybrids on average produced 2.31 t/ha yield and they expressed 19.7% heterosis over the pure line cultivars (1.93 t/ha). On the other hand, under high disease pressure, the overall productivity was relatively low (Table 4.4) but the mean heterosis (68%) was three-fold greater than that of disease-free conditions. This data showed that besides genetic resistance to the two diseases, the hybrids also demonstrated additional resilience that enabled plants to produce more grains than inbred varieties under severe disease pressure. This high level of resilience expressed by the hybrids will help in producing more grains than inbred verities under endemic situations and contribute positively toward negotiating the ill effects of diseases.

4.5.2.3 Traits Related to Recovery from Stress Damages

Pigeonpea embraces a tremendous genetic variation for crop maturity extending from < 90 (super early) to >275 (long) days. A perusal of germplasm data shows that this variation is almost continuous in nature, and for this reason, it finds adaptation in diverse environments and cropping systems. The medium (160–190 days) and late (>200 days) maturing types are invariably cultivated under rainfed conditions with short-aged intercrops or mixed crops. In this crop combination, the cereal is harvested in <100 days and pigeonpea is retained in the field for another 100 (medium genotypes) to 150 (late genotypes) days. During this extended duration the plants invariably encounter intermittent and/or terminal stresses. Among the abiotic stresses, drought is the most common yield reducer. Because the *Cajanus* is a perennial species, its roots grow 2–3 m deep into the soil, the plants use underground water for their survival and later produce a few new branches, leaves, flowers, pods, and seeds.

Since pigeonpea takes a long time to complete its growth cycle, many times the crop canopy is damaged by stray animals. Due to its perennial growth habit, the pigeonpea plants produce new branches with good canopy and produce grain at maturity. This inherent recovery mechanism of pigeonpea compensates for loss and also help in tolerating other stresses. For example, if the crop is severely damaged by bod borers, then after two to three weeks, a new flush of flowers emerges to produce new pods, and thus compensates for the potential yield loss. Sharma et al. (1978) demonstrated that there are genetic differences for the recovery from physical damages to the crop. For example, variety AS 71-37 recorded the highest (64.32%) recovery, while in HY 4 (21.7%) had the least recovery in terms of yield of ratooned crop. This trait can be incorporated in the breeding program to provide resilience to crop production.

4.6 POTENTIAL APPROACHES TO ENHANCE YIELD AND ADAPTATION OF PIGEONPEA HYBRIDS

4.6.1 Enhancing Resilience by Incorporating Dominant Wilt Resistance Genes in CMS Lines

In pigeonpea, the resistance to Fusarium wilt disease is generally controlled by recessive alleles with variable effects. Under this scenario, to breed wilt resistant hybrids, it would be necessary to have such recessive alleles in both male and female parents. Saxena et al. (2012) while studying inheritance of Fusarium wilt in pigeonpea, reported the presence of a single dominant gene that conferred high level of resistance to Fusarium wilt.

This unique wilt resistance gene has high importance in breeding hybrids. To achieve this, the dominant gene need to be transferred to the elite female (A-) and their maintainer (B-) lines. This would ease the breeding of wilt resistant pigeonpea hybrids. Such resistant female parents, carrying dominant resistant nuclear gene, when crossed to either wilt resistant or susceptible fertility restorer (male parent), the resultant hybrid(s) will be resistant to Fusarium wilt. This will allow the selection of more fertility restorers and selection of more high-yielding combinations rather easily. Among these hybrids further selection for yielding and wide adaptation could be performed.

4.6.2 Enhancing Resilience in Hybrids through New Variability

For the long-term sustainability of hybrid breeding programs, it is important that new variability is introduced from time to time. This would not only protect the breeding products from the new variants of different biotic and abiotic stresses but also enhance the general conditioning of resilience. In the case of hybrids, the new variability would be quite useful to infuse variability in both parents.

4.6.2.1 Cytoplasmic Diversification

Cytoplasmic or mitochondrial uniformity in the hybrid breeding programs of a crop always carry the entrenched threats of mass crop losses. This may happen due to the presence of the susceptibility gene(s) for certain diseases in mitochondria of a specific cytoplasm. This fear is based on the incidence of en masse outbreak of leaf blight disease in the hybrids sown in the corn-belt of USA, and its susceptible genes were found to be located in its mitochondrial genome (Tatum, 1971). Therefore, to protect pigeonpea hybrids from such potential catastrophes, the diversification of cytoplasm was taken up on priority, and nine CMS-inducing cytoplasms derived from different wild species representing secondary and tertiary gene pools (Table 4.5) were bred (Saxena et al., 2010; Saxena, 2014). This will help in developing a strong broad-based hybrid breeding program.

4.6.2.2 Nuclear Diversification

Induction of new genetic variability is necessary to breed high-yielding hybrids with greater stability and adaptation. In the hybrid breeding programs, the availability of genetically diverse male (R-) and female (A-lines) parents with high general combining ability is important. In this context, the gene banks situated at ICRISAT and ICAR centers are rich resources of diversity with about 20,000 accessions conserved. Such gene pools contain tremendous variability for a range of quantitative and qualitative traits (Table 4.6) and these could be exploited in the breeding of new cultivars. Studies conducted to quantify the diversity at the molecular level in pigeonpea germplasm showed that the genetic diversity available within the primary gene pool of pigeonpea is not large enough for the genetic enhancement of yield. The genetic diversity among crossable wild species, however, is relatively high (Yang et al. 2006, Yadav et al. 2014), and it could be used to diversify the genetic variation within the primary gene pool. The genetic variability among the hybrid parents can also be enhanced by incorporating the favorable genes available in the wild species from the secondary gene pool. These include *Cajanus scarabaeoides*, *C. cajanifolius*, *C. lineatus*, *C. albicans*, *C. sericeous*, etc.

TABLE 4.5
Different CMS Systems Derived from Secondary and Tertiary Gene Pools of Genus Cajanus

CMS ID	*Cytoplasm* Donor Species	Gene pool
A_1	*C. sericeus*	Secondary
A_2	*C. scarabaeoides*	Secondary
A_3	*C. volubilis*	Tertiary
A_4	*C. cajanifolius*	Secondary
A_5	*C. cajan*	Secondary
A_6	*C. lineatus*	Secondary
A_7	*C. platycarpus*	Tertiary
A_8	*C. acutifolius*	Secondary
A_9	*C. cajan*	Secondary

TABLE 4.6
Phenotypic Variation Conserved for Different Key Traits in Pigeonpea Germplasm

Trait	Minimum	Maximum
Days to 50% flowering	45.0	237.0
Days to 75% maturity	85.0	299.0
Plant height (cm)	39.0	385.0
Primary branches (number)	2.0	66.0
Secondary branches (number)	0.3	146.0
Racemes (number)	6.0	915.0
Seed/pod (number)	1.6	9.0
100-seed weight (g)	2.8	25.8
Seed protein (%)	12.4	32.0

Source: Gene Bank, ICRISAT.

4.6.2.3 Application of Heterotic Groups

The concept of "heterotic groups" involves clustering of genotypes on the basis of their performance in F_1 hybrid generation or genetic diversity. According to this concept, crosses of lines within a cluster (group) are not expected to produce superior hybrids, while the crosses between the lines of two diverse groups are likely to yield highly heterotic crosses. In pigeonpea, Saxena and Sawargaonkar (2014) formed seven heterotic groups. To overcome the effects of the environment in such exercises, use of genomics tools has been suggested to discriminate the parental lines into heterotic groups. Use of such heterotic groups in breeding is likely to lift the performance level of hybrids with reasonable resources (Saxena et al. 2021). They also reported that heterosis for seed yield was much greater when the parental lines representing two diverse heterotic groups were crossed. It is believed that the use of genetic markers provide more reliability to diversity data by eliminating environmental and genotype x environment effects in the formation of heterotic groups. This information generated on potential male and female parents would help in synthesizing heterotic combinations, which can impart a greater degree of resilience to the hybrids.

4.6.2.4 Use of Temperature-Sensitive Male Sterility

Saxena (2014) bred a temperature-sensitive male-sterile (TGMS) line from an interspecific population. This selection remains male sterile with no pollen grain in the temperature under high (>25° C) temperature regime. Interestingly this same male-sterile genotype, when exposed to low (<24°C) temperatures produced fertile pollen grains and self-pollinated seeds (Table 4.7). This type of male sterility can be used in breeding single cross hybrids, but under strict monitoring of temperature. A high temperature site is used for the production of hybrid seeds and therefore, it should be undertaken in rainy season when the ambient temperatures across the season generally remain well over 26°C. This production environment would allow

TABLE 4.7
Number of Fertile and Sterile Plants in TGMS Lines Under Two Temperature Regimes

	September (>25°C)		November (<24°C)	
Selection Number	**Sterile**	**Fertile**	**Sterile**	**Fertile**
Selection -1	37	0	0	37
Selection -2	32	0	0	32
Selection -3	27	0	0	27
Selection -5	23	0	0	22

Source: Saxena (2014).

full expression of the gene(s) responsible for male sterility. Such genotypes, when male sterile, can be used to produce hybrid (A x R) seeds with the help of pollinating insects present in nature. Under low temperature environments, these plants will produce self-pollinated seeds like a pure line variety and thus exclude the need of the maintainer (B-) line. This TGMS-based hybrid breeding program is popularly called the "two-parent hybrid" system. The main advantage of the two-parent hybrid system is that it allows development and testing of a greater number of hybrid combinations to cover wider genetic diversity.

4.6.2.5 Judicious Use of Wild *Cajanus* Species

The genetic diversity in the primary gene pool of pigeonpea is limited (Yang et al., 2006), and there is a need to diversify the germplasm through the introgression of genetic materials from its secondary gene pool (Bohra et al., 2010). The wild relatives of cultivated species are well-established resources of new genes. Pundir and Singh (1985) reported that the success in the inter-specific hybridizations varied considerably from <5% to 35%. The widening of genetic diversity in pigeonpea can be achieved by exploiting secondary gene pool through hybridization and precise marker-based selection. The transfer of genes from their wild relatives is a resource-intensive operation and often suffers from unwanted linkage drag (Bohra et al. 2022a, 2022b). Therefore, selection of the candidate trait and is breeding approach should be given very careful consideration. In overcoming the problems associated with linkage drag, the emerging technologies of genomics can be a boon to breeders. In the last half century, conventional research has failed to deliver the products most required by small-holder farmers. Major issues of plateauing yield and pod borer damage still persist. We believe that the answers to these issues can come through the intelligent use of the genetic materials from the secondary gene pool and advanced genomics science in pigeonpea breeding programs.

4.6.3 Enhancing Adaptation through Early Maturing Genotypes

Pigeonpea cultivars used in the past were late-maturing and highly photo-sensitive. These two vital traits limit the adaptation of the crop in the areas placed between

TABLE 4.8
Seed Yield (t/ha-) of Early Lines at Different Latitudes (°N)

ICPL No.	7°	17°	29°	32°	46°
83015	2.32	2.35	1.06	3.73	2.06
83019	2.21	1.46	1.00	3.58	1.76
84023	2.34	1.42	1.37	2.99	1.59
85010	2.79	1.59	1.17	3.16	2.15
83006	3.09	2.22	1.28	3.38	1.43
Mean	2.17	1.65	1.16	3.19	1.77
SE±	0.28	0.15	0.10	0.04	NA*
CV%	22.8	14.5	12.9	17.1	NA*

equator and 30° latitudes. According to Wallis et al. (1981), the adaptation of pigeonpea can be enhanced only if photo-insensitivity is associated with early flowering (Saxena et al., 2019). In order to widen the adaptation of pigeonpea and take the crop to new production areas and to diversify the existing cropping systems, new high-yielding extra early maturing cultivars were developed under the aegis of ICAR and ICRISAT. To test their adaptation, the promising lines were tested at locations ranging from 7 to 46 °N latitudes in an International Nursery in five countries. The data from these assessments (Table 4.8) showed that extra early genotypes such as ICPL 83015 and ICPL 85010 produced over 2 t /ha of grain even at 46° N. The enhanced adaptation of early maturing pigeonpea has been documented very well in literature. These include introduction of pigeonpea in (i) rice-wheat production areas (Dahiya et al., 2002) of Haryana, Punjab, and Uttar Predesh; (ii) abandoned and rainfed areas of mid- and low hills (Saxena et al., 2012) of Uttarakhand; (iii) and semiarid dry areas of Rajasthan (Singh et al., 2005); and (iv) rainfed rice fallows of Sri Lanka (Saxena, 1999; Chauhan et al., 1999).

4.7 GENOMICS IN BREEDING FOR STRESS TOLERANCE

The development of climate-resilient crop varieties is crucial for mitigating the adverse effects of environmental stresses such as drought, heat, and salinity while ensuring stable or increased yields. These improved varieties help sustain agricultural productivity amid climate change challenges (Ravikiran et al., 2022). Abiotic stresses alone can reduce crop yields by over 50%, underscoring the urgency of breeding stress-tolerant cultivars (Hossain et al., 2021). Advancements in genomics have transformed crop breeding, with over 1,000 plant genomes sequenced in the past two decades (Sun et al., 2022). Large-scale re-sequencing projects, including the 3K Rice Genomes (Wang et al., 2018), Chickpea 3K Genomes (Varshney et al., 2021), and the Gene Bank Genomics Program in barley (Milner et al., 2019) have provided critical insights into genetic diversity among crop wild relatives, landraces, and elite breeding lines. Recently, haplotype-based breeding has emerged as a powerful strategy for precisely selecting genetic variants linked to enhanced resistance against biotic and abiotic stresses (Abbai et al., 2019; Sinha et al., 2020). Refining breeding strategies using superior haplotypes accelerates genetic gains in major crops by integrating high-throughput sequencing, genomic selection, and precision breeding techniques. This approach deepens our understanding of stress tolerance genetics, enabling the precise selection of beneficial allelic combinations to develop high-yielding, resilient cultivars and strengthen global food security (Bevan et al., 2017).

4.7.1 From Traits to Haplotypes: A Roadmap for Stress Breeding

Genome-wide association studies (GWAS) have emerged as a powerful tool for identifying genetic variants associated with key agronomic traits in crops. The advent of whole-genome sequencing has significantly enhanced the efficiency of GWAS, enabling the rapid detection of significant marker-trait associations (MTAs), candidate genes, and superior haplotypes. Once MTAs are identified, the next crucial step is pinpointing candidate genes within the associated genomic regions. These genes, often identified based on their biological functions or roles in key pathways, provide deeper insights into the molecular mechanisms regulating trait expression.

Beyond identifying candidate genes, haplotype analysis further refines breeding strategies by precisely targeting genetic variations linked to improved crop performance. Mapping haplotypes to specific traits allows researchers to integrate superior haplotypes into breeding programs, ultimately enhancing stress resilience and yield potential. For instance, Sinha et al. (2020) investigated drought tolerance in pigeonpea by analyzing ten drought-responsive candidate genes, validating the identified haplotypes under drought-stress conditions. Similarly, soil conditions significantly impact crop productivity. Sharma et al. (2022) conducted a pioneering GWAS study on wheat grown in sodic-dispersive soils, identifying several superior haplotypes that improved stress tolerance and yield. Additionally, Wang et al. (2022) highlighted the role of superior haplotypes in addressing phosphorus (P) deficiency in soybeans. Their study identified *GmPHR14-Hap2* and *GmPHR32-Hap2*, two elite haplotypes linked to P-use efficiency, emphasizing the importance of haplotype-based breeding in developing resilient, high-performing crop varieties.

4.7.2 Introgression of Superior Stress-Tolerant Haplotypes into Elite Genetic Backgrounds

Haplotype-based breeding can be broadly classified into forward and backcross breeding, each serving distinct objectives in crop improvement. Forward breeding focuses on integrating and stacking superior haplotypes from well-adapted sources into elite lines using haplotype-specific markers. In contrast, backcross breeding facilitates the transfer of beneficial haplotypes from non-adapted sources, such as landraces or wild relatives, into widely cultivated varieties with consumer-preferred traits and strong environmental adaptability. A key strategy in this approach is the development of Haplo-NILs (Near Isogenic Lines) for specific traits, followed by stacking multiple superior haplotypes to enhance overall crop performance (Abbai et al., 2019).

The successful integration of superior haplotypes into breeding programs accelerates the development of high-yielding, stress-resilient varieties. For instance,

Tong et al. (2022), explored the impact of the *ZmCCT* haplotype H5 on yield, stalk-rot resistance, and drought tolerance in maize. By introgressing nine *ZmCCT* haplotypes into seven elite maize inbred lines, they developed NILs and evaluated their performance under field conditions. The results demonstrated that NILs carrying the H5 haplotype exhibited improved disease resistance, reduced photoperiod sensitivity, and significantly enhanced yield potential. Similarly, Thakro et al. (2023) introduced two superior haplotypes linked to the *CabHLH10* gene—responsible for drought tolerance and yield—from wild chickpea germplasm into the elite cultivar JG 11. Through repeated backcrossing and selection, they developed drought-tolerant chickpea lines with improved productivity. These studies highlight haplotype introgression as a transformative approach for developing stress-tolerant, high-yielding crop varieties.

4.7.3 Challenges and Future Direction

Leveraging superior haplotypes offers significant potential for tailoring crop varieties to specific product profiles. However, several challenges must be addressed. Efficient identification of these haplotypes demands advanced computational resources and algorithms capable of analyzing vast genetic datasets. Many superior haplotypes originate from underutilized sources like wild relatives or landraces, posing risks of linkage drag. To mitigate this, breeders can transfer haplotypes into elite backgrounds or use gene editing to retain beneficial traits while eliminating undesirable genetic associations. To fully harness the benefits of haplotype-based breeding, several key challenges must be addressed. These include a deeper understanding of the functional roles of key genes linked to target traits, managing undesirable epistatic interactions when selecting optimal haplotype combinations, and accounting for all genetic effects—additive, dominance, and epistatic. Integrating artificial intelligence (AI) and systems biology with precise phenotyping is crucial for maximizing the vast genetic and phenotypic data generated by high-throughput technologies.

4.8 GENERAL DISCUSSION

Pigeonpea is predominantly cultivated as a rainy-season rainfed crop at subsistence level. It is an important pulse with wide adaptation. The potential yield of pigeonpea is over 3000 kg/ha but its on-farm realized productivity is at around 700 kg/ha. There may be various production and social factors responsible for the undesirable huge yield gap. Some of them can be addressed rather easily but others cannot, at least at the present time and in the near future.

Pigeonpea is embedded with a special trait, i.e., "recovery from stress damage." This is a very important trait for pigeonpea due to its various yield reducing factors operating both from above and below ground surface. The recovery from damages primarily depends on the deep root system that provides sufficient water to the plant to generate canopy and flowers and pods. The breeding of these critical traits (Table 4.9) is cumbersome and would need more resources. However, the breeding activities can be substantiated by integrating genomics science and speed breeding.

TABLE 4.9
Some Resilience and Adaptation Parameters Identified in Pigeonpea

Parameter	Source	Genetic Control	Genetic Mapping	Breeding Activity	Remarks
Rapid germination	ICPH 2671	No	No	No	In hybrids
Radicle length and weight	ICPH 2740	No	No	No	In hybrids
Plumule length and weight	ICPH 3762	No	No	No	
White seed color	HY 3C	Rec.	No	Yes	In all groups
Aerenchyma cells	ICPH 2671	Dom	No	No	Only under
Lenticels	ICP 5028	–	No	No	waterlogging
Adventitious roots	ICP 5028	–	No	No	
Seedling shoot weight	ICPL 2671	Fusarium No	No	No	In hybrids
Seedling root weight	ICPH 2671	No	No	No	In hybrids
Drought tolerance	ICPH 2671	Susc.	No	No	In late types
Wilt resistance	ICP 8863	Susc.	No	No	In hybrids
Stress recovery	AS71-37	No	No	No	In late types
Photo-insensitivity	MN 1	Susc	No	Yes	Earliness
Early maturity	ICPL 88039	Susc	Yes	Yes	Photo-insens

Tolerance to initiate marker-aided selection for traits associated with drought tolerance. The genomic and molecular breeding approaches (e.g., marker-assisted selection, QTL mapping) can further enhance selection efficiency. Utilization of wild pigeonpea relatives with high osmotic adjustment and deep root systems can improve the developing drought-resilient hybrids with high seed yield, superior root traits, and stress-adaptive physiology can significantly enhance pigeonpea production in semiarid regions.

4.9 CONCLUSIONS

To enhance yield in pigeonpea, the breeders successfully bred hybrid pigeonpea with significant yield advantages. It has been observed that hybrids are also not free from the threat of biotic and abiotic stresses. The major abiotic stresses in pigeonpea (with which this discussion is dealing) are drought, waterlogging, and soil salinity. In order to understand the potential ways to overcome these, the crop physiologists identified some major and minor parameters that are directly or indirectly associated with tolerance to these stresses. However, the breeding for resistance could not be undertaken due to various reasons, such as lack of screening technologies, inadequate understanding of the genetic nature of the trait or poor heritability, and genetic advance. In this context, we believe that a beginning can be made by incorporating the simply inherited traits in the hybrid parents. These may include earliness, photo-insensitivity, desirable branching (compact or spreading), height (tall or dwarf), growth habit (determinate or non-determinate), rapid seedling growth, market preferred traits, and incorporation of dominant gene for resistance to wilt, soil salinity, and waterlogging in the female parents of the hybrids.

ACKNOWLEDGMENTS

Authors of this manuscript acknowledge the support received from the Pigeonpea Breeding Unit of ICRISAT for providing access to various progress reports and photographs.

REFERENCES

Abbai R, Singh VK, Nachimuthu VV, Sinha P, Selvaraj R, Vipparla AK, Kumar A (2019) Haplotype analysis of key genes governing grain yield and quality traits across 3K RG panel reveals scope for the development of tailor-made rice with enhanced genetic gains. *Plant Biotechnol J* 17(8): 1612–1622.

Bevan MW, Uauy C, Wulff BB, Zhou J, Krasileva K, Clark MD (2017) Genomic innovation for crop improvement. *Nature* 543(7645): 346–354.

Bidinger PD, Nag B (1981) *The role of pigeonpeas in village diets.* In: *Proceedings of International Workshop on Pigeonpeas. Volume 1. 15–19 December 1980. International Crops Research Institute for the Semi-arid Tropics*, Patancheru, 502324 India, Pages 357–364.

Bohra A, Jha UC, Premkumar A, Bisht D, Singh NP (2016) Cytoplasmic male sterility (CMS) in hybrid breeding in field crops. *Plant Cell Reports* 35: 967–993

Bohra A, Kilian B, Sivasankar S, Caccamo M, Mba C, McCouch SR, Varshney RK (2022a) Reap the crop wild relatives for breeding future crops. *Trends in Biotechnology* 40: 412–431.

Bohra A, Mallikarjuna N, Saxena KB, Upadhyaya HD, Vales MI, Varshney R (2010) Harnessing the potential of crop wild relatives through genomics tool for pigeonpea improvement. *J Plant Biol* 37: 83–98.

Bohra A, Tiwari A, Kaur P, Ganie S, Raza A, Roorkiwal M, Mir R, Fernie A, Smykal P, Varshney R (2022b) The key to the future lies in the past: Insights from grain legume domestication and improvement should inform future breeding strategies. *Plant and Cell Physiology* 63(11):1554–1572.

Bohra A, Tiwari A, Pareek S, Joshi R, Satheesh Naik SJ, Kumari K, Verma RL, Parihar AK, Patil PG, Dixit GP (2025) Past and future of cytoplasmic male sterility and heterosis breeding in crop plants. *Plant Cell Rep* 22: 33.

Bond DA (1987) Recent developments in breeding field beans (*Vicia faba* L.). *Plant Beed* 99: 1–2.

Chauhan RM, Parmar LD, Patel PT, Tikka SBS (2004) Fertility restoration in cytoplasmic male sterile lines in pigeonpea. (*Cajanus cajan* (L.) Millsp.). *Indian J Genet* 64: 112–114.

Chauhan YS, Atukorala WD, Perera KDA, Joseph KDSM, Saxena KB, Johansen C (1999) Adaptation of extra-short-duration pigeonpea in the short rainy season of a tropical bimodal rainfall environment. *Exp Agric* 35: 87–100.

Dahiya SS, Chauhan YS, Johansen C, Waldia RS, Sekhon HS, Nandal JK (2002) Extra short duration pigeonpea for diversifying wheat-based cropping systems in sub-tropics. *Exp Agric* 38: 1–11.

Daniel VA, Narayanaswamy D, Desai BLM, Kurien S, Swaminathan M, Paripia HAB (1970) Supplementary value of varying levels of red gram (*Cajanus cajan*) to poor diets based on rice and ragi. *Indian J Nutr Diet* 7 358–362.

FAO. (2021) http://faostat.fao.org

Hossain A, Ahmad Z, Moulik D, Maitra S, Bhadra P, Ahmad A, Aftab T (2021) Jasmonates and salicylates: Mechanisms, transport and signaling during abiotic stress in plants. *Jasmonates* and salicylates signaling in plants 1–29.

Hulse JH (1977) Problems of nutritional quality of pigeonpea and chickpea and prospects of research. *In nutritional standards and methods of evaluation for food legume breeders* (eds. Hulse J. H. et al.). IDRC, Ottawa, Canada, pp 88–100.

Loomis RS, Connor DJ (1992) *Crop ecology: Productivity and management in agricultural systems*. Cambridge University Press, Cambridge, 32–59.

Lopez FB, Chauhan YS, Johansen C (1996) Effects of timing of drought stress on phenology, yield and yield components of short duration pigeonpea. *J Agron Crop Sci* 177: 311–320.

Milner SG, Jost M, Taketa S, Mazón ER, Himmelbach A, Oppermann M, Stein N (2019) Genebank genomics highlights the diversity of a global barley collection. *Nat Genet* 51(2): 319–326.

Mudaraddi B, Saxena KB (2012) A comparative study of hybrid and inbred cultivars for germination and other related traits of pigeonpea. *J Food Legumes* 25: 351–354.

NIN (2010) *Dietary guidelines for Indians - a manual*. National Institute of Nutrition, Hyderabad.

Pal RK (1939) A review of the literature on the nutritive value of pulses. *Indian J Agric Sci* 9: 133–137.

Palmer RG, Gai J, Sun H, Burton JW (2001) Production and evaluation of hybrid soybean. *Plant Breed Rev* 21: 263–307.

Pundir RPS, Singh RB (1985) Crossabilty relationships among *Cajanus, Atylosia* and *Rhynchosia* species and detection of crossing barriers. *Euphytica* 34: 303–308.

Rajjou L, Duval M, Gallardo K, Catusse J (2012) Seed germination and vigor. *Annu Rev Plant Physiol* 63: 507–533.

Ramanujam S, Singh SP (1981) *Pigeonpea breeding in the All India Coordinated Programme*. In: *Proc. Intl. Workshop on Pigeonpeas. Vol. 1. International Crops Research Institute for the Semi-arid Tropics (ICRISAT)*, Patancheru, 502 324. India.

Ravikiran KT, Gopala Krishnan S, Abhijith KP, Bollinedi H, Nagarajan M, Vinod KK, Singh AK (2022) Genome-Wide Association Mapping Reveals Novel Putative Gene Candidates Governing Reproductive Stage Heat Stress Tolerance in Rice. *Front genet* 13: 876522.

Reddy S, Virmani SM (1981) *Pigeonpea and its climatic environments*. In: *Proc. Intl. Workshop on Pigeonpeas. Vol. 1. International Crops Research Institute for the Semi-arid Tropics (ICRISAT)*, Patancheru, 502 324. India, Pages 259–270.

Romagnoti S, Maddaloni M, Livini C, Motto M (1990) Relationship between gene expression and hybrid vigor in primary root tips of young maize (*Zea mays* L.) plantlets. *Theor Appl Genet* 80: 769–775.

Saeki K, Aburai N, Aratani S, Miyashita H, Abe K (2016) Salt-stress and planthormone-likeresponses for selective reaction of esterified xanthophylls in the aerial microalga *Coelastrella* sp. KGU-Y002. *J Applied Phycol*. DOI: 10.1007/s10811-016-0911-7

Saxena KB (1999) Pigeonpea in Sri Lanka. Monograph. International Crops Research Institute for the Semi-Arid Tropics, Patancheru, 502 324, A.P., India.

Saxena KB (2014) Temperature-sensitive male sterility system in pigeonpea. *Curr Sci* 107: 277–281.

Saxena KB (2015) From concept to field: evolution of hybrid pigeonpea technology in India. *Indian J Genet* 75: 279–293.

Saxena KB, Chauhan YS, Johansen C, Singh L (1992) *Recent developments in hybrid pigeonpea research*. In: *New Frontiers in Pulses Research and Development. Proc. National Symposium*, 10–12 Nov. 1989, Directorate of Pulses Research, Kanpur 208 024, India, Pages 58–69.

Saxena KB, Choudhary AK, Dalvi VA, Saxena RK, Ghosh J, Singh S, Verma P, Kumar S (2021) Pigeonpea is significantly more than just a delicious pulse. *J Agri Search* 8: 177–187.

Saxena KB, Choudhary AK, Saxena RK, Chauhan YS (2020) Can pigeonpea hybrids negotiate stresses better than inbred cultivars? *Breed Sci* 70: 423–429.

Saxena KB, Choudhary AK, Srivastava RK, Bohra A, Saxena RK, Varshney RK (2019) Origin of early maturing pigeonpea germplasm and its impact on adaptation and cropping systems. *Plant Breed* 138: 243–251.

Saxena KB, Kumar RV, Saxena RK, Sharma M, Srivastava RK, Sultana R, Varshney RK, Vales MI, Pande S (2012) Identification of dominant and recessive genes for resistance to Fusarium wilt in pigeonpea and their implication in breeding hybrids. *Euphytica* 188: 221–227.

Saxena KB, Kumar RV, Srivastava N, Shiying B (2005) A cytoplasmic-nuclear male-sterility system derived from a cross between *Cajanus cajanifolius* and *Cajanus cajan. Euphytica* 145: 289–294.

Saxena KB, Sawargaonkar SL (2014) First information on heterotic groups in pigeonpea (*Cajanus cajanus* (L.) Millsp.). *Euphytica* 200: 187–196.

Saxena KB, Sharma D, Vales MI (2018) Development and commercialization of CMS pigeonpea hybrids. *Plant Breed Rev* 41: 103–167.

Saxena KB, Singh L, Gupta MD (1990) Variation for natural out-crossing in pigeonpea. *Euphytica* 46: 143–148.

Saxena KB, Sultana R, Mallikarjuna N, Saxena RK, Kumar RV, Sawargaonkar KL (2010) Male- sterility systems in pigeonpea and their role in enhancing yield. *Plant Breed* 129: 125–134.

Saxena KB, Tikle AN (2015) Believe it or not, hybrid technology is the onl6y way to enhance pigeonpea yields. *Int J Sci Res* 5: 1–7.

Shalendra Gummagolmath KC, Sharma P, Patil SM (2013) Role of pulses in the food and nutritional security in India. *J Food Legumes* 26: 124–129.

Sharma D, Saxena KB, Green JM (1978) Potential of ratooning in pigeonpea. *Field Crops Res* 1: 165–172.

Sharma DL, Bhoite R, Reeves K, Forrest K, Smith R, Dowla MA (2022) Genome-wide superior alleles, haplotypes and candidate genes associated with tolerance on sodic-dispersive soils in wheat (*Triticum aestivum* L.). *Theore Appl Genet* 135(3): 1113–1128.

Shimamura S, Yamamoto R, Nakamura T, Shimada S, Komatsu S (2010) Stem hypertrophic lenticels and secondary aerenchyma enable oxygen transport to roots of soybean in flooded soil. *Ann Bot* 106: 277–284.

Singh SJ, Chand U, & Saxena KB. (2005) Pigeonpea in Rajasthan. (*Unpublished Report*). ICRISAT. Patancheru, A.P.

Sinha P, Singh VK, Saxena RK, Khan AW, Abbai R, Chitikineni A, Varshney RK (2020) Superior haplotypes for haplotype-based breeding for drought tolerance in pigeonpea (*Cajanus cajan* L.). *Plant Biotechnol J* 18(12): 2482–2490.

Subbarao GV, Johansen C, Jana MK, Rao JVDKK (1991) Comparative salinity responses among pigeonpea genotypes and their wild relatives. *Crop Sci* 31: 415–418.

Sultana R, Vales MI, Saxena KB, Rathore A, Rao S, Rao SK, Mula MG, Kumar RV (2013) Water-logging tolerance in pigeonpea (*Cajanus cajan* (L.) Millsp.): genotypic variability and identification of tolerant genotypes. *J Agric Sci* 151: 659–671.

Sun Y, Shang L, Zhu QH, Fan L, Guo L (2022) Twenty years of plant genome sequencing: achievements and challenges. *Trends Plant Sci* 27: 391–401.

Tatum LA (1971) The southern corn leaf blight epidemic. *Sci* 171: 1113–1116.

Thakre DP, Mula MG, Mehtre SP, Saxena KB, Subbarao GV, Johansen C, Jana MK, Subbarao GV, Johansen C, Jana MK, Rao JVDKK (1991) Comparative salinity responses among pigeonpea genotypes and their wild relatives. *Crop Sci* 31: 415–418.

Thakro V, Malik N, Basu U, Srivastava R, Narnoliya L, Daware A, Parida SK (2023) A superior gene allele involved in abscisic acid signaling enhances drought tolerance and yield in chickpea. *Plant Physiol* 191(3): 1884–1912.

Tikka SBS, Pawar LD, Subbarao GV, Johansen C, Jana MK, Subbarao GV, Johansen C, Jana MK, Rao JVDKK (1991) Comparative salinity responses among pigeonpea genotypes and their wild relatives. *Crop Sci* 31: 415–418.

Tong L, Yan M, Zhu M, Yang J, Li Y, Xu M (2022) ZmCCT haplotype H5 improves yield, stalk-rot resistance, and drought tolerance in maize. *Front Plant Sci* 13: 984527.
Troyer AP (1991) Breeding corn for export market. 46th Annual Corn and Sorghum Research Conference. *Seed Trade Association (USA)*, 46: 165–176.
Varshney RK, Roorkiwal M, Sun S, et al. (2021) A chickpea genetic variation map based on the sequencing of 3,366 genomes. *Nature* 599: 622–627.
Wallis E S, Byth DE, Saxena KB (1981) *Flowering responses by thirty-seven early maturing lines of pigeonpea*. In: *Proc. Int. Workshop on Pigeonpeas, Vol. 2, International Crops Research Institute for the Semi-Arid Tropics*, Patancheru, 502 324, India, Pages 143–159.
Wang Q, Du W, Zhang S, Yu W, Wang J, Zhang C, Yu D (2022) Functional study and elite haplotype identification of soybean phosphate starvation response transcription factors GmPHR14 and GmPHR32. *Mol Breed* 42(5): 29.
Wang, W, Mauleon R, Hu Z, et al. (2018) Genomic variation in 3,010 diverse accessions of Asian cultivated rice. *Nat* 557: 43–49.
Yadava K, Yadava SK, Yadava A, Pandey VP, Dwivedi UN (2014) Comparative analysis of genetic diversity among cultivated (*Cajanus cajan*) and its wild relatives (*C. albicans* and *C. lineatus*) using randomly amplified polymorphic DNA (RAPD) and inter single sequence repeats (ISSR) finger printing. *Am J Plant Sci* 5: 1668–1678.
Yang S, Ash G, Harper J, Varling J, Wenzl P, Huttner E (2006) Low level of genetic diversity in cultivated pigeonpea compared to its wild relatives by diversity arrays technology. *Theor Appl Genet* 113: 585–595.
Yu J, Tuinstra MR (2001) Genetic analysis of seedling growth under cold temperature stress in grain sorghum. *Crop Sci* 41: 1438–1443.

5 Understanding Trait Architectures to Accelerate Pigeonpea (*Cajanus Cajan*) Improvement

Insights from Breeding and Multi-omics

Rajib Roychowdhury, Sagar K Rangari, Abhinav Moghiya, Sunil Gangurde, Prakash Gangashetty, Jagdish Jaba, Sunita Choudhary, Mamta Sharma, Sean Mayes, Girish Prasad Dixit, and Manish K Pandey

5.1 INTRODUCTION

Pigeonpea (*Cajanus cajan*) is an important leguminous crop of tropical and semiarid regions and is largely cultivated in South Asia, Africa, and the Caribbean regions (Sarkar et al. 2020). It is the primary source of protein for millions of people, as it can fix atmospheric nitrogen grow under drought-prone conditions, and this makes it an important crop for sustainable agriculture. Cultivated and wild relatives of pigeonpea are members of the legume family Fabaceae with considerable genetic diversity (Smýkal et al. 2015; Athanas, 2022). Presently cultivated pigeonpea varieties are characterized by the absence of the genes responsible for the potent traits, which are conversely characterized in wild relatives such as *Cajanus scarabaeoides*, *Cajanus albicans*, and *Cajanus platycarpus* that confer both disease resistance and abiotic stress tolerance, in addition to favorable agronomic traits (Singh and Abhilash 2017). Trait enhancement is also highly relevant in pigeonpea breeding programs, since the introgression of wild species to cultivated species can potentially improve crop resilience, along with the yield (Pratibha et al., 2015). Traditionally, pigeonpea breeding research targeted yield potential, tolerance for biotic and abiotic stresses, flowering

DOI: 10.1201/9781003603092-5

time, maturation period, and seed quality. Due to a lack of genetic and genomic resources, pigeonpea is considered an "orphan crop," although it is one of the most important food legume crops. Despite the extensive use of conventional breeding practices, its long generation time and complex nature of trait inheritance have hampered its further progress (Bohra et al. 2020; Saxena 2008). To exploit quantitative genetics for rapid breeding efforts and to ensure food security, the genetic architecture of traits is highly essential. In this crop, complex genotype-environment interactions (G×E) control the expression of different traits under varied climatic conditions (Bomma et al. 2024). Wide variation in performance across numerous agro-ecological zones is a consequence of the strong influence of environmental variables such as soil type, temperature, and precipitation on trait expression. The variations in genotype performance for pigeonpea under locations and seasons are due to different expressions of genes caused by environmental signals, which hinder trait selection in breeding programs and require genomic prediction models and multilocation experiments to identify stable and adaptable genotypes. A number of pigeonpea traits express pleiotropy and linkage disequilibrium (LD), where a single gene or groups of genes influence multiple traits (Kumar et al. 2022). For example, days to flowering and maturity are commonly correlated traits that form the basis of adaptation and yield capacity. Understanding these interactions at the genetic level will enable us to choose better genotypes and to disrupt unwanted linkage between the traits. With current progress the multi-omics approach, such as genomics, transcriptomics, metabolomics, and proteomics, or even whole cell study, has revolutionized our ability to explore complex traits and thereby enhance crop performance (Roychowdhury et al. 2023a). These omics tools act as a whole new technology for understanding the genomics of pigeonpea and eventually enhancing breeding productivity. Recent renaissance of modern genomic technologies such as whole genome sequence (WGS), genome-wide association study (GWAS), and quantitative trait loci (QTL) has expedited pigeonpea research (Naik et al. 2024). The release of the pigeonpea genome has facilitated the identification of genes controlling disease resistance, stress tolerance, and yield (Bohra et al. 2020). Today, high-throughput genomic analysis is being performed by means of single-nucleotide polymorphism (SNP) arrays or genotyping-by-sequencing (GBS), thus making marker-assisted selection (MAS) possible for valuable traits. Across a range of environmental conditions and developmental stages, transcriptome sequencing (RNA-Seq) provides information on gene expression profiles. The DEGs in response to pod development and resistance mechanism, drought, and pod development using transcriptomics have been reported. It also allows for the description of alternative splicing events, noncoding RNAs (ncRNAs), gene regulatory networks and other characteristics that can inform a more accurate understanding of trait architecture (Jha et al. 2023). Metabolomics is the detailed study of certain metabolites associated with stress and physiological responses. Studying metabolite changes in pigeonpea under pathogen infestation or drought assists in identifying key metabolic pathways and biomarkers linked to nutritional quality and stress tolerance. The integration of genomics, transcriptomics and metabolomics increases the ability to develop climate-resilient pigeonpea varieties. Mass spectrometry (MS)-based proteomics conducted in pigeonpea has resulted in stress-responsive proteins, enzymes involved in lignin synthesis and seed storage proteins. Functional validation of these proteins using gene-editing tools like CRISPR-Cas9 may facilitate the development of better cultivars (Rasheed et al. 2022). While individual omics techniques give some information, the integration of these technologies using multi-omics analysis leads to a complete understanding of a well-regulated trait. To identify candidate genes and regulatory networks controlling important agronomic traits, to understand G×E interactions at a molecular level to develop climate-resilient pigeonpea varieties, to develop functional markers for MAS and genomic selection (GS), and to improve nutritional quality and stress tolerance through targeted breeding strategies, a systems biology approach combining genomics, transcriptomics, proteomics, and metabolomics is indispensable (Figure 5.1).

5.2 TOLERANCE OF ABIOTIC STRESSES IN PIGEONPEA

5.2.1 Drought Stress

In semiarid and rainfed agricultural environments, drought stress tolerance in pigeonpea is an essential trait to improve productivity. Though pigeonpea is renowned for its adaptation to water-limited conditions, a more prolonged drought period causes a reduction in plant growth and yield (Dwivedi et al. 2023). Drought-resistant traits of this crop are due to physiological and morphological modifications. Pigeonpea has an extensive root system, which allows deep penetration into the soil to avail maximum moisture. In addition, the control for stomatal aperture allows a reduction in water loss while permitting photosynthesis. Moreover, the thick and small leaves with waxy, glossy cuticle (sometimes can be seen in wild species) may reduce transpiration and increase WUE. Large numbers of QTL and putative genes were revealed for drought tolerance by the genetic basis (Sinha et al. 2020). The response to drought is influenced by genes associated with osmotic adjustment and oxidative stress management, and ABA hormonal signaling. Its wild relatives with high drought tolerance, such as *C. scarabaeoides* become valuable sources of genetic material for pre-breeders (Khoury et al. 2015). In this context, omics approaches could contribute to enhancing drought tolerance and to phenotyping for drought-tolerant alleles. Transcriptomics enables the identification of differentially expressed genes upon drought stress, allowing for to search for crucial regulating genes. Meanwhile, proteomics and metabolomics contribute to elucidating the stress-responsive proteins and metabolites for drought acclimation. From the agronomic point of view, early planting, mulching, soil water conservation, and biostimulant application may enhance drought tolerance in pigeonpea will then serve to prevent evaporation and protect groundwater levels (Melo-Sabogal et al. 2024). Through the absence of nutrition and hormone sprays, the application of foliar biostimulants increases drought tolerance. Enhancement of drought tolerance was made in pigeonpea through breeding and biotechnological approaches, including introgression of QTLs for drought tolerance using MAS, which have been exploited for developing crop varieties (Joshi et al. 2024). Such as CRISPR-Cas9, genome-editing tools offer an effective strategy for the precise engineering of drought-responsive genes to enhance water stress tolerance in crops. However, hybridization and introgression approaches enable the incorporation of drought-tolerant traits from wild relatives into improved cultivars, thus providing increased adaptation and yield stability under stressful environments

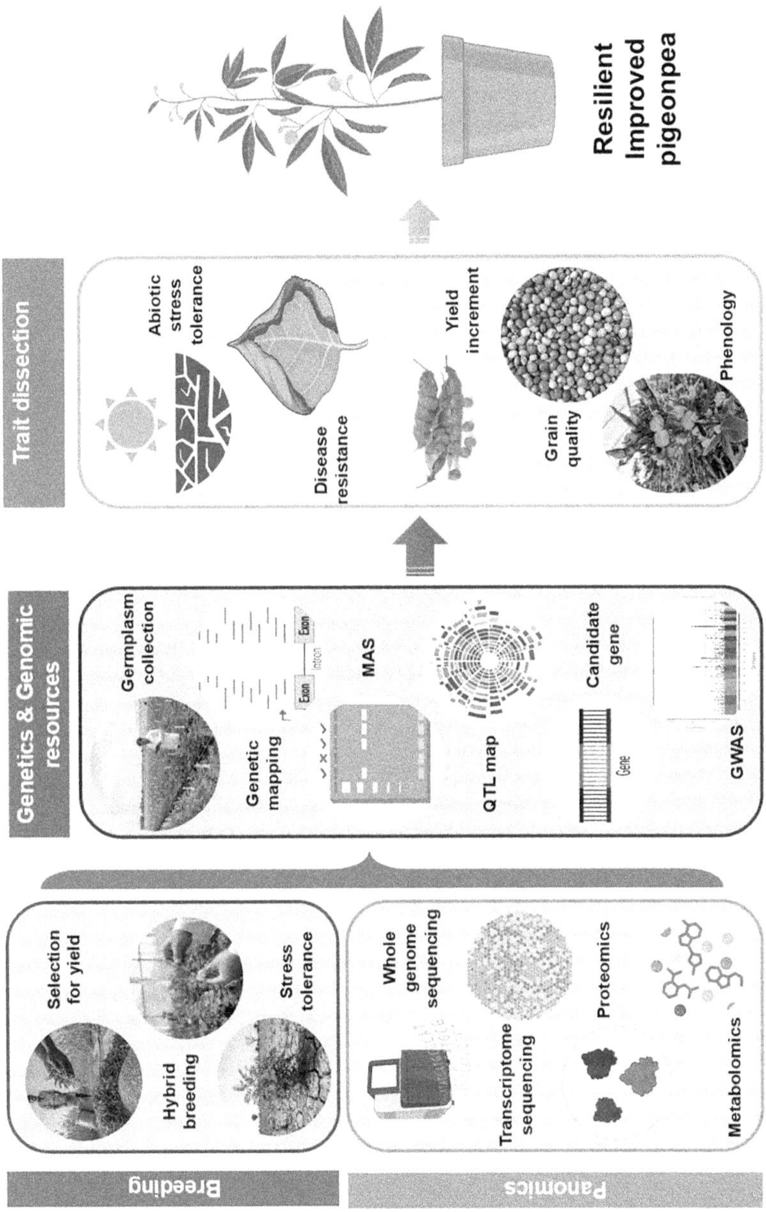

FIGURE 5.1 Schematic diagram representing the connecting pipeline between breeding and multi-omics for pigeonpea trait architecture and improvement.

(Bakala et al. 2024). Food security in climate-vulnerable regions can be sustained through the development of drought-tolerant pigeonpea varieties using diverse modern technologies and sustainable agricultural approaches.

5.2.2 Salinity Stress

Soil salinization is a developing issue, particularly in dry and semiarid environments, and salinity stress constitutes a significant abiotic constraint on pigeonpea production. The elevated soil salinity can disrupt pigeonpea's water absorption, induce ion toxicity, and create osmotic-oxidative stress, thus diminishing crop development, yield, and quality. All stressed saline regions could enhance pigeonpea productivity solely through the creation of salinity-tolerant genotypes. Pigeonpea employs diverse physiological processes (Awana et al. 2019), for example, osmotic adjustment that facilitates the accumulation of osmolytes (e.g., proline, glycine betaine, and soluble carbohydrates) to sustain cellular water equilibrium and avert dehydration, which is essential for osmotic shock tolerance under salinity. An additional critical mechanism concerning osmoregulation and salt tolerance is ion homeostasis, which involves the regulation of sodium (Na^+) uptake and distribution, alongside the compartmentalization of both Na^+ and chloride (Cl^-), thereby limiting their concentrations to non-toxic levels while preserving the essential potassium (K^+) content necessary for normal cellular function (Hasanuzzaman et al. 2015). Pigeonpea activates its antioxidant defense mechanism to mitigate oxidative damage caused by salinity by inducing the activity of essential enzymes such as superoxide dismutase (SOD), catalase (CAT), and peroxidase (POD), thereby neutralizing reactive oxygen species (ROS) and safeguarding cellular structural integrity. It can regulate the chlorophyll concentration and stomatal conductance ratio, allowing the plant to sustain high photosynthetic efficiency and continue photosynthesis under saltwater conditions. These collective adaptive mechanisms allow pigeonpea to withstand salt stress while maintaining its physiological activities under harsh conditions. Enhancing salinity tolerance while maximizing yields to sustain production potential, the development of salt-tolerant pigeonpea varieties necessitates the integration of breeding and molecular methods (Kumar et al. 2019). Inherent resistant genotypes from various germplasm sources are assessed and selected by traditional breeding methods, and thereafter, they are utilized in hybridization to generate an enhanced variety. By expediting the cross-breeding process, MAS can facilitate the identification and introgression of significant QTL for salinity tolerance into superior breeding lines. Utilizing high-throughput SNPs, GWAS enables the mapping of salinity-responsive genes and the identification of novel genetic loci that regulate stress adaptation (Pandey et al. 2016). Transcriptomics and proteomics technologies are utilized to examine differential gene expression and protein profiles, thereby finding essential regulatory genes involved in salt stress adaptation and enhancing the understanding of molecular responses. Moreover, targeted point mutations in salinity response genes via CRISPR-Cas9 mediated genome editing can serve as a precise method to augment resistance without compromising production. By utilizing these technologies collectively, a comprehensive extension for the development of pigeonpea lines with improved salinity tolerance was accomplished and necessitating a collaborative strategy that integrates advanced biotechnological methods, contemporary genomics, and traditional breeding practices. Furthermore, agronomic management measures such as regulated irrigation, biofertilizers, and soil amendments are also effective in alleviating the impacts of salinity stress (Sangma 2020). Future studies must prioritize the identification of novel stress-responsive genes, the enhancement of root systems for effective ion exclusion, and the application of multi-omics techniques to elucidate the intricate mechanisms of salinity tolerance in pigeonpea.

5.2.3 Water Logging Stress

Waterlogging or flooding stress significantly impairs pigeonpea development, yield, and overall productivity by inducing hypoxic conditions that diminish root function and nutrient absorption from the soil. Pigeonpea is primarily cultivated in semiarid areas and is highly sensitive to excessive soil moisture; therefore, resistant cultivars must be developed by physiological, biochemical, and genetic methods. Pigeonpea adapts to waterlogging stress via many physiological and biochemical means that mitigate cellular damage and promote growth in hypoxic environments (Duhan and Sheokand 2020). A primary adaptive response is the development of aerenchyma tissue, which involves the production of specialized air pockets inside root tissues to enhance oxygen transport to submerged roots and mitigate stress-related damage. Fundamental to the above-ground morphological variations, crucial for adaptive responses, are modifications in root architecture; adventitious roots can enhance oxygen availability and nutrient absorption in anaerobic soils (Zhang et al. 2025). The enhancement of ethylene and ABA disrupts this equilibrium, resulting in essential activities like as stomatal closure, root elongation, and metabolic alterations in response to hypoxia; hormonal modulation is fundamental to tolerance. Moreover, under waterlogging conditions, the antioxidant defense system is enhanced, exhibiting increased activity of SOD, CAT, and POD concurrently to neutralize ROS and mitigate oxidative damage to plant cells. The integration of these systems improves waterlogging tolerance in pigeonpea, leading to increased survival and productivity in moisture-stressed situations. Creating waterlogging-tolerant pigeonpea genotypes necessitates the integration of traditional breeding methods and genomics technologies (Belliappa et al, 2024). Ultimately, traditional breeding is crucial for choosing naturally resistant genotypes from various germplasm banks, which can be utilized in hybridization programs aimed at enhancing resistance. The identification and introgression of QTLs linked to waterlogging tolerance into high-yielding lines via MAS further expedite the breeding process. Transcriptomics and proteomics technologies provide significant molecular insights into altered gene expression in hypoxic environments, ultimately revealing crucial regulatory genes associated with oxygen sensing, root adaptation, and metabolic switching (Basit et al. 2024). Furthermore, contemporary CRISPR-Cas9 targets specific genes associated with stress response mechanisms, enhancing tolerance and leading to enhanced root system architecture and metabolic robustness. The amalgamation of genetic and genomic methodologies enables breeders to create pigeonpea cultivars with improved tolerance to waterlogging, hence providing consistent yields under conditions of excessive moisture. The amalgamation of these physiological, biochemical, and molecular methodologies is anticipated to enhance pigeonpea breeding programs in the development of high-yielding, waterlogging-tolerant hybrids and cultivars, ensuring output in moisture-saturated situations.

5.3 PEST AND DISEASE RESISTANCE IN PIGEONPEA

Due to climate change and resistance breakdown, diseases in plants have become one of the important threats to food security and sustainability. The traditional chemical methods are not effective enough and also deteriorate environmental health (Upadhyay et al. 2024). Several genomic approaches, like genome-wide association studies, QTL-sequencing, and genomic selection, help to identify the potential genomic regions and resistant donors for plant disease resistance.

5.3.1 Fusarium Wilt (FW) Resistance

FW is the most dangerous biotic threat to pigeonpea production throughout the world. *Fusarium udum* enters the pigeonpea plant through the roots and starts infecting seedlings, causing significant damage at the flowering and podding stage. The occurrence of infection has been recorded up to 60%, particularly at the flowering and crop maturity stages of the plant (Sharma et al. 2016), nevertheless, 100% yield losses were also reported in susceptible cultivars (Sharma et al. 2012). Interveinal chlorosis, internal browning of xylem vessels, loss of turgidity, yellowing and drooping of leaves, and withering and wilting of plants are key symptoms of FW. Environmental effects on pigeonpea *Fusarium udum* interactions and genotype stability to FW were reported by a group of scientists led by Sharma et al. (2016). This study shows that different isolates respond to hosts differently. Severe yield losses due to FW across the major pigeonpea-producing regions have been reported. This study shows the occurrence of *F. udum* isolates into nine variants (races). Among these, the two and three variants are found to be most predominant in most pigeonpea-growing areas (Ravikumara et al. 2022).

5.3.2 Phytophthora Blight (PB) Resistance

PB is the third potentially important disease of red gram (pigeonpea) caused by the fungus *Phytophthora drechsleri* f. sp. *cajani*. It can affect the crops at any stage of growth. PB is an economically important disease of pigeonpea, especially when extreme rains fall within a short time, and hot and humid environmental conditions stay during the crop season (Pande et al. 2011). Optimal conditions create frequent outbreaks of PB epidemics due to unfavorable climatic changes, their unpredictable occurrence, and changes in rainy day frequency. Very little knowledge about the resistance mechanism of pigeonpea is available.

5.3.3 Sterility Mosaic Disease (SMD) Resistance

For the first time, SMD was identified in 1931 at Pusa, Bihar, and became endemic in the pigeonpea cultivated regions of India and its nearby countries. Approximately over US$300 million in yield losses were reported in India alone due to SMD

(Sharma et al. 2020). The characterization of SMD can be done by observing bushy and stunted plants whose leaves have been reduced in size with chlorosis on them and reduced flower production (Patil and Kumar 2015). Two variants of SMD virus were observed: one is pigeonpea sterility mosaic virus 1 (PPSMV-1), having five negative-sense RNA segments, and another is pigeonpea sterility mosaic virus-2 (PPSMV-2), having six negative-sense RNA segments (Patil et al., 2017). Some cultivated and wild species of pigeonpea were found to support a vector of this virus, *Aceria cajani*, an eriophyid mite. This scenario was only found in the infected plant, which shows the strong relation between these infected plants and virus-carrying vectors (Singh et al. 2021). Many popular cultivars, such as ICP8863 (Maruti) and JKM-189, are susceptible to SMD and become a host to spread this disease in the nearby region. Hence, the identification of resistant sources is highly vital.

5.3.4 Resistance to Pod Borer, Pod Fly, Buchids, and Maruca

The pigeonpea crop is affected by about 300 species of insect pests, out of which *Helicoverpa armigera* (Hubner) or pod borer is a main pest causing 20%–30% yield losses (Sreekanth et al. 2015; Moghiya et al. 2025). The primary impact on crops occurs during pod maturation and subsequent harvesting, particularly due to pod borers, resulting in annual production losses exceeding $1,000 (Sharma et al. 2001). The newly developed larvae initially consume tender leaves, followed by the flower bud. During the podding stage, the newly emerging larvae eat soft leaves, prolifically reproducing by chewing before progressing to the flower bud. During the pod stage, the larvae enter the pods and consume the contents within (Chavan et al. 2014). Pod borers cause significant output losses in crops. The damage attributed only to *H. armigera* is reported to range from 13.2 to 36.4% across several regions in India. The newly released eggs are hemispherical in shape with a flat base, initially remain yellowish white in color, subsequently change to profound yellow to dark or gray-black before one day to hatching and shrivel after three to four days. The average hatching success rate of eggs is 93.57 ± 2.67%. The larval period of *H. armigera* passed through six distinct instars. The freshly emerging first instar larvae are pellucid and yellowish white in appearance with a black head, while second instar larvae are yellowish green with black thoracic limbs. The pupa is a structure form that is broadly rounded frontally with a tapering posterior end; the mean of pupal duration is 14.80 ± 1.30 days. However, the back wings exhibit a deeper coloration with a dark spot near the anterior end. The male moth is smaller than the female moth and is easily identifiable due to the spike of hairs at the top of the body (Chavan et al. 2014).

The pod fly, *Melanagromyza obtusa* (Malloch; Diptera, Agromyzidae), is a highly detrimental pest, causing grain losses of around 20 and 80% due to its damaging effects (Sreekanth et al. 2020). The infection causes considerable loss of pigeonpea pods and seeds, leading to a diminished rate of germination and making these unfit for human food or other purposes . Pod fly egg formation transpires in soft pods and along the inside of the pods. Larvae consume the seeds and undergo pupation within the pods (Nair et al. 2017). The pod fly produced a reduced number of eggs in December and January due to low temperatures. The population of pod flies develops with rising temperatures (Chiranjeevi and Patange 2018). The female releases up to

80 eggs into developing tender pods. These pods remain normal before the arrival of the larvae, which then create shot-holes in the pod walls following maturation. Generally, a single maggot needs only one seed for its maturation (Yadav et al. 2016). After hatching, a larva excavates into the pods and consumes the tender seeds, making them unsuitable for consumption by humans and future reproduction. This insect, as an inside feeder, resides within the pod wall during both larval and pupal stages, resulting in a delicate translucent membrane. The larvae within the pod ingest the developing seeds, subsequently undergoing pupation (Patange et al. 2017). This incision functions as an exit path for the adult flies developed inside the pod (Kumar et al. 2015). The egg phase generally lasts from 3 to 5 days, succeeded by a larval growth period of 6 to 11 days, and subsequently, the pupal stage spans around 9 to 23 days' duration (Yadav et al. 2020). It was also found that mature insects have a lifespan of approximately 6 days deprived of sustenance, which extends to 12 days when food is available. The newly emerged larvae are small and black in color. Dried pigeonpea pods display one or several apertures on the superior surface. The seeds within contaminated pods seem dry, wrinkled, and half-consumed (Sharma and Keval 2021). Several studies have determined that over 20 Hymenopteran parasites target this pest (Yadav et al. 2011). The prevalence of parasitism in *M. obtusa* has been examined by many studies (Yadav et al. 2012).

The biggest challenge in pulses storage is the infestation of storage pests, among them, bruchid is considered the most devastating post-harvest insect pest of tropical regions (Ndong et al. 2012). Bruchid, commonly known as seed beetle or seed weevil, belongs to *Coleoptera* order and the *chysomelidae (Bruchidae*) family is a serious insect pest of grain legumes during storage (Southgate, 1979). The Bruchidae family harbors about 1,300 species of seed beetles, of which 20 are acknowledged as pests for stored seeds of legume crops (Talekar 1988). Bruchids a cosmopolitan insect pests of legume crops and are distributed across the continents (Rees 2004). It is an infestation that causes seed weight loss, reduces germination and nutritional quality of seed, which ultimately leads to reduced market value, being unsuitable for further utilization in any form (Duan et al. 2014; Rizvi et al., 2022). This pest can cause up to 25%–30% loss in the field itself, while during storage, up to 80% loss in a six- to eight-month time span. Among the 20 species, two species, i.e., *C. maculatus* and *C. chinensis* are predominant species in legume crops and have similar life cycle and ecology. The infestation started in the field itself with egg laying on the pods and seeds. Normally, it lays 1–3 eggs/seed, but it varies based on the seed size. Bruchid usually has a preference for yellow color seeds for oviposition and development rather than green and black seeds. After 6 days of oviposition, hatching of the egg takes place (Devi and Devi 2014). Later on, larvae enter the seed and initiate feeding of the seed and subsequently mine a tunnel toward the seed coat for adult emergence, which appears as a translucent opening on the seed coat. The adult male and female average life span is seven days under controlled conditions; however, a few can live >2 weeks (Fatima et al. 2016). Females lay a large number of eggs depending on the substrate quality that the female perceives. A good quality substrate will lead to good larval fitness and a successful population buildup. Female bruchid can check the host appropriateness accordingly, deposit proper egg load, and also discriminate the eggs laid by other females on a seed (Guedes et al. 2007).

The spotted caterpillar (*M. vitrata*), an occasional pest of pigeonpea, often damages this short crop. Varieties exhibiting spreading characteristics and determinate growth are the most susceptible. The larvae inflict significant damage on flowers and floral buds. The characteristic symptom includes the webbing of flowers, pods, and leaves, together with frass present on the terminals of pods and stems. This is a notable pest in early maturing cultivars. They contaminate flowers, resulting in their wilting and discoloration. Damage is evidenced by small, darker entry pores on the surface and borers within the compromised pods. Webbing generally connects leaves and pods while exhibiting evidence of surface feeding. Young larvae can attain a length of 18 mm and exhibit drab to yellow-white coloration. Each segment features longitudinal black patches arranged in a distinct series. This design is really prominent on the upper surface. The larval head is either black or dark brown. The mature moth possesses three brown forewings adorned with white spots and a grayish-white hind wing featuring brown patterns at its distal edge. The eggs are positioned singularly on the pods, within the flower's buds, sepals, or petals. The caterpillar devotes three weeks post-hatching to consuming the tender leaves, flower buds, and developing seeds within the pods (Mishra et al. 2023). *M. vitrata* is responsible for almost 80% of annual yield losses in pigeonpea (Chauhan et al. 2025).

5.4 IMPROVEMENT OF AGRONOMIC BREEDING TRAITS IN PIGEONPEA

5.4.1 Determinacy

Determinant growth habit in pigeonpea significantly influences its plant architecture, reproductive behavior, and overall agronomic performance. The determinate lines have a distinct developmental pattern in which the terminal buds retard vegetative growth and transform floral structures. The incorporation of determinate growth habit in pigeonpea breeding has been acknowledged for its capacity to improve agronomic efficiency, yield stability, and adaptability to contemporary farming systems. One advantage of determinate pigeonpea lines is their uniform maturity, which ensures synchronous pod setting and blooming (Mir et al. 2014). This characteristic is particularly advantageous for mechanical harvesting, as it diminishes post-harvest deterioration and labor expenses (Gayacharan Parida et al. 2023). Furthermore, by allocating additional resources to reproductive development at the cost of perpetual vegetative growth, determinate habit enhances the harvest index. In intensive farming settings, determinate pigeonpea lines are preferable because of their enhanced assimilate allocation to the pods, which improves total production efficiency and grain setting. Another significant advantage is the appropriateness of determinate pigeonpea for short-duration cropping. Determinate cultivars assist growers in optimizing land use and attaining elevated yields by enabling crop intensification and various cropping systems through early synchronized maturity (Bohra et al. 2020). Determinate genotypes possess a drought escape mechanism in rainfed and marginal settings where moisture availability is problematic, enabling them to complete their life cycle before experiencing detrimental dry matter stress. This adaptation may be especially beneficial in mitigating the adverse impacts of climate change and enhancing the resilience of pigeonpea agricultural systems in semiarid regions (Bakala et al. 2024). Although the manifestation of the determinacy characteristic in pigeonpea is influenced by environmental factors, its genetic foundation has been well investigated and is primarily ascribed to one or a few key genes. Significant genomic areas linked to determinacy have been predominantly found via QTL mapping and GWAS, facilitating the development of molecular markers for MAS (Roychowdhury et al. 2014; Roychowdhury et al. 2023b). Furthermore, investigations into the molecular regulation of determinacy, particularly those concerning floral transition and meristem identity genes like *Terminal Flower 1* (*TFL1*) homologs, have used genomic and transcriptomic methodologies. These genes are recognized for governing the transition from indeterminate to determinate development, leading to changes in flowering time and plant architecture (Goretti et al. 2020). Identifying genes that specifically regulate flowering time and growth habit can expedite genetic improvement in pigeonpea (Kumar et al. 2025). Contemporary genomic tools provide novel prospects for the swift advancement of enhanced, determinate cultivars and can be utilized alongside conventional breeding methods. To effectively utilize the determinacy trait in pigeonpea enhancement, integrated breeding strategies must be implemented. These stable, high-yielding genotypes with exceptional agronomic characteristics are produced by combining determinate lines with heterotic indeterminate cultivars using hybridization and selection programs (Saxena et al. 2021). Furthermore, the conjugation of traditional selection methods with genomics-assisted breeding may substantially improve the efficacy of these breeding programs, resulting in the creation of early maturing, high-yielding, and climate-resilient pigeonpea cultivars (Sheoran et al., 2022). The extensive adoption of determinate varieties provides substantial advantages for breeders, researchers, and, particularly, farmers. The generation of successful determinate pigeonpea varieties would facilitate sustainable and profitable global production, as breeding efforts increasingly combine contemporary genomic technologies with classical selection methods.

5.4.2 Days to Flowering and Maturity

Critical agronomic factors influencing crop adaptation, production potential, and appropriateness for various agro-climatic zones are the days to flowering and maturity duration in pigeonpea (Saxena et al. 2019, 2021). While maturation duration influences the general cropping cycle, resource use, and stress tolerance, flowering time defines the change from vegetative to reproductive development. Depending on their blooming and maturation period, pigeonpea cultivars fall generally into extra-early (90–100 days), early (120–140 days), medium (150–180 days), and late-maturing (>180 days) categories (Tripathi et al. 2025). Although these features are genetically regulated, environmental elements—especially photoperiod and temperature sensitivity—also affect them. In pigeonpea, the control of blooming and maturity relies on intricate interactions among environmental, hormonal, and genetic elements. Important for the photoperiodic response of the plant, several qualitative loci control the expression of essential flowering-related genes, including homologs of *FLOWERING LOCUS T (FT) and CONSTANS (CO)*

(Varshney et al., 2012). Furthermore, revealing the role of hormonal pathways in regulating flowering initiation and maturity is a transcriptome and proteomic analysis of gibberellins and abscisic acid (Bohra et al., 2020). Breeding high-yielding, climate-resilient pigeonpea cultivars requires the capacity to control flowering and maturity length (Lohithaswa et al. 2025). Integration of the multi-omics toolkit helps pigeonpea improvement projects create long-duration varieties for stress-sensitive areas and short-duration varieties for crop intensification, therefore ensuring improved yield and better resilience to climate change (Saxena et al., 2021).

5.4.3 Cytoplasmic Male Sterility (CMS)

CMS is a condition wherein plants do not generate viable pollen necessary to produce hybrid seeds (Bohra et al. 2025). Female parents must incorporate specific nuclear genes known as fertility restorer (*rf*) genes, which counteract the effects of CMS and enable these plants to generate viable pollen. The genetic composition of hybrids governs fertility restoration in pigeonpea (Lay & Saxena, 2011). Hybrid seed production relies on sporophytic CMS, commonly referred to as the three-line hybrid system, which comprises a male-sterile line (A-line), a maintainer line (B-line), and a fertility restoring line (R-line). Several articles regarding the regeneration of CMS from *Cajanus* sp. have been released (Mallikarjuna et al. 2006). The cytoplasm-isolated (CMS) system, originating from the interspecific hybrid of *C. cajanifolius* (A4-Cytoplasm), is now essential for the commercial use of hybrid cultivars in pigeonpea (Saxena et al. 2005).

5.4.4 Seed Weight

The yield, market acceptance, and seedling vigor in pigeonpea are correlated with the seed weight characteristic. It is a quantitative trait regulated by multiple genes and significantly affected by environmental, physiological, and genetic influences. With seed weight directly influencing grain productivity and consumer demand in various places, pigeonpea shows a broad variety of seed sizes from small-seeded (5–7 g per 100 seeds) to large-seeded (15–25 g per 100 seeds). Important candidate genes implicated in seed growth and filling, including homologs of *GW2*, *GIF1*, and *SHB1*, help to define ultimate seed size and weight (Varshney et al. 2012). Hormonal mechanisms, including auxins, cytokinins, and gibberellins, which control cell division and expansion during seed development, have been shown by transcriptomics and proteomics studies (Bohra et al. 2020). Environmental factors, including soil nutrients, water availability, and temperature fluctuations, may influence final seed weight by impacting the duration and efficiency of seed filling (Chakraborty et al. 2014; Roychowdhury 2014; Roychowdhury et al. 2020). The integration of advantageous alleles from various germplasm sources by MAS and GS enhances seed weight in pigeonpea breeding programs. Precision breeding tactics have benefited from the identification of loci associated with seed weight, facilitated by advancements in QTL mapping and GWAS that assist routine breeding (Saxena et al. 2021).

A viable method for enhancing seed yield and quality is CRISPR-Cas9 mediated genome editing, which enables the exact modification of genes that regulate seed weight and size (Roychowdhury et al. 2025). The integration of genomics, transcriptomics, and metabolomics, alongside other multi-omics methodologies, will facilitate a comprehensive understanding of seed weight regulation, ultimately resulting in the development of high-yielding pigeonpea varieties that exhibit superior adaptability to various agro-climatic conditions (Varshney et al. 2020).

5.4.5 Photo-Insensitivity

The day-neutral trait in pigeonpea is a crucial agronomic characteristic that allows plants to flower and mature regardless of day length, hence promoting its cultivation across diverse latitudes, geographical regions, and cropping systems. Photoperiod-insensitive pigeonpea genotypes can flower under both short and long day conditions, thereby enabling broad adaptability and flexibility in planting schedules, unlike photoperiod-sensitive varieties that necessitate specific day lengths for flowering induction in diverse growing locations (Sennhenn 2016). This trait is crucial for expanding pigeonpea cultivation into non-traditional areas and for intercropping systems that require precise flowering and maturity durations. Photoperiod sensitivity is primarily governed by genes associated with the flowering pathway, including *CO*, *FT*, and *GIGANTEA* (*GI*) homologs, which serve as sensors for day length and are essential for inducing flower production (Soni et al. 2024). QTL mapping and GWAS have pinpointed genetic regions associated with photoperiod sensitivity, aiding in the development of molecular markers for the selection of day-neutral varieties. Furthermore, a substantial quantity of differentially expressed genes was identified between photoperiod-sensitive and insensitive genotypes using transcriptome analyses, which may elucidate their regulatory processes. The creation of photoperiod-insensitive genotypes has been a significant breeding goal focused on enhancing adaptability, broadening cultivation range, and ensuring consistent yields across various agro-ecological conditions. Various tactics, including classical breeding, MAS, genome editing, and multi-omics approaches, are employed to achieve that objective. These methods effectively influence blooming time, hence expanding pigeonpea cultivation across various latitudes and facilitating its incorporation into diverse cropping scenarios with varying photoperiods (Saxena 2008). Traditional breeding methodology involves screening natural germplasm collections to identify photoperiod-insensitive progenitors, which are subsequently crossed with elite cultivars. This technique can facilitate photoperiod insensitivity in flowering time during the creation of high-yield varieties, while preserving other desirable agronomic features. Heterosis combined with selection facilitates the development of stable pigeonpea lines that can flower and mature regardless of photoperiodic conditions. Nevertheless, conventional breeding can be a protracted process, requiring multiple generations of selection and backcrossing to stabilize a trait in new lines. The implementation of MAS has revolutionized pigeonpea breeding by enabling the discovery and selection of genes associated with photoperiod insensitivity. QTL mapping and GWAS have identified genetic markers linked to flowering timing and photoperiod responsiveness. The integration of these markers into the breeding program will enable breeders to expedite and reduce the time necessary for generating photoperiod-insensitive cultivars. This approach enhances efficiency by focusing breeding material on certain genetic loci linked to the trait, rather than relying solely on phenotypic selection. Progressive enhancement of flowering genetics for photoperiod insensitivity in wheat by gene-editing technologies provides a novel method to alter the flowering-related genes that regulate photoperiod (Naik et al. 2025). Similarly, researchers can also translate it into pigeonpea backbone through genetically targeting a set of priority genes associated with flowering, such as *CO*, *FT*, and *GI* to alter photoperiod sensitivity. This would facilitate the development of early maturing, high-yielding pigeonpea types without hazardous genetic alterations. Integrating new holistic genomes, transcriptomics, proteomics, and metabolomics can provide a comprehensive understanding of the molecular events associated with photoperiod insensitivity. Genomic investigations elucidate genetic variants that influence flowering time, while transcriptome analysis reveals gene expression under varying photoperiod conditions. Proteomic and metabolomic techniques can enhance the comprehension of the roles of proteins and metabolites in photoperiod response, facilitating targeted breeding tactics. This multi-omics approach improves the effectiveness and accuracy of breeding schemes, allowing breeders to develop pigeonpea genotypes with consistent photoperiod insensitivity having high-yield potential with stability and adaptability to a changing environment (Mangal et al. 2024). A photoperiod-insensitive pigeonpea is essential for climate-resilient agriculture, facilitating consistent production amid seasonal fluctuations. These novel varieties may be cultivated with adaptable sowing schedules, integrate more effectively into crop rotation, and expand the cultivation of pigeonpeas to new regions. Advanced research on genomics-assisted breeding will focus toward fine-tuning of flowering responses coupled with high-yield, disease resistance, and abiotic stress tolerance for sustainable pigeonpea production and food security worldwide.

5.4.6 Cleistogamy

Cleistogamy, a form of sexual reproduction in which pigeonpea plants self-pollinate flowers before their opening, holds significant implications for breeding and genetic improvement (Suso et al. 2015). Being primarily a partial outcrosser, pigeonpea may benefit from cleistogamy, which can enhance genetic purity, increase hybrid seed production efficiency, and diminish the likelihood of inadvertent cross-pollination. The genetics and physiology of cleistogamy in pigeonpea are crucial for the development of stable, high-yielding cultivars with a regulated pollination mechanism, and are crucial for enhancing efficiency and stability. This characteristic restricts genetic contamination and offers numerous benefits for the preservation of superior breeding lines, hybrid seed generation, and ecological adaptation. The regulated pollination mechanism linked to cleistogamy has the potential to improve genetic purity, decrease hybrid seed production costs, stabilize yields, and promote reproductive success, thus becoming a significant trait for pigeonpea breeding. Economic and environmental importance of cleistogamy in pigeonpea is genetic purity, i.e., the emphasis on genetic purity is vital in seed production to uphold quality, yield potential, and quantity. As cleistogamous organisms, they undergo self-pollination devoid of interference from external pollen sources (Tawaha et al. 2022). This attribute is particularly beneficial for the preservation of parental seed supplies, inbred lines, and elite breeding materials, preventing the dilution of favorable features through genetic interference. Consequently, breeders can cultivate and distribute true-to-type varieties exhibiting enhanced uniformity for more consistent and superior seeds. Cleistogamy offers a benefit for hybrid seed production, as it allows for controlled pollination, eliminating the need for artificial emasculation. In traditional hybrid breeding programs, manual anther removal is a labor-intensive and time-consuming task. By including cleistogamous traits, breeders can mitigate pollen contamination, hence enhance seed purity, and improve the production efficiency of hybrid seeds. This trait is vital in the large-scale propagation of hybrid seeds, where seed integrity is crucial for consistent seed performance and market acceptance. A second significant advantage of cleistogamy is less dependence on pollinators, as it enhances seed production in environments lacking pollinators, perhaps contributing to elevated and more consistent yields. This contributes to greater stability in pigeonpea propagation and reduces dependence on external environmental variables for effective reproduction. In addition to its benefits for breeding and pollination, cleistogamy provides an adaptive advantage under unfavorable environmental conditions. Closed flowers are shielded from adverse weather conditions (intense rain, strong winds, humidity) that may diminish pollination success and the survival of pollen in chasmogamous flowers (Murphy 2022). Cleistogamous flowers enhance fertilization and seed production by protecting reproductive organs from external conditions, hence increasing productivity in challenging agro-climatic settings. In conclusion, the integration of cleistogamy serves as an alternative method to improve genetic purity, hybrid seed production, pollination, and adaptability to various conditions in pigeonpea. Research on cleistogamy in legumes suggests that a limited number of genes regulate the transition between cleistogamous and chasmogamous phases. QTL mapping and GWAS in pigeonpea may facilitate the identification of loci affecting floral development and self-pollination traits. Transcriptomic and proteomic investigations can elucidate gene expression disparities between cleistogamous and chasmogamous flowers, enabling breeders to develop molecular markers for focused selection. One method to accomplish this is through screening and selection, whereby naturally occurring cleistogamous mutants are identified from diverse germplasm sources, specifically pigeonpea lines. These lines are subsequently utilized in breeding programs to generate high-yielding, genetically stable cleistogamous pigeonpea cultivars. To prevent the loss of agronomic features during the selection of promising genotypes in breeding programs, meticulous evaluation and selection are essential. An alternative method is MAS, which facilitates the swift identification of cleistogamous plants at an early stage, independent of exclusive reliance on phenotypic evaluation (Cobb et al. 2019). It provides a chance to modify the floral structure and improve the adaptation of cleistogamous pigeonpea across various agro-climatic situations. Moreover, hybridization breeding is essential for utilizing cleistogamy in hybrid seed production. Breeders can attain controlled hybridization with excellent seed purity by developing cleistogamous male-sterile lines. This plan ensures that hybrid development is exceptionally efficient, reduces the chance of incidental cross-pollination, and enhances the productivity of hybrid seed production. By employing breeding techniques such as screening and selection, MAS, genome editing, and hybrid breeding, breeders could effectively integrate the cleistogamy trait into pigeonpea enhancement programs. These tactics facilitate the development of genetically pure, high-yielding, and stress-resilient pigeonpea hybrids and varieties, ultimately benefiting both farmers and seed producers across many agro-ecological regions. Notwithstanding its potential benefits, the adoption of cleistogamy for pigeonpea variety development may encounter challenges, such as reduced genetic recombination due to complete self-fertilization, and potential yield reductions if the intrinsic floral structures are not favorable for efficient self-pollination. Future research must ascertain the ideal equilibrium between cleistogamy and limited outcrossing to maintain hybrid vigor and genetic advancement. Supplemental feeding of adults significantly attracted them but did not enhance the efficacy of an artificial sugar meal. Utilizing contemporary genomics and molecular breeding techniques, we want to leverage cleistogamy to enhance the stability of pigeonpea yields, optimize seed production efficiency, and ensure genetic purity for the benefit of farmers and breeders.

5.5 QUALITY ENHANCEMENT IN PIGEONPEA GRAINS

5.5.1 Protein Content

Pigeonpea has the same kinds of nutrients as other legumes. It has a low amount of fat and a high amount of protein in the grain. It has an average of 20%–22% protein, 65% carbohydrates, 3.8% ash, and 1.2% fat. However, the protein level can be anywhere from 18% to 25%, and it can even reach 32% (Aggarwal et al. 2015). Sarkar et al. (2020) showed that pigeonpea can be eaten in a number of different ways, such as fresh pods, whole seeds, dal (split seeds without seed coats), and green seeds as veggies (Jeevarathinam and Chelladurai 2020; Kuraz Abebe 2022). Today, pigeonpea flour is used as an interesting ingredient in many foods, like pasta, noodles, biscuits, and sausages. This is because it has a lot of fiber and protein, is gluten-free, has a low glycemic index, a lot of antioxidants, and useful properties like being able to absorb and hold water (Singh 2022). Saxena et al. (2010a) highlighted that this legume is a great way to get dietary fiber, iron, sulfur, calcium, potassium, manganese, and water-soluble vitamins, especially thiamine, riboflavin, and niacin.

5.5.2 Iron and Zinc Content

Pigeonpea is a valuable source of important minerals, including magnesium, calcium, iron, potassium, sodium, manganese, chromium, and zinc. Metals such as Fe, Zn, Cu, Cr, and Mn are important for both animals and humans, as they are crucial for various metabolic processes, enzymatic functions, receptor sites, hormonal activities, and protein transport at certain concentrations (Mekassa and Chandravanshi 2015). The cotyledon, which is the primary part of the pigeonpea grain, contributes significantly to the grain's nutritious content. On average, cotyledon comprises a significant proportion of protein (97.58% ± 0.45%), phosphorus (99.08% ± 0.33%), potassium (91.48% ± 1.61%), magnesium (77.11 ± 3.50%), iron (87.40% ± 1.61%), zinc (90.05% ±

TABLE 5.1
Instances of Using Genomic Strategies for Trait Improvement in Pigeonpea

Traits	Genomic Strategy Employed	References
Drought tolerance	Identification of drought-responsive genes through transcriptome analysis	Saxena et al. (2011)
Salinity tolerance	Functional validation of candidate genes conferring salinity tolerance	Sekhar et al. (2010)
FW resistance	Development of mapping populations and identification of associated QTLs	Saxena et al. (2010b)
Sterility mosaic disease	SSR-based diversity analysis and mapping for resistance genes	Saxena et al. (2010b, 2010c)
Pod borer resistance	Introgression of resistance genes from wild relatives and development of transgenic lines	Jadhav et al. (2012), Ramu et al. (2012), and Moghiya et al. (2025)
Determinate growth habit	Mapping of the determinacy gene and development of molecular markers for selection	Mir et al. (2014)
Earliness	Identification of QTLs associated with early flowering and maturity	Kumawat et al. (2012)
Fertility restoration	Identification of markers linked to fertility restoration genes in hybrid breeding	Bohra et al. (2012)
Seed protein content	QTL mapping and identification of genomic regions associated with protein content	Pazhamala et al. (2015)

2.49%), copper (88.68% ± 3.09%), and manganese (86.59% ± 4.17%), while the seed coat contains the highest level of calcium (66.78% ± 3.33%; Susmitha et al. 2022). The nutritional supplementation of cereals with protein-dense legumes is regarded as one of the most effective strategies to mitigate protein-calorie malnutrition in developing nations (Odeny 2007). Hunger and malnutrition are increasing due to population growth, a shortage of agricultural land, and high food prices. It is the primary cause of death and disease among millions of pregnant women and young children worldwide. The development of nutrient-dense food products from an economical and inexpensive source, such as pigeonpea, is an effective alternative technique to combat malnutrition (Pelletier et al. 1995). Thus, dedicated efforts have been made toward trait improvement in pigeonpea using different genomic strategies and identified genes/QTLs associated with target traits as presented in Table 5.1. The identified genes/QTLs enhance the efficiency and precision of breeding programs, enabling breeders to develop climate-resilient, nutrient-dense, and high-yielding pigeonpea varieties.

5.6 TRAIT DRAGGING USING WILD ACCESSIONS OF PIGEONPEA

Although wild pigeonpea accessions serve as a key source of beneficial traits, their introgression through conventional breeding practice leads to undesirable trait dragging that results in the co-inheritance of deleterious and useful attributes

(Bohra et al. 2010). These closed genetic associations are induced by the tight linkage among the target loci and adverse alleles from wild relatives (Bohra et al. 2022). The following are some critical factors of trait drag when imposed in pigeonpea:

1. **Yield penalty:** Many wild pigeonpea species have lower yield potential compared to elite cultivars. Incorporating stress-resistance traits from wild relatives may inadvertently reduce overall productivity.
2. **Delayed flowering and maturity:** Some wild accessions exhibit late flowering and extended growth periods, which may not be desirable in breeding programs aimed at developing early maturing varieties.
3. **Reduced seed quality:** Wild pigeonpea species may carry alleles associated with undesirable seed characteristics, such as harder seed coats, lower palatability, and altered nutritional composition.
4. **Cross-incompatibility:** Hybridization between cultivated and wild pigeonpea is often challenging due to reproductive barriers, limiting the efficiency of introgression breeding.

In parallel, the following innovation strategies are established to minimize such trait dragging:

1. **Advanced backcross breeding:** Successive backcrossing with elite cultivars while selecting for the desired traits can help dilute unwanted wild alleles.
2. **MAS:** Genomic markers linked to beneficial traits allow breeders to track and retain favorable alleles while eliminating linked undesirable genes.
3. **GS:** By using genome-wide marker data, GS enables the prediction of breeding values and the selection of superior genotypes with minimal linkage drag.
4. **Gene editing (CRISPR-Cas9):** Precise genome-editing techniques can be used to eliminate deleterious alleles while retaining beneficial genes from wild accessions.
5. **Speed breeding:** Accelerating generation cycles through controlled environment facilities allows breeders to quickly break linkages between undesirable and desirable alleles.

The era of pangenomics has provided an opportunity to scan wider genetic diversity in pigeonpea species, which shall help discover new naturally recombined genotypes with almost zero linkage drag. Furthermore, further development of haplotype-assisted breeding and genomic-assisted introgression pipelines is likely to streamline wild relative utilizations in the pigeonpea breeding improvement program (Figure 5.2). Using these contemporary genomic and breeding methodologies will enable pigeonpea researchers to more fully capitalize on the advantages of wild germplasm, while minimizing the challenges associated with trait dragging. This will enhance the designing of high-yielding, stress-tolerant and

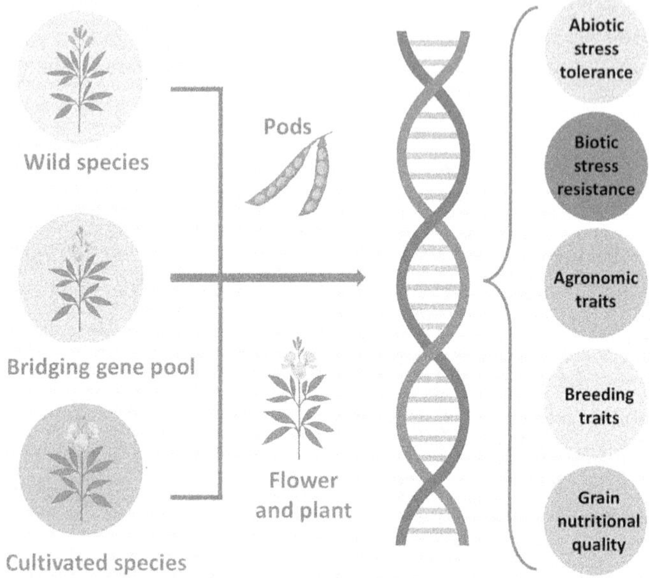

FIGURE 5.2 Utilization of wild and cultivated pigeonpea to improve omics-mediated improvement of the desirable traits.

nutrition-rich pigeonpea cultivars, resulting in global food security under sustainable agriculture.

5.7 CONCLUSIONS

The rapid advancement of omics technology is generating novel prospects for the enhancement of pigeonpea. Clarifying the genetic architecture of a trait is essential for plant breeding, enabling breeders to create pigeonpea varieties with improved yields, stress resilience, and environmental adaptability. The integration of omics like genomics, transcriptomics, metabolomics, and proteomics can provide robust methodologies for accelerating trait identification and improving breeding efficacy in pigeonpea. The application of these techniques for predictive breeding will present the next hurdle, as artificial intelligence (AI) and machine learning (ML) offer potential utilities that may inform optimal selection for specific traits in a sustainable manner. Integrated multi-omics breeding strategies will be essential for transforming pigeonpea from an "orphan crop" into a climate-resilient legume necessary for global food security.

ACKNOWLEDGMENTS

RR is thankful to the Department of Biotechnology (DBT, Govt. of India) for the funding grant in the form of Research Associateship (DBT-RA/2023-24/Call-I/RA/01). All the authors acknowledge CGIAR-ICRISAT (HQ, Hyderabad, India) for providing the infrastructure and research opportunity.

REFERENCES

Aggarwal A, Nautiyal U, Negi D (2015) Characterization and evaluation of antioxidant activity of *Cajanus cajan* and *Pisum sativum*. *Int J Rec Adv Sci Tech* 2(1): 21–26.

Athanas K (2022) *Morphological and genetic characterization of pigeon pea (Cajanus cajan L) using simple sequence repeat markers* (Doctoral dissertation, Kenyatta University).

Awana M, Yadav K, Rani K, et al. (2019) Insights into salt stress-induced biochemical, molecular and epigenetic regulation of spatial responses in pigeonpea (*Cajanus cajan* L.). *J Plant Growth Regul* 38: 1545–1561.

Bakala HS, Devi J, Singh G, Singh I (2024) Drought and heat stress: insights into tolerance mechanisms and breeding strategies for pigeonpea improvement. *Planta* 259(5): 123.

Basit F, Khalid M, El-Keblawy A et al. (2024) Hypoxia stress: plant's sensing, responses, and tolerance mechanisms. *Environ Sci Pollut Res* 31(55): 63458–63472.

Belliappa SH, Bomma N, Pranati J et al. (2024) Breeding for water-logging tolerance in pigeonpea: current status and future prospects. *CABI Agric Biosci* 5(1): 98.

Bohra A, Mallikarjuna N, Saxena KB, Upadhyaya H, Vales I, Varshney R (2010) Harnessing the Potential of Crop Wild Relatives Through Genomics Tools for Pigeonpea Improvement. *J. Plant Biology*, 37: 8–98.

Bohra A, Saxena KB, Varshney RK, Saxena RK (2020) Genomics-assisted breeding for pigeonpea improvement. *Theor Appl Genet* 133(5):1721–1737.

Bohra A, Saxena RK, Varma P, et al. (2012) Analysis of BAC-end sequences (BESs) and development of BES-SSR markers for genetic mapping and hybrid purity assessment in pigeonpea (*Cajanus cajan* L. Millsp.). *BMC Plant Biol* 11: 56.

Bohra A, Tiwari A, Kaur P, Ganie S, Raza A, Roorkiwal M, Mir R, Fernie A, Smykal P, Varshney R (2022) The key to the future lies in the past: Insights from grain legume domestication and improvement should inform future breeding strategies. *Plant and Cell Physiology* 63(11):1554–1572.

Bohra A, Tiwari A, Pareek S, Joshi R, Satheesh Naik SJ, Kumari K, Verma RL, Parihar AK, Patil PG, Dixit GP (2025) Past and future of cytoplasmic male sterility and heterosis breeding in crop plants. *Plant Cell Rep* 22: 33.

Bomma N, Shruthi HB, Soregaon CD, et al. (2024) Multi-environment testing for G×E interactions and identification of high-yielding, stable, medium-duration pigeonpea genotypes employing AMMI, GGE biplot, and YREM analyses. *Front Plant Sci* 15: 1396826.

Chakraborty S, Pattanayak A, Mandal S, et al. (2014) *An overview of climate change: causes, trends and implications. Crop improvement in the era of climate change*. IK International Publishing House, New Delhi, pp. 1–29.

Chauhan N, Singh I, Taggar GK, Vij S, (2025) Evaluation of pigeonpea interspecific derivatives for resistance to spotted pod borer, *Maruca vitrata* (Fabricius). *Plant Arch 25*(1):167–175.

Chavan AP, Patil SK, Tilekar VB, et al. (2014) Biology of *Helicoverpa armigera* Hubner on Pigeonpea. *J Agric Res Technol* 39(2):256.

Chiranjeevi B, Patange NR (2018) Field biology of pigeonpea pod fly, *Melanagromyza obtusa* (Malloch). *J Entomol Zool Stud* 6(1): 717–719.

Cobb JN, Biswas PS, Platten JD (2019) Back to the future: revisiting MAS as a tool for modern plant breeding. *Theor Appl Genet* 132: 647–667.

Devi MB, Devi NV, (2014) Biology and morphometric measurement of cowpea weevil, *Callosobruchus maculatus* fabr. (Coleoptera: Chrysomelidae) in green gram. *J Entomol Zool Stud* 2(3): 74–76.

Duan CX, Zhu ZD, Ren GX, Wang XM, Li DD (2014) Resistance of faba bean and pea germplasm to *Callosobruchus chinensis* (Coleoptera: Bruchidae) and its relationship with quality components. *J Econ Entomol* 107(5):1992–1999.

Duhan S, Sheokand S (2020) Mechanism of waterlogging stress tolerance in pigeonpea plants: Biochemical and anatomical adaptation under waterlogging. In *Plant ecophysiology and adaptation under climate change: mechanisms and perspectives II: mechanisms of adaptation and stress amelioration*, pp. 89–105.

Dwivedi SL, Chapman MA, Abberton MT, et al. (2023) Exploiting genetic and genomic resources to enhance productivity and abiotic stress adaptation of underutilized pulses. *Front Genet* 14: 1193780.

Fatima, S.M., Usman, A., Sohail, K., et al. (2016) Rearing and identification of *Callosobruchus maculatus* (Bruchidae: Coleoptera) in Chickpea. *J Entomol Zool Stud* 4: 264–266.

Gayacharan Parida, SK, Mondal N, Yadav R, Vishwakarma H, Rana JC (2023) Mining legume germplasm for genetic gains: An Indian perspective. *Front Genetics* 14: 996828.

Goretti D, Silvestre M, Collani S, et al. (2020) *TERMINAL FLOWER1* functions as a mobile transcriptional cofactor in the shoot apical meristem. *Plant Physiol* 182(4): 2081–2095.

Guedes RN, Guedes NM, Smith RH (2007) Larval competition within seeds: from the behaviour process to the ecological outcome in the seed beetle *Callosobruchus maculatus*. *Austral Ecology* 32(6): 697–707.

Hasanuzzaman M, Roychowdhury R, Karmakar J, et al. (2015) Recent advances in biotechnology and genomic approaches for abiotic stress tolerance in crop plants. In *Genomics and proteomics: concepts, technologies and applications*. Apple Academic Press, Burlington, pp. 333–366.

Jadhav DR, Mallikarjuna N, Sharma HC, Saxena KB (2012) Introgression of Helicoverpa armigera resistance from *Cajanus acutifolius* - a wild relative from secondary gene pool of pigeonpea (*Cajanus cajan*). *Asian J Agric Sci* 4(4): 242–248.

Jeevarathinam G, Chelladurai V (2020) *Pigeon pea. Pulses*: *Processing and Product Development*, pp.275–296.

Jha UC, Nayyar H, Roychowdhury R, et al. (2023) Non-coding RNAs (ncRNAs) in plant: Master regulators for adapting to extreme temperature conditions. *Plant Physiol Biochem* 205: 108164.

Joshi A, Kumar A, Adhikari S, Saroj R (2024) Marker-assisted selection: strategy for biotic and abiotic stress tolerance in plants. In *Climate-resilient agriculture* (pp. 137–191). Apple Academic Press.

Khoury CK, Castañeda-Alvarez NP, Achicanoy HA, et al. (2015) Crop wild relatives of pigeonpea [*Cajanus cajan* (L.) Millsp.]: Distributions, ex situ conservation status, and potential genetic resources for abiotic stress tolerance. *Biol Conserv* 184: 259–270.

Kumar J, Choudhary AK, Gupta DS, Kumar S (2019) Towards exploitation of adaptive traits for climate-resilient smart pulses. *Int J Mol Sci* 20(12): 2971.

Kumar K, Anjoy P, Sahu S, et al. (2022) Single trait versus principal component based association analysis for flowering related traits in pigeonpea. *Sci Rep* 12(1): 10453.

Kumar K, Durgesh K, Anjoy P, et al. (2025). Transcriptional reprogramming and allelic variation in pleiotropic QTL regulates days to flowering and growth habit in pigeonpea. *Plant Cell Env* 48(4): 2783–2803.

Kumar R, Keval R, Yadav A, Srivastava CP, (2015) Field screening and evaluation of long duration pigeonpea genotypes against the infestation of pod fly (*Melanagromyza obtusa* Malloch). *The Ecoscan*, 7: 291–295.

Kumawat G, Raje RS, Bhutani S, et al. (2012) Molecular mapping of QTLs for plant type and earliness traits in pigeonpea (*Cajanus cajan* L. Millsp.). *BMC Genet* 13: 84. https://doi.org/10.1186/1471-2156-13-84

Kuraz Abebe B (2022) The dietary use of pigeon pea for human and animal diets. *The Sci World J* (1): 4873008.

Lay KK, Saxena KB (2011) Inheritance of fertility restoration in pigeonpea. *J Food Leg* 24(4): 273–276.

Lohithaswa HC, Shreekanth SM, Banakara S, Rajanna LT, Kashyap GSS, Hosakoti S, Gonal B, Mallikarjuna MG, Gangashetty P (2025). Breeding Climate-Resilient Pigeonpea in Climate Change Era: Current Breeding Strategies and Prospects. In: *Breeding Climate Resilient and Future Ready Pulse Crops* (pp. 255–288). Singapore: Springer Nature Singapore.

Mallikarjuna N, Jadhav D, Reddy P (2006) Introgression of *Cajanus platycarpus* genome into cultivated pigeonpea, *Cajanus cajan*. *Euphytica* 149: 161–167.
Mangal V, Verma LK, Singh SK, et al. (2024) Triumphs of genomic-assisted breeding in crop improvement. *Heliyon* 10(15).
Mekassa B, Chandravanshi BS (2015) Levels of selected essential and non-essential metals in seeds of korarima (*Aframomum corrorima*) cultivated in Ethiopia. *Brazilian Journal of Food Technology*, 18(2): 102–111.
Melo-Sabogal DV, Contreras-Medina LM (2024) Elicitors and biostimulants to mitigate water stress in vegetables. *Horticulturae 10*(8): 837.
Mir RR, Kudapa H, Srikanth S, Saxena RK, et al. (2014). Candidate gene analysis for determinacy in pigeonpea (*Cajanus* spp.). *Theor Applied Genet* 127(12): 2663–2678.
Mishra AK, Kurre D, Mishra SK, Tiwari B (2023) Insect Pests of Pigeon Pea (*Cajanus cajan*) and Integrated Management Approaches. *Vigyan Varta* 4(12): 19–23.
Moghiya A, Munghate RS, Sharma V, et al. (2025) Dissecting genomic regions and candidate genes for pod borer resistance and component traits in pigeonpea minicore collection. *Front Plant Sci* 16: 1630435.
Murphy S (2022) The role of wind pollination in crop plants. *Promoting Pollination and Pollinators in Farming*, 126: 120.
Naik SSJ, Padmaja G, Tiwari A, et al. (2024) Genomics-Aided Breeding Strategies for Biotic Stress in Pigeonpea. In *Genomics-aided Breeding Strategies for Biotic Stress in Grain Legumes* (pp. 49–84). Singapore: Springer Nature Singapore.
Naik YD, Bahuguna RN, Garcia-Caparros P, et al. (2025) Exploring the multifaceted dynamics of flowering time regulation in field crops: Insight and intervention approaches. *The Plant Genome* 18(2): e70017.
Nair N, Shah SK, Thangjam B, et al. (2017) Insect pest complex of Pigeon pea (*Cajanus cajan*) in agro ecosystem of Tripura, *NE India. J Entomol Zool Stud* 5(4): 765–771.
Ndong A, Kébé K, Thiaw C, et al. (2012) Genetic distribution of the Cowpea (*Vigna unguiculata* (L.) Walp) bruchid (*Callosobruchus maculatus* F., Coleoptera, Bruchidae) populations in different agro-ecological areas of West Africa. *J Anim Sci Adv 2*(7): 616–630.
Odeny DA (2007) The potential of pigeonpea (*Cajanus cajan* (L.) Millsp.) in Africa. *Natural Resources Forum* 31(4): 297–305. Oxford, UK: Blackwell Publishing Ltd.
Pande S, Sharma M, Naga Mangla U, Ghosh R, Sundaresan G (2011) Phytophthora blight of Pigeonpea [*Cajanus cajan* (L.) Millsp.]: An updating review of biology, pathogenicity and disease management. *Crop Protect* 30(8): 951–957. https://doi.org/10.1016/j.cropro.2011.03.031
Pandey MK, Roorkiwal M, Singh VK, et al. (2016) Emerging genomic tools for legume breeding: current status and future prospects. *Front Plant Sci 7*: 455.
Patange NR, Sharma OP, Chiranjeevi B (2017) Population dynamics of *Melanagromyza obtusa* (Malloch) (Diptera: Agromyzidae) and its natural parasitization in pigeonpea. *Entomon* 42(3): 201–206.
Patil BL, Dangwal M, Mishra R (2017) Variability of emaravirus species associated with sterility mosaic disease of pigeonpea in india provides evidence of segment reassortment. *Viruses* 9(7):183. https://doi.org/10.3390/v9070183
Patil BL, Kumar PL (2015) Pigeonpea sterility mosaic virus: a legume-infecting Emaravirus from South *Asia. Mol Plant Pathol* 16: 775–786. https://doi.org/10.1111/mpp.12238
Pazhamala LT, Purohit S, Saxena RK et al. (2015) Gene expression atlas of pigeonpea and its application to gain insights into seed development and pod filling. *J Exp Bot* 68(8): 2037–2054. https://doi.org/10.1093/jxb/erx010
Pelletier DL, Frongillo Jr EA, Schroeder DG, Habicht JP (1995) The effects of malnutrition on child mortality in developing countries. *Bulletin of the World Health Organization* 73(4): 443.

Pratibha G, Srinivas I, Rao KA, et al. (2015) Impact of conservation agriculture practices on energy use efficiency and global warming potential in rainfed pigeonpea–castor systems. *European J Agron* 66: 30–40.

Ramu SV, Rohini S, Keshavareddy G, et al. (2012) Expression of a synthetic *cry1AcF* gene in transgenic Pigeon pea confers resistance to *Helicoverpa armigera. J Appl Entomol* 136(9): 675–687.

Rasheed A, Barqawi AA, Mahmood A, et al. (2022) CRISPR/Cas9 is a powerful tool for precise genome editing of legume crops: a review. *Mol Biol Rep* 49(6): 5595–5609.

Ravikumara BM, Naik MK, Telangre R, Sharma M (2022) Distribution and pathogenic diversity in *Fusarium udum* Butler isolates: the causal agent of pigeonpea Fusarium wilt. *BMC Plant Biol* 22(1): 147.

Rees DP (2004) *Insects of stored products*. CSIRO Publishing.

Rizvi Q, Kumar K, Ahmed N, et al. (2022) Influence of soaking and germination treatments on the nutritional, anti-nutritional, and bioactive composition of pigeon pea (*Cajanus cajan* L.*). J Appl Biol Biotechnol 10*(3): 127–134.

Roychowdhury R (2014) *Crop improvement in the era of climate change*. IK International Publish, New Delhi, India.

Roychowdhury R, Ballén-Taborda C, Chaturvedi P (2023b) Characterizing and improving traits for resilient crop development. *Front Plant Sci* 14: 1307327.

Roychowdhury R, Choudhury S, Hasanuzzaman M, Srivastava S (2020) *Sustainable agriculture in the era of climate change*. Springer-Nature Switzerland.

Roychowdhury R, Das SP, Das S, et al. (2025) Advancing vegetable genetics with gene editing: a pathway to food security and nutritional resilience in climate-shifted environments. *Funct Integr Genom* 25(1): 31.

Roychowdhury R, Das SP, Gupta A, et al. (2023a) Multi-omics pipeline and omics-integration approach to decipher plant's abiotic stress tolerance responses. *Genes 14*(6):1281.

Roychowdhury R, Taoutaou A, Hakeem KR, Gawwad MRA, Tah J (2014) *Molecular marker-assisted technologies for crop improvement. Crop improvement in the era of climate change*. IK International Publishing House, New Delhi, pp.241–258.

Sangma CB (2020) Soil and crop health management for the cultivation of pigeon pea: an overview of management practices. In: *Management of Fungal Pathogens in Pulses: Current Status and Future Challenges*, pp.143–167.

Sarkar S, Pand S, Yadav KK, Kandasamy P (2020) Pigeonpea (*Cajanus cajan*) an important food legume in Indian scenario-A review. *Legume Res* 43(5): 601–610.

Saxena KB (2008) Genetic improvement of pigeonpea—a review. *Tropical Plant Biol* 1: 159–178.

Saxena KB, Bohra A, Choudhary AK, Sultana R, Pazhamala L, Saxena RK (2021) The alternative breeding approaches for improving yield gains and stress response in pigeonpea [*Cajanus cajan* (L.) Millsp.]. *Plant Breed* 140: 76–86.

Saxena KB, Choudhary A, Srivastava R, Bohra A, Saxena R, Varshney RK (2019) Origin of early maturing pigeonpea germplasm and its impact on adaptation and cropping systems. *Plant Breed* 138: 243–251.

Saxena KB, Kumar RV, Gowda CLL (2010a) Vegetable pigeonpea–a review. *J Food Leg* 23(2): 91–98.

Saxena KB, Kumar RV, Srivastava N, Shiying B (2005) A cytoplasmic-genic male-sterility system derived from a crosses between *Cajanus cajanifolius* and *Cajanus cajan. Euphytica* 145: 291–296.

Saxena, RK, Cui X, Thakur V, Walter B, Close TJ, Varshney RK (2011) Single feature polymorphisms (SFPs) for drought tolerance in pigeonpea (*Cajanus* spp.). *Funct Integr Genomics* 11(4): 651–657.

Saxena RK, Prathima C, Saxena KB, Hoisington DA, Singh NK, Varshney RK (2010b) Novel SSR markers for polymorphism detection in pigeonpea (*Cajanus* spp.). *Plant Breeding* 129(2): 142–148.

Saxena RK, Saxena KB, Kumar RV, Hoisington DA, Varshney RK (2010c) Simple sequence repeat-based diversity in elite pigeonpea genotypes for developing mapping populations to map resistance to *Fusarium* wilt and sterility mosaic disease. *Plant Breeding* 129(2): 135–141.

Sekhar K, Priyanka B, Reddy VD, Rao KV (2010) Isolation and characterization of a pigeonpea cyclophilin (CcCYP) gene, and its over-expression in *Arabidopsis* confers multiple abiotic stress tolerance. *Plant Cell Environ* 33(8): 1324–1338.

Sennhenn A (2016) *Exploring niches for short-season grain legumes in semi-arid Eastern Kenya* (Doctoral dissertation, Niedersächsische Staats-und Universitätsbibliothek Göttingen).

Sharma HC, Stevenson PC, Simmonds MSJ, Green PWC (2001) *Identification of Helicoverpa armigera (Hübner) feeding stimulants and the location of their production on the pod-surface of pigeonpea [Cajanus cajan (L.) Millsp.]*. Final Technical Report. International Crops Research Institute for the Semi-Arid Tropics (ICRISAT), Patancheru, Andhra Pradesh, India.

Sharma M, Ghosh R, Telangre R, et al. (2016). Environmental influences on pigeonpea-*Fusarium udum* interactions and stability of genotypes to *Fusarium* wilt. *Frontiers Plant Science* 7: 253.

Sharma M, Rathore A, Mangala UN (2012) New sources of resistance to Fusarium wilt and sterility mosaic disease in a mini-core collection of pigeonpea germplasm. *European J Plant Pathol* 133(3): 707–714. https://doi.org/10.1007/s10658-012-9949-9

Sharma R, Keval R (2021) Screening of pigeonpea [*Cajanus cajan* (L.) Millsp.] Against Tur Pod fly, *Melanagromyza obtusa* (Malloch)(Diptera: Agromyzidae) in long duration pigeonpea genotypes. *J Entomol Zool Stud* 9(3): 198–202.

Sharma S, Paul PJ, Sameer Kumar, Nimje C (2020) Utilizing Wild *Cajanus platycarpus*, a Tertiary Genepool Species for Enriching Variability in the Primary Genepool for Pigeonpea Improvement. *Front Plant Sci* 11: 525517. https://doi.org/10.3389/fpls.2020.01055

Sheoran S, Thribhuvan R, Saini M, et al. (2022) Conventional, genomics, and post-genomics era of pulses breeding: Current status and future prospects. *Advances in legumes for sustainable intensification*, pp.553–574.

Singh AK, Abhilash PC (2017) *Wild Relatives of Cultivated Plants in India*. Springer Nature, Singapore, 1, p.308.

Singh N, Narula B, Ujinwal M, Langyan S (2021) Pigeonpea sterility mosaic virus a green plague-Current status of available drug and new potential targets. *Ann Proteom Bioinform* 5: 008–026. https://doi.org/10.29328/journal.apb.1001013

Singh YP (2022) Impact of climatic abnormities and mitigating technologies on pigeonpea (*Cajanus cajan*) in central India. *Indian Journal of Agricultural Sciences* 92(6): 779–784.

Sinha P, Singh VK, Saxena RK, et al. (2020) Superior haplotypes for haplotype-based breeding for drought tolerance in pigeonpea (*Cajanus cajan* L.). *Plant Biotechnol J* 18(12): 2482–2490.

Smýkal P, Coyne CJ, Ambrose MJ, et al. (2015) Legume crops phylogeny and genetic diversity for science and breeding. *Crit Rev Plant Sci* 34(1–3): 43–104.

Soni K, Durgesh K, Anjoy P, et al. (2024) Transcriptional reprogramming and allelic variation in pleiotropic QTL regulates days to flowering and growth habit in pigeonpea. *Authorea Preprints*.

Southgate BJ (1979) Biology of the Bruchidae. *Annu Rev Entomol* 24(1): 449–473.

Sreekanth M, Lakshmi MSM, Rao YK (2015) Efficacy and economics of certain new generation novel insecticides against legume pod borer, *Maruca vitrata* (Geyer) on pigeonpea (*Cajanus cajan* L.). *J Appl Biol Biotechnol* 3(3): 7–10.

Sreekanth M, Seshamahalakshmi M, Ramana MV (2020) Management of Pod fly, *Melanagromyza obtusa* on Pigeonpea (*Cajanus cajan* L.). *Agric Sci Digest Res J 40*(4): 382–386.

Susmitha D, Kalaimagal T, Senthil R, et al. (2022) Calcium-Rich pigeonpea seed coat: a potential byproduct for food and pharmaceutical industries. *Sustainability* 14(9): 4918.

Suso MJ, Bebeli PJ, Palmer RG (2015) Reproductive biology of grain legumes. *Grain Legumes* 365–399.

Talekar NS (1988) *Biology, damage, and control of bruchid pests of mungbean* (No. AVRDC/SP). AVRDC.

Tawaha ARM, Kozuharova E, Valchev H, et al. (2022) Pollinators ecology and management. In *Climate Change and Ecosystems* (pp. 245–260). CRC Press.

Tripathi K, Thendral US, Panda RR, Gore PG, Pamarthi R, Sahi VP, Singh GP (2025) Underutilized grain legumes breeding. In *Fundamentals of Legume Breeding: A Text for Students and Practitioners* (pp. 257–282). Singapore: Springer Nature Singapore.

Upadhyaya SR, Danilevicz MF, Dolatabadian A, et al. (2024) Genomics-based plant disease resistance prediction using machine learning. *Plant Pathol 73*(9): 2298–2309. https://doi.org/10.1111/ppa.13988

Varshney RK, Bohra A, Roorkiwal M, et al. (2020) Translational genomics in agriculture: Some examples in grain legumes. *Plant Biotechnol J* 18(1): 39–57.

Varshney RK, Chen W, Li Y, et al. (2012). Draft genome sequence of pigeonpea (*Cajanus cajan*), an orphan legume crop of resource-poor farmers. *Nature Biotechnol* 30(1): 83.

Yadav AK, Yadav S, Singh, MK (2012) Effect of temperature on the population of parasitoids and their impact on the pest, *Melanagromyza obtusa* (Diptera: Agromyzidae). *The Ecoscan* 1: 45–50.

Yadav ANK, Yadav S, Singh MK (2011) Impact of climatic variation on the parasitoid-complex of *Melanagromyza obtusa* (Diptera: Agromyzidae). *J Ecobiol* 29(3): 161.

Yadav DK, Sachan SK, Singh G, Singh DV (2016) Insect pests associated with pigeonpea variety UPAS 120 in Western Uttar Pradesh, India.

Yadav SK, Kumar P, Singh G, et al (2020) Study the natural enemies associated with pigeonpea pod fly. *Int J Chem Stud* 8(5): 97–99.

Zhang Y, Chen X, Geng S, Zhang X (2025) A review of soil waterlogging impacts, mechanisms, and adaptive strategies. *Front Plant Sci* 16: 1545912.

6 Application and Prospects of Modern Genomics to Deliver Climate-Smart Pigeonpea

Gurjeet Singh, Inderjit Singh, Suruchi Vij, Harmeet Singh Bakala, Navjot Singh Sidhu, T R Likhithashree, Amandeep, Harshavardan J. Hilli, Mehakpreet Kaur, Abhishek Bohra, and Rachit K. Saxena

6.1 INTRODUCTION

Pigeonpea (*Cajanus cajan* (L.) Millspaugh) is an important food legume mainly grown in tropical and subtropical regions, valued for its drought tolerance and versatile cultivation methods (Bohra et al., 2020b). As a diploid species (2n = 22) with a genome size of 833 Mbp, pigeonpea shows a wide range of maturity durations, from extra short (90 days) to long (300 days) growing periods (Saxena et al., 2019, 2021a). Pigeonpea, the fifth most important pulse crop globally, is predominantly grown in tropical regions, especially in Asia and Africa, where it plays a key role in the livelihoods of resource-poor farmers in drought-prone and degraded areas (Bakala et al., 2024). It is cultivated in both sole cropping or intercropping systems with cereals, other legumes, and commercial crops like cotton, etc. Globally, pigeonpea grows across 5.38 million hectares, yielding 4.5 million tons annually, with an average productivity of 851 kg/ha (FAOSTAT 2023). India, with 4.06 million hectares under pigeonpea cultivation leads production of 3.3 million tons. Besides India, pigeonpea is cultivated in Malawi, Myanmar, Tanzania, and Kenya (FAOSTAT 2023). India, which accounts for over 70% of global pigeonpea production and area, primarily consumes it as dehusked splits or dhal. On the other hand, countries in Africa and Latin America commonly consume it as canned peas. The nutritional value and versatility in cooking makes pigeonpea a crucial component of food security in these regions. Additionally, the plant's fallen leaves enrich the soil, and its ability to fix atmospheric nitrogen through symbiosis improves soil fertility (Singh et al., 2012). Pigeonpea is a vital source of

DOI: 10.1201/9781003603092-6

protein (20%–22%), carbohydrates, vitamin B complex, and essential minerals, making it an important dietary staple in many developing regions (Singh et al., 2022a).

Pigeonpea belongs to the subtribe *Cajaninae* within the tribe Phaseoleae of subfamily Papilionoideae, which is part of the Leguminosae family. *C. cajan* is the only domesticated species within the subtribe *Cajaninae*. This subtribe is notably characterized by the presence of vesicular glands on various parts of the plant, including the leaves, calyx, and pods. Currently, 12 genera are included in the subtribe *Cajaninae* (Saxena et al., 2018a). Based on their crossability with cultivated pigeonpea, all accessions were classified into three gene pools (GPs; Bohra et al., 2010). The genus *Cajanus* includes a total of 32 species divided into three groups: 1 species is classified as GP-1, 12 species as GP-2, and 20 species as GP-3 (Maesen, 1990). GP-2 and GP-3 are considered the rich reservoir of useful genes for many useful traits which can be used to broaden the narrow genetic base of cultivated pigeonpea to develop climate-resilience pigeonpea (Mallikarjuna et al., 2006). Multiple traits inherent in wild species contribute to the development of climate-resilient pigeonpea (Table 6.1).

Limited or stagnant growth in productivity has led to the declining per capita availability of pigeonpea in major producing regions. Several factors including low

TABLE 6.1
Genetic Resources from Wild Pigeonpea Species to Breed Climate-smart Cultivars

Traits	Important Traits	Wild Relatives	References
Plant type	Extra-early flowering and maturity	*C. platycarpus*	Ariyanayagam and Spence (1978)
	Photo-insensitivity	*C. platycarpus*	Mallikarjuna and Moss (1995)
	Cleistogamy	*C. lineatus*	Saxena et al. (1992)
	Fruiting branches and pods per plant	*C. scarabaeoides*	Sandhu et al.(2009) and Singh et al.(2020)
Quality	High protein content	*C. cajanifolius, C. sericeus, C. albicans, C. lineatus, C. scarabaeoides*	Saxena et al. (1987) and Reddy et al. (1997)
	High seed weight	*C. acutifolius*	Jadhav et al.(2012)
Resistance to pests	Pod borer	*C. scarabaeoides, C. acutifolius, C. platycarpus, C. reticulates, C. sericeus, C. albicans*	Sujana et al. (2008) and Singh et al. (2020)
	Pod fly	*C. reticulates, C. acutifolius*	Sharma et al. (2003)
Resistance to diseases	Bruchids	*C. scarabaeoides, C. acutifolius, C. platycarpus*	Mallikarjuna et al. (2011)
	Cyst nematode	*C. platycarpus*	Saxena et al. (1996)
	Fusarium wilt	*C. platycarpus*	Mallikarjuna et al. (2011)
	Sterility mosaic disease	*C. scarabaeoides*	Kulkarni et al. (2003)
	Phytophthora blight	*C. platycarpus, C. sericeus, C. acutifolius*	Mallikarjuna et al. (2005)

(*Continued*)

TABLE 6.1 (Continued)
Genetic Resources from Wild Pigeonpea Species to Breed Climate-smart Cultivars

Traits	Important Traits	Wild Relatives	References
Tolerance to abiotic stresses	Salinity	*C. platycarpus, C. albicans*	Subbarao et al. (1990) and Srivastava et al. (2006)
	Drought	*C. sericeus, C. albicans, C. lineatus, C. scarabaeoides*	
	Heat	*C. cajanifolius, C. platycarpus, C. scarabaeoides*	Khoury et al. (2015) and Basu et al. (2016)
	Waterlogging	*C. scarabaeoides, C. acutifolius*	Krishnamurthy et al. (2012)
Cytoplasmic male sterility	A1–A9 cytoplasm for hybrid seed production	*C. sericeus* (A1), *C. scarabaeoides* (A2), *C. volubilis* (A3), *C. cajanifolius* (A4), *C. cajan* (A5), *C. lineatus* (A6), *C. platycarpus* (A7), *C. reticulates* (A8), *C. lanceolatus* (A9)	Sharma et al. (2019)

level of genetic diversity in the active breeding material, biotic and abiotic stresses, inadequate management practices, limited exploitation of heterosis breeding, and lack of long-term planning and structured approach in policy framework are contributing to the low productivity and declining availability. Furthermore, breeding pigeonpea has been more challenging compared to other food legumes due to its unique floral biology. Pigeonpea is predominantly cross-pollinated, with natural outcrossing (20%–70%), which complicates the use of efficient selection and mating strategies commonly applied in self-pollinating crops. Traditional breeding methods such as pure line breeding, population breeding, mutation breeding, and wide hybridization have resulted in gradual improvements in yield potential. To overcome these challenges, genetic male-sterility (GMS) and cytoplasmic-genetic male-sterility (CMS) systems were discovered, offering a means to produce hybrids with a yield advantage of up to 30% to 50% (Bohra et al. 2025). However, the seed production technology is yet to be optimized to reap the full potential of hybrid breeding in pigeonpea. It is important to mention that first CMS-based pigeonpea hybrid, GTH-1, was released for cultivation in India in 2007 and more recently IPH 15-03 and IPH 09-5 in 2020 and 2021, respectively. Subsequently, several CMS-based hybrids have released in India (Singh et al., 2024a).

While wild relatives of pigeonpea harbor a rich pool of genes for resistance to biotic and abiotic stresses, as well as traits that enhance yield such as number of pods per plant, fruiting branch length, and the number of primary branches, utilization of these wild species for pigeonpea improvement has been limited so far (Sharma et al., 2019; Singh et al., 2020, 2021). This is primarily due to the poor crossability of *C. cajan* with its wild species relatives. Successful crosses could only be made between the cultivated pigeonpea and closely related species like

C. cajanifolius and *C. scarabaeoides* (Bohra et al., 2010; Singh et al., 2020). However, biotechnological approaches, such as in vitro rescue and propagation of hybrids from wide crosses, alongside the use of bridge crosses, may facilitate the transfer of valuable genes from a broader range of germplasm within and outside the *Cajanus* genus. Ongoing research using molecular tools to study the taxonomic and phylogenetic relationships within the subtribe *Cajaninae* could provide insights into potential routes for efficient trait introgression and enhance pigeonpea improvement.

Conventional breeding and hybrid technology for pigeonpea improvement continue to progress; molecular breeding is expected to accelerate the utilization of the considerable variability found in pigeonpea landraces and germplasm lines. However, the genetic basis of key traits in pigeonpea remains largely unknown, researchers need to explore through modern genomic tools and technologies to develop climate-smart pigeonpea cultivars. The efforts aimed at developing genome-wide sets of genetic markers have got an impetus with the shift from simple sequence repeats (SSRs) to high automated single nucleotide polymorphisms (SNPs)(Varshney et al., 2010; Bohra et al., 2020b). Subsequently, several advancements were made in pigeonpea genomics research through various multi-institutional programs. These high-density genetic markers will help build a saturated genetic map to facilitate mapping of disease resistance genes, identification of quantitative trait loci (QTLs) for economically important traits and enabling the integration of phenotypic data across diverse experimental populations and germplasm sets. Additionally, the development of mutant lines and large DNA insert libraries, such as bacterial artificial chromosome (BAC) libraries, will facilitate map-based cloning and functional analysis of traits in pigeonpea.

Climate change and associated changes are presenting a great challenge to global crop production (Benitez-Alfonso et al., 2023, Saeed et al., 2023). Advances in genomics remain crucial for addressing the challenges posed by climate change, which affects pigeonpea farming through rising temperatures, erratic rainfall, and increased pest and disease pressures (Bohra et al., 2020b). Extreme heat can harm plants during flowering and pod setting, while droughts and floods can reduce yields (Saeed et al., 2023; Bakala et al., 2024). Shifting seasons and changing weather patterns further complicate the growing cycle, potentially leading to poor pollination and shorter growing periods, which affect the yield potential of pigeonpea. Keeping these vital points in consideration, the present chapter covered the genomics advancements in pigeonpea and their potential role in addressing the challenging climatic scenarios. Additionally, it is also focused on the important climate-smart traits for future pigeonpea breeding programs and the basic role of genomic databases as well as bioinformatics tools for climate-smart pigeonpea.

6.2 FUNDAMENTAL GENOMICS IN PIGEONPEA

Pigeonpea has seen significant advancements in genomics over the last two decades that have bolstered its research and breeding efforts. For instance, multi-omics resources have been developed and effectively used in molecular breeding and functional genomics research (Kudapa et al., 2012; Pazhamala et al., 2015;

Bohra et al. 2021a, b, c). Transcriptomics studies have further enhanced gene annotation and identification of candidate genes for traits such as drought tolerance and pest resistance. The rapid evolution of genetic marker technologies has been pivotal in advancing marker-assisted selection (MAS) and genomic selection (GS), accelerating the development of superior pigeonpea cultivars with enhanced productivity and climate-resilience.

6.2.1 Genomics Tools in Pigeonpea Improvement

Next-generation genomics tools have accelerated pigeonpea research and breeding, which helped in the development of superior cultivars with improved yield, and multiple stress tolerance (Figure 6.1). Next-generation sequencing (NGS) and third generation sequencing (TGS) techniques have been used for sequencing the reference pigeonpea genome and further improving the fragmented genome assembly (Varshney et al., 2012; Garg et al., 2022). Latest research in pigeonpea genome has provided insights into genome organization, repetitive elements, and stress-responsive genes, establishing a foundation for genomics-assisted breeding (GAB). Application of NGS and TGS has also facilitated discovery of genome-wide genetic markers and large-scale genetic mapping, which are essential for identifying markers/genomic segments associated with agronomic traits. Integration of sequence data with traditional QTL mapping has enabled precise identification of loci governing important traits like resistance to Fusarium wilt (FW), *Maruca vitrata* and *Helicoverpa armigera* (Satheesh Naik et al., 2024).

Advances in multi-omics science in pigeonpea has further enhanced genetic improvement efforts. For instance, transcriptome sequencing has been employed to study gene expression under stress conditions, helping identify genes related to drought tolerance and disease resistance (Njaci et al., 2020). Additionally, metabolomics has provided insights into the biochemical pathways associated with multiple stress adaptation, offering a direct link between genotypes and phenotypes (Singh et al., 2024b). Together, these multi-omics tools provide a systems-level understanding of gene function, enabling breeders to target specific pathways for crop improvement (Saini et al., 2024). GWAS (genome-wide association study) has complemented these efforts by taking advantage of natural genetic diversity to identify novel loci associated with key traits, such as pod borer resistance and high protein content in pigeonpea (Satheesh Naik et al., 2024).

Gene editing has emerged as a versatile tool to introduce precise and rapid genetic changes in the genome in a flexible and cost-effective manner (Kumar et al., 2023). When sequence and functional information are available for these genes, their sequence can be modified using programmable nuclease enzymes-based gene-editing tools. This precise method of altering target DNA is widely used in precision breeding. Different gene-editing tools including Zinc Finger Nucleases (ZFNs), Transcription Activator-Like Effector Nucleases (TALENs), and CRISPR/Cas9 can be used for precision breeding of climate-resilience in pigeonpea. Although it looks quite rewarding to use precision breeding, there is a need to develop efficient tissue culture protocols and plant transformation methods in pigeonpea to get full potential of this technique to develop climate-resilient cultivars.

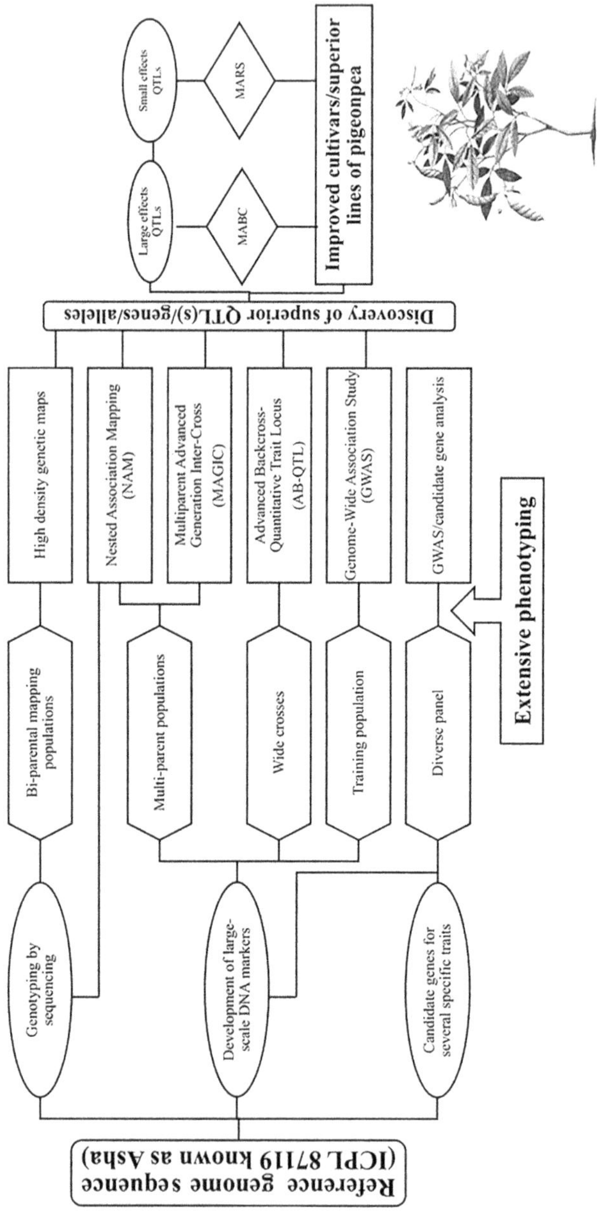

FIGURE 6.1 Modern genomics tools to understand the complex trait and development of superior pigeonpea cultivars.

In this context, clustered regularly interspaced palindromic repeats/(CRISPR)/CRISPR-associated protein 9 (Cas9) has been increasingly applied in pigeonpea to induce targeted modification in the genome. CRISPR/Cas9 has been explored to develop pest-resistant and drought-tolerant crop varieties, addressing the climate change challenges (Jaganathan et al., 2018). By integrating NGS/TGS, QTL mapping, GWAS, transcriptomics, metabolomics, genome-editing, and pigeonpea breeding programs can meet the growing demands of sustainable agriculture (Figure 6.1).

6.2.2 Current Genomic Landscape of Pigeonpea

The first draft of the genome of pigeonpea was published by the International Crops Research Institute for the Semi-Arid Tropics (ICRISAT), marking a milestone in legume genomics (Varshney et al., 2012). This sequencing project revealed the genome's size of approximately 833 Mb, containing 48,680 protein-coding genes, and provided insights into its genome architecture, including repetitive elements and GC content. Subsequent efforts have focused on refining the genome assembly and developing a pangenome to capture the genetic diversity across pigeonpea cultivars and landraces (Zhao et al., 2018). Garg et al. (2022) developed high-quality chromosome-length genome assemblies by Hi-C sequencing technique for six legumes including pigeonpea, enhancing the understanding of their genetics and evolution, and providing resources to accelerate genetic improvement in legume crops. The availability of high-quality genome assemblies of different wild species creates exciting opportunities to elucidate the full spectrum of genetic diversity of the crop in the form of a 'super-pangenome' (Khan et al. 2024). These genomic resources have laid the foundation to advance trait discovery and GAB.

Advances in DNA sequencing technologies have given a significant impetus to gene discovery methods such GWAS to enable identification of future breeding targets (Bohra et al., 2020a). For instance, QTLs for drought tolerance have been mapped to regions harboring genes related to root architecture and osmotic adjustment (see Satheesh Naik et al., 2024). Similarly, GWAS has identified SNPs associated with early maturity, disease resistance, and seed protein content (Obala et al., 2020; Megha et al., 2024). Transcriptomic analyses under drought stress have highlighted genes involved in ABA (abscisic acid) signaling, proline biosynthesis, and aquaporins, such as *CcDREB2A* and *CcLEA*. For heat tolerance, heat-shock proteins (HSPs), and transcription factors like *CcHSF* have been identified as critical players in cellular thermotolerance (see Bohra et al., 2020b). Resistance to major pests, such as *H. armigera*, has been linked to genes involved in secondary metabolite biosynthesis, including protease inhibitors and lectins (Malhotra et al., 2022). The latest discoveries have provided pigeonpea breeders with modern tools to support development of stress-tolerant pigeonpea varieties. An integration of these genomic resources with genome-editing tools has further enhanced the ability to edit genes of interest, accelerating the development of resilient pigeonpea cultivars suited for challenging agro-climatic conditions.

6.2.3 Molecular Markers and Breeding in Pigeonpea

Molecular markers have become a cornerstone of modern crop breeding, offering an efficient and precise way to accelerate the improvement of important traits in crops including pigeonpea. Traditional breeding methods, which rely on phenotypic selection, are often slow and imprecise, especially for traits governed by multiple genes or those that are difficult to assess due to strong environmental influences (Singh et al. 2018). Molecular markers, on the other hand, provide a direct link between genotype and phenotype, allowing breeders to select desirable traits without the need for labor-intensive field trials (Bohra et al., 2020b). The modern DNA markers technology is particularly valuable in pigeonpea breeding, where traits such as drought tolerance, disease resistance, and pest resistance are crucial for developing climate-resilient pigeonpea, especially under marginal growing conditions (Bohra et al., 2020b).

6.2.3.1 Role of Molecular Markers in Pigeonpea Breeding

Molecular markers are DNA sequences with known locations in the genome that can be used to identify specific genes/alleles/genetic variations associated with desirable traits (Bohra 2013; Bohra et al., 2020a). In pigeonpea, various types of molecular markers, such as SNPs, SSRs, and DArT markers, have been developed and applied to map genes related to key agronomic traits (Bohra et al., 2020b). For example, using SSR markers, significant QTLs for drought tolerance have been identified on chromosomes 6 and 11, providing valuable genetic resources for breeding programs targeting climate-resilient pigeonpea varieties (Saxena et al., 2014a). Additionally, markers linked to pest resistance have been identified, enabling the selection of pigeonpea varieties with enhanced resistance to the *H. armigera* pest, one of the major threats to pigeonpea production. By using molecular markers, pigeonpea breeders can precisely track these QTLs during the breeding process, even in early generations. This reduces the time and cost involved in traditional phenotypic screening, making breeding programs more efficient. Molecular markers also help in understanding the genetic architecture of complex traits, which can be used to design better crosses and improve the genetic gain in pigeonpea breeding programs.

6.2.3.2 Marker-Assisted Selection in Pigeonpea

MAS is a powerful tool that has revolutionized crop breeding by allowing for the selection of plants based on their genotype rather than phenotype. For instance, MAS has been used to introgress genes for sterility mosaic disease (SMD) and FW resistance into elite pigeonpea lines, thereby reducing the risk of yield loss due to this devastating disease (Saxena et al., 2021b). In addition to disease resistance and seed quality, MAS has been instrumental in developing and managing CMS systems and fertility restoration in pigeonpea. The CMS system is essential for hybrid breeding, as it facilitates the production of hybrids without the need for manual emasculation (Bohra et al. 2015, 2017). MAS has enabled the identification and incorporation of fertility restorer genes, ensuring the successful restoration of fertility in CMS-based hybrids. This advancement has led to the development of commercial pigeonpea hybrids with significant yield advantages over traditional cultivars (Saxena et al., 2018b). MAS also aids in the improvement of seed quality traits, such as protein content and cooking time, which are critical for enhancing the nutritional value of pigeonpea. By selecting for these traits at the molecular level, breeders can expedite development of varieties with improved agronomic performance without extensive field testing. MAS is particularly advantageous when working with traits that are difficult to evaluate phenotypically due to environmental variability or late expression, such as resistance to biotic stresses or tolerance to abiotic stresses like drought and heat. By incorporating molecular markers linked to these traits, breeders can speed up the selection process while ensuring that desirable genes are efficiently incorporated into the varietal improvement programs.

6.2.3.3 Genomic Selection in Pigeonpea Breeding Programs

GS is relatively a recent plant breeding approach that utilizes genome-wide genetic marker information to predict the breeding value of an individual in the absence of phenotype data (Bohra et al., 2020a). In contrast to MAS, which targets specific QTLs/genes, GS leverages the comprehensive genetic data obtained from a genome-wide scan to estimate the genetic potential for a specific trait. GS can be quite beneficial in pigeonpea breeding, where multiple genes influence complex traits like yield, drought tolerance, and pest resistance. By using GS, breeders can predict the performance of breeding lines more accurately, even before they are tested in field conditions. This predictive power allows for faster and more efficient breeding cycles to predict traits such as yield potential and drought tolerance in pigeonpea. The use of GS to predict yield-related traits in pigeonpea could be used to improve breeding efficiency by reducing the number of generations required to achieve desired genetic gains (Ghosh et al., 2023). Furthermore, by combining GS with high-throughput phenotyping technologies (such as drone-based imaging or soil moisture sensors), breeders can continuously update their genomic models to improve the accuracy of trait prediction.

6.3 ADDRESSING CLIMATE CHALLENGES WITH GENOMICS TOOL IN PIGEONPEA

6.3.1 Drought Stress Tolerance in Pigeonpea

Owing to its deep-rooted system, pigeonpea behaves as a drought resilient crop and holds significant importance in semiarid and tropical regions. Water availability is a key limiting factor in the sustainable management of water resources in rural areas, which are integral to both agricultural production and environmental preservation (Yadav et al., 2024). The cultivation of pigeonpea can help improve water use efficiency (WUE), as it requires minimal irrigation compared to other crops, supporting both agricultural output and conservation of water resources in rural ecosystems. Traditionally, breeding for drought tolerance in pigeonpea has focused on selecting phenotypic traits such as deep root systems, osmotic adjustment, reduced transpiration rate and high WUE (Varshney et al., 2012). These traits contribute to its survival under prolonged periods of water scarcity and ensure yield stability. However, conventional breeding has inherent limitations (low heritability of drought tolerance traits) and challenges in accurate phenotyping and strong influence of G × E (genotype × environment) interactions.

Recent advances in genomics have transformed the research aimed at improving drought tolerance in pigeonpea. With the availability of pigeonpea reference genome and high-throughput genotyping, it has become feasible to identify drought responsive genes, QTLs and regulatory pathways that regulate stress adaptation (Varshney et al., 2012). Several candidate genes, such as *CcHyPRP*, *CcCYP*, and *CcCDR* have been linked to drought and salinity tolerance through over expression and homology-based analyses (Priyanka et al., 2010; Tamirisa et al., 2014). GWAS and QTL mapping have revealed significant genomic loci associated with root traits, osmotic adjustments, and photosynthetic efficiency, which are critical for drought adaptation in pigeonpea (Saxena et al., 2021a). Recent studies using haplotype-based breeding have identified superior drought tolerance haplotypes associated with traits such as plant height, shoot and root biomass, and relative water content (Sinha et al., 2020). Additionally, genomic regions linked to determinacy and early maturity traits such as *CcTFL1* and *CcLG03* have been mapped using GWAS and QTL sequencing (Mir et al., 2014; Singh et al., 2022b). These findings offer valuable resources for improving drought tolerance in pigeonpea. The advent of genomic tools like MAS, GS and gene editing have opened new avenues for enhancing WUE in pigeonpea. Targeted introgression of QTLs linked to root traits and drought responsive physiological phenotypes into elite germplasm have shown promising results in other food legume crops like chickpea (Varshney et al., 2014). The deployment of drought resilient pigeonpea cultivars such as ICPH 2671 and ICPL 15011 shows the success of breeding programs that combine conventional and molecular breeding approaches (Saxena et al., 2013). Pigeonpea hybrid ICPH 2671 has demonstrated superior WUE and yield stability under water limited conditions. A recent study in pigeonpea used whole genome resequencing (WGRS)-based approach to identify the superior haplotypes associated with drought tolerance traits, thus paving the way for obtaining optimal combinations of haplotypes in future cultivars (Sinha et al., 2020). A haplotype-led approach has immense potential to devise efficient breeding strategies to develop pigeonpea cultivars that could withstand extreme drought conditions and yield better.

6.3.2 Heat Stress Tolerance

Pigeonpea is often exposed to high temperature stress during critical growth stages, particularly flowering and pod setting, which leads to flower drop, reduced pod filling, and overall seed yield loss (Kumar et al., 2016). Physiologically, heat stress disrupts cellular homeostasis by causing protein denaturation, membrane instability, and oxidative stress due to accumulation of reactive oxygen species (ROS). High temperatures adversely impact pollen viability and fertilization, compromising reproductive success by reducing pollen function and impairing fertilization processes in pigeonpea (Bakala et al., 2024). To mitigate heat stress, pigeonpea employs mechanisms such as synthesis of HSPs and chaperons that stabilize proteins and cellular structures. Antioxidant enzymes like superoxide dismutase (SOD) and catalase (CAT) help in scavenging ROS, while osmoprotectants such as glycine betaine and proline preserve cellular integrity under heat stress (Møller et al., 2007). Genomics has provided a deeper understanding of molecular responses to heat stress in pigeonpea (Ramakrishna et al., 2021). Transcriptomics analyses in legumes have identified heat responsive genes including HSPs, ROS scavenging enzymes and transcription factors like HSFs (heat-shock factors; Zhang et al., 2015).

Despite some progress, breeding heat- and drought-tolerant pigeonpea varieties remains challenging due to the complex and polygenic nature of stress tolerance. Limited genetic diversity for heat tolerance within the cultivated germplasm pool is a major bottleneck (Saxena et al., 2021b). Additionally, high temperature period often coincides with drought stress, necessitating development of dual tolerant varieties. To address these challenges, collaborative efforts combining genomics, phenomics and system biology approaches are being undertaken (Bakala et al., 2020). High-throughput phenotyping platform, coupled with advanced genomic tools are enabling precise selection for heat tolerance traits. Nevertheless, scaling these technologies in resource-limited regions remains a critical hurdle. Future research must focus on leveraging pigeonpea wild relatives (Table 6.1), GPs, and landraces to broaden the genetic base for heat tolerance while integrating GAB strategies for sustainable crop improvement.

6.3.3 Pest and Disease Resistance

In the modern era, genetics and genomics research plays a pivotal role in enhancing pest and disease resistance in major and minor crops (Sharma et al., 2022; Bakala et al., 2021; Saxena et al., 2021b). The pod borer, *H. armigera* and *M. vitrata* are the most significant pests of pigeonpea. It causes substantial yield losses valued at approximately $2 billion annually (Mapfumo et al., 2013). Host plant resistance is a critical component of sustainable pest management, but conventional breeding has shown limited success due to the narrow genetic base of cultivated pigeonpea. To overcome this limitation, researchers have explored wild relatives such as *C. scarbaeoides* and *C. acutifolius*, to introduce resistance traits, including antibiosis, antixenosis and trichome based pest repellence (Sharma et al., 2022).

Introgression lines (ILs) developed through crosses with wild relatives have shown promise in enhancing resistance to pod borer. For instance, introgressions for *C. scarabaeoides* into elite lines significantly reduced pod damage and pesticide dependency (Singh et al., 2020). Recent studies have also demonstrated the importance of specific resistance mechanisms. Vishal et al. (2023) identified genotypes such as ICPH 3461 and BSMR 853 with strong antixenosis for oviposition and high antibiosis under bioassay conditions, making them valuable sources for breeding programs targeting pest resistance. Transgenic approaches using bacterial *Cry* proteins have also been explored to provide durable resistance to pod borer in pigeonpea. However, development of transgenic pigeonpea is hindered by recalcitrant tissue culture responses, poor regeneration efficiency, and limited reproducibility. These barriers necessitate further innovation in the strategies. The need for insect-resistant pigeonpea varieties is particularly urgent in the regions, where short-duration varieties are prevailing. Additionally, the impact of climate change on pest-resistant genotypes must be addressed through interdisciplinary collaborations. By leveraging genetic diversity and advanced genomic technologies, pigeonpea breeding can deliver insect-resistant cultivars, reducing pesticidal use and ensuring sustainable production in resource-limited regions (Bakala et al., 2024).

Pigeonpea is also susceptible to several diseases, including FW, SMD, and *Phytophthora* blight (PB), which collectively cause significant yield losses. Among these, FW is considered the most destructive with yield losses ranging from 30% to 100% in susceptible genotypes (Patilet al., 2017). SMD and stem blight also severely impact pigeonpea productivity, emphasizing the urgent need to develop resistant varieties. Recent advances in pigeonpea genomics have revolutionized disease resistance breeding. The availability of molecular markers, transcript sequences and genetic maps has facilitated mapping resistance loci for key diseases such as FW and SMD (Pazhamala et al., 2015). Notably, Saxena et al. (2021b) identified 19 robust markers, among them nine for FW and 10 for SMD resistance and discovered two genes for FW resistance (*C.cajan_03691* and *C.cajan_18888*) and four genes for SMD resistance (*C.cajan_20995*, *C.cajan_07858*, *C.cajan_21801*, and *C.cajan_17341*). These genes encode proteins such as "*Pumilio homolog 6* (APUM-6)" "*Allene oxide cyclase*" "*Myb protein 1* (DdTom1)" and "*ABC transporter G family*" which are pivotal in disease resistance mechanisms. For example, the FW resistance gene *C.cajan_03691* on *CcLG11* plays a critical role in imparting resistance, while the SMD resistance gene *C.cajan_20995* on *CcLG04* encodes a Myb-related protein, emphasizing their potential for functional analysis and resistance breeding applications.

The availability of whole genome assembly has further accelerated GAB approaches, enabling more efficient introgression of resistance traits in legume crops (Bohra et al., 2014; Bohra et al., 2022c). Concerning disease resistance, significant progress has been made in identifying resistant germplasm in pigeonpea. For example, lines such as IPA8F, IPA 9F, and IPA 16F show resistance to both FW and SMD, while ICPL 366 is resistant to SMD, and PB 9 exhibited resistance to phytophthora blight. Landmark varieties like ICPL 87119 (Asha), resistant to FW and SMD, and Malviya Chamatkar (MAL 13) resistant to FW, SMD and pod borer, have demonstrated durability over decades (Saxena et al., 2021b). However, effective resistance breeding has primarily focused on FW and SMD with limited success against other diseases such as blight. A recent Meta-QTL analysis (2011–2020 data) have identified 34 meta-QTLs associated with agronomic traits, disease resistance, seed quality, and fertility restoration traits (Halladakeri et al., 2023). This comprehensive approach highlights 53 gene models containing potential candidate genes, including calmodulin-binding proteins and P-loop-containing nucleoside triphosphate hydrolases, which are critical in disease resistance signaling pathways. Despite these advances, the durability of pigeonpea disease resistance remains a challenge. Factors such as pathogen diversity, genetic contamination and extreme environmental conditions often lead to resistance breakdown. Sharma et al. (2016) evaluated genotypes across diverse locations, revealing significant G x E interactions and variability in pathogen virulence. Similarly, three geographically specific strains of SMD were identified, complicating breeding efforts (Mediga et al., 2023). To sustain productivity in diverse environments, future breeding programs must focus on stable donor germplasm, multi-disease resistance, and omics-driven resilience to changing pathogen populations and environmental stresses.

6.3.4 Nitrogen Fixation and Soil Adaptation

Nitrogen fixation in pigeonpea is a genetically regulated process that occurs through symbiotic interactions at the host plant-rhizobium interface. Pigeonpea forms nodules with rhizobia from cowpea miscellany group, shared with various legumes such as cowpea and groundnut. Despite this promiscuous association with indigenous rhizobia, variability exists in the symbiotic efficiency among strains. Biological N_2 fixation (BNF) is mediated by key bacterial genes: *nod*, *nif*, and *fix* that regulate nodulation, nitrogenase activity and the symbiosis process, with the genes often encoded on transferable genetic elements like plasmids and chromids, enabling horizontal gene transfer within bacterial species (Oldroyd et al., 2011; Wang et al., 2018).

Recent genomic research has provided new insights into rhizobial compatibility with pigeonpea. For example, Bopape et al. (2023) recently sequenced genome of *Rhizobium tropici* strain "10ap3," a robust symbiont compatible with pigeonpea. The genome consists of a large circular chromosome (~6.3 Mbp) harboring 6013 genes, including those linked to N, P and Fe metabolism as well as stress responses. Notably, genome lacked canonical *nod* and *nif* genes, suggesting that an alternative symbiotic pathway might involve purine derivatives rather than typical lipochito-oligosaccharidic nod factors encoded by *nodABC* genes. Similar findings were reported in *R. tropici* strain CIAT899 and photosynthetic *Bradyrhizobia* strains, which demonstrated variations in the presence of these nodulation genes (Ormeño-Orrillo et al., 2012). These discoveries highlight the potential for unique biochemical pathways enabling symbiosis in pigeonpea. The unique strain "10ap3" exhibit efficient N_2 fixation and genetic traits for pigeonpea abiotic stress tolerance and nutrient acquisition, making it a promising candidate for commercial inoculant development, particularly under challenging environmental conditions.

Pigeonpea's promiscuity with diverse rhizobia populations supports its adaptability in sustainable agriculture. Bopape et al. (2021) reported effective nodulation in pigeonpea by indigenous soil rhizobia across South Africa, even in nontraditional pigeonpea growing regions, minimizing reliance on commercial inoculants. However, variability in host-rhizobial compatibility underscores the need for molecular characterization such as 16S rRNA and *recA* gene analysis, to identify strains with high symbiotic potential. Similarly, Jain et al. (2020) explored molecular diversity of pigeonpea-nodulating rhizobia in semiarid regions of India. Of the 55 isolated identified, 37 demonstrated effective nodulation under controlled conditions, with five exhibiting high symbiotic efficiency and multiple plant growth-promoting (PGP) activities, such as tolerance to pH, salinity, temperature, and drought. Molecular characterization revealed significant genetic diversity, attributed to mixed legume cropping systems. The diversity underscores the importance of identifying efficient strains for enhancing pigeonpea nodulation and productivity under stress conditions. Effective symbiosis in pigeonpea can enhance N_2 fixation, improving seed quality and yield. However, legume-rhizobium interactions are sensitive to drought, limiting N_2 fixation efficiency (Varshney et al., 2012). Globally, studies have revealed diverse *Bradyrhizobium* species nodulating pigeonpea, including *B. elkanii*, *B. yuanmingense*, *B. ivorense*, and *B. cajani*, while *Ensifer spp.* are rarely reported. Jorrin et al. (2021) isolated 19 native and eight commercial pigeonpea endosymbionts from

Indian soils for genome sequencing. Phylogenomic analysis identified *B. yuanmingense* harboring nod genes resembling *B. cajani* and *B. zhanjiangense*, as dominant symbiont. Strain BRP05 exhibited superior nodulation and growth promotion, alongside *Ensiferalkalisoli* NBAIM29, effective in alkali soils (Choudhary et al., 2020). These findings suggest that *B. yuanmingense* BRP05 is a promising inoculant, with potential to enhance pigeonpea yields across South Asia, specifically India.

6.4 CLIMATE-SMART TRAITS FOR FUTURE PIGEONPEA BREEDING

Integrating the climate-smart traits like drought and heat tolerance, disease, and pest resistance with modern genomic tools such as GS, MAS, and genome editing, can accelerate the development of climate-resilient pigeonpea cultivars (Figure 6.2). By using advanced sequencing (NGS) and high-throughput phenotyping, breeders can develop pigeonpea cultivars for low-input agriculture that thrive under the changing climate conditions. Low-input crops improve sustainability by utilization of proper natural resources (Jyoti et al., 2024).

6.4.1 Breeding for Multiple Stress Tolerances

The change in climate can expose the crop plants to many extreme biotic and abiotic stresses. Therefore, breeding climate-resilient pigeonpea cultivars that can maintain productivity under changing climate are the need of the hour. The development of climate-resilient cultivars can be achieved by introducing climate-smart traits (agro-morphological, physiological, or biochemical) into the breeding pipelines (Kumar et al., 2019). The cleistogamous flower is one such climate-smart trait that aids in preventing desiccation of the anthers and stigma under dry climatic conditions. It also promotes cross pollination, thereby encouraging heterozygosity in the population, which can increase the buffering ability under various climatic conditions. Pod filling ability is another important agro-morphological trait which is adversely affected when plants are exposed to moisture deficit conditions at various stages of its life cycle (Pang et al., 2017). Other traits like leaf area, leaf weight, leaf growth rate should be considered while breeding for climate change since these traits are reported to be associated with drought stress (Quan et al., 2016).

Physiological traits like pod abscisic acid concentration can aid in breeding for pest resistance, whereas sucrose infusion has been identified to be associated with salt tolerance (Khan et al., 2017). Another key climate-smart trait (Figure 6.2) that can lend a hand in breeding for drought stress, moisture stress, nutrient use efficiency, and biomass accumulation is root system architecture (RSA). RSA can be improved by breeding for traits like xylem plasticity (Prince et al., 2017), nodulating ability, and active nodules percentage. The leaf physiological traits such as total chlorophyll content, photosynthesis rate, transpiring ability, stomatal conductance and source to sink relationship are reported to be linked with climate resilience (Mohanty et al., 2007). The frost tolerance can be bred by altering the biochemicals such as sugars, prolines, and glycine betaine (Nayyar et al., 2005) at the cellular level. Nevertheless, the quantification of these organic compounds is still a challenging

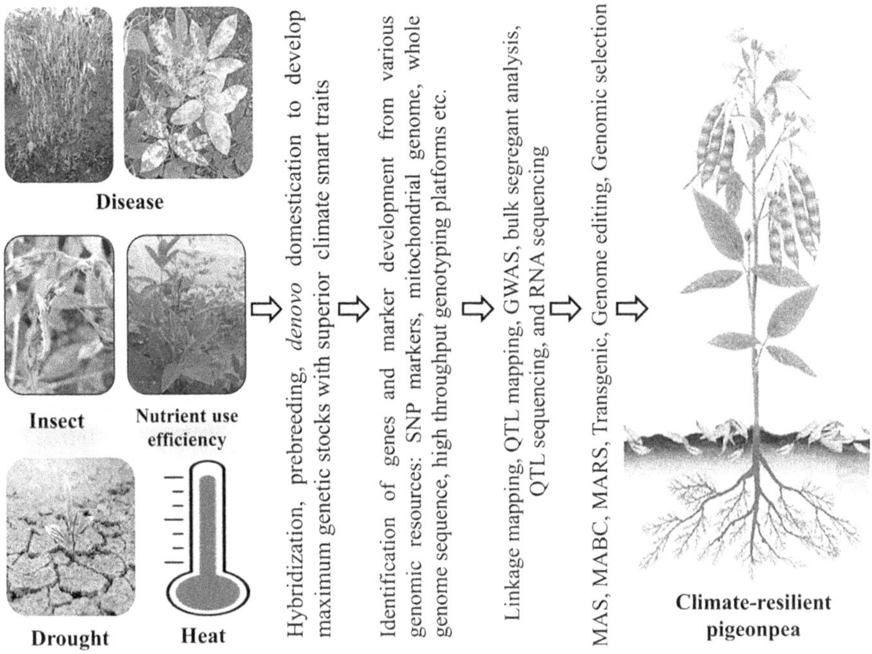

FIGURE 6.2 Climate-smart traits to integrate with modern genomic resources and tools to deliver climate-resilient pigeonpea cultivars.

task for physiologists since these are cell, tissue, and organ specific. Even though many climate-smart traits have been studied in different crop plants, the reports from pigeonpea are still lacking. Hence, there is an urgent need to design and conduct experiments pertaining to climate-smart traits which can be further introduced to breeding pipelines to breed climate-resilient pigeonpea.

Identification of climate-smart traits is the first step toward the development of climate-resilient cultivars. The next step in the process is identification of donors for the traits through which the traits can be introduced into the developing cultivars. The cultivated GPs of pigeonpea have a narrow genetic base to withstand climate change (Bohra et al. 2020b; Singh et al., 2022a). The crop wild relatives (CWRs) and landraces are the most important genetic resources (Amandeep et al., 2024), which can aid to search for climate-smart traits and alleles, since the evolution would have favored those plants that harbor genes for climate-resilient traits to sustain the extreme climatic conditions (Bohra et al., 2022a, 2022b). Interspecific and intergeneric hybridization between elite cultivars and germplasm lines are the methods to bring climate-smart traits into the cultivated pool.

To bring CWRs and landraces into breeding pipeline, prebreeding, and more recently, *de novo* plant domestication plays a vital role (Figure 6.2). Prebreeding is the link between genetic resources and cultivar development. Prebreeding develops the genetic stocks with the climate-smart traits from different GPs, which can be utilized as parents in the crop improvement. On the other hand, *de novo* domestication and re-domestication utilizes gene editing, whole genome sequence information, and phenotyping platforms to modify wild germplasm to render them amenable to modern agricultural settings (Bohra et al., 2022a, 2022b; Pisias et al., 2022). This technology has been adopted by Yu et al. (2021) to develop new allotetraploid rice by editing the *Oryza alta* for multiple traits—namely, seed shattering, awn length, plant height, grain length, stem thickness, and heading date. Likewise, these technologies can be adopted in pigeonpea breeding in order to develop climate-resilient cultivars.

6.4.2 Speeding Up Pigeonpea Improvement with GAB Approaches

Traditional plant breeding has made significant contributions to crop improvement. However, it was based on phenotypic selection, which has reduced the crop improvement efficiency. To increase the pace of varietal development for changing climate, modern plant breeders have adopted GAB. In contrast to phenotypic selection, GAB tools are extremely beneficial to boost breeding efficiency by reducing the breeding cycle and improving the selection efficiency (Bohra et al., 2020b). First of all, different QTL/gene mapping techniques such as QTL mapping, GWAS, and bulk segregant analysis can be utilized to identify markers linked to QTLs/genes responsible for climate-smart traits. Thereafter, marker-assisted backcrossing (MABC) can be used for introgression of QTLs/genes that are associated with climate-smart traits into high-yielding, widely accepted cultivated varieties. GS is another revolutionizing methodology that accounts for minor effect QTLs for traits. All these methods can be used effectively in pigeonpea by taking advantage of widely available genomic resources including different types of molecular markers, genome sequence, genetic maps, and different genotyping platforms.

Though many genomic resources and tools are available for pigeonpea improvement, in order to take their full advantage, standardized protocols for rapid generation advancement are needed. For this, speed breeding is one such modern plant breeding tool that accelerates generation time of crops by enabling their growth in controlled environmental conditions such as greenhouse, glasshouse (Aggarwal et al. 2025). When combined with techniques like MABC and GS, it can be effective in reducing length of breeding cycle further accelerating rate of pigeonpea improvement. Saxena et al. (2017), presents a technology for rapid generation advancement (RGA) in pigeonpea by using early-maturing cultivars and germinating 35-day-old immature seeds. This method enables the growth of three to four generations per year by following the single seed decent (SSD) method in RGA technology, which accelerates the breeding and genomics research in pigeonpea.

More recently, speed breeding protocol to hasten the breeding cycle of pigeonpea was optimized by Gangashetty et al. (2024) at ICRISAT, Hyderabad, India to reduce the total length of breeding cycles. They concluded that a photoperiod of 13h:8h:13h at vegetative: flowering: pod filling stage that was able to reduce the breeding cycle. Broad spectrum light accelerated early vegetative and pod formation phase, whereas far light was effective in accelerating early flowering. Their RGA protocol could reduce the breeding cycle to two to four years, in place of around seven to eight years taken by a conventional breeding to develop a stable line. Additionally, they suggested that integration of this technique with seed/pod chip-based genotyping, marker-assisted breeding, and SSD method resulted in enhanced efficiency of selections. This speed breeding facility is a boon for development of climate-resilient cultivars in pigeonpea.

6.4.3 Genomic Resources for Imparting Climate-resilience in Hybrid Pigeonpea

Hybrids have inherent qualities such as higher yield, stable tolerance to multiple stresses, faster growth, better adaptability to changing environment due to hybrid vigor that make them suitable candidates for production of climate-resilient crops (Bohra et al., 2016, 2025). For hybrid seed production at commercial scale, the hybridization process should be efficient and cost-effective. Pigeonpea, an often cross-pollinated crop, has wider scope to harness the advantages of hybrid breeding. Additionally, male-sterility system is also crucial for wider success of hybrid varieties. In par with this, pigeonpea has nine (Table 6.1) CMS sources among them A2 (*C. scarabaeoides*) and A4 ((*C. cajanifolius*) Maesen) have been used for hybrid development in India (Sharma et al., 2019). ICPH 2671 is a successful example of climate-resilient hybrid developed in pigeonpea production (Saxena et al., 2013). More recently, two pigeonpea hybrids IPH 15-03 and IPH 09-5 have been released in India using A2-CMS at ICAR-Indian Institute of Pulses Research (IIPR) for cultivation in the North-Western Plain Zone (Bohra et al., 2025). The hybrid IPH 15-03 showed 28.3% superiority over the popular variety Pusa 992, while the other hybrid IPH 09-5 offers 30.5% yield advantage over the national check variety UPAS 120. The hybrid IPH 15-03 matures in 150 days, whereas the maturity duration of IPH 09-5 is 140-145 days.

Genomics tools can be used for effective hybrid production in pigeonpea crop. Since the first reports of pigeonpea rough draft genome (Singh et al. 2012;Varshney et al. 2012), significant progress has been made toward pigeonpea genomic resources. Some of the developments include mitochondrial genome sequence (Tuteja et al., 2013), high-throughput genotyping platforms (Varshney et al., 2010), large collection of molecular markers (Saxena et al., 2014b), and genetic maps (Bohra et al., 2012). These genomic resources can be effectively used in various steps involved in hybrid seed production such as identification of true hybrids, determination of genetic diversity, genetic purity test of hybrids, and in understanding molecular basis of male sterility. These genomic resources can be of great help to conduct these tasks that in turn will make it possible to produce climate-resilient hybrids in pigeonpea.

Hybrids have improved traits, as compared to their parents, due to the phenomenon of heterosis. For production of climate-resilient pigeonpea hybrids, heterosis for traits imparting drought and heat stress tolerance, soil adaptation, pest, and disease resistance is required. In a review by Saxena et al. (2020), it was highlighted that their hybrid showed greater heterosis than pure lines for stress related traits. To better use the heterosis phenomenon, researchers need to identify the heterotic patterns for a crop species. For this, Saxena et al. (2021c) conducted an experiment to characterize heterosis and establish genomic prediction based heterotic patterns in pigeonpea. In this study whole genome resequencing of 104 parental lines and 292 germplasm lines was used to study the mid-parent heterosis (MPH) and identify heterotic patterns for yield-related traits. The analysis revealed the positive MPH values in 78.6% of hybrids. Additionally, genomic prediction and association mapping identified 52 copy number variations and 129 SNPs with significant heterotic effects, establishing high-yielding heterotic patterns in pigeonpea. However, most heterosis studies in pigeonpea have focused on yield and its component traits. Given the need for climate-resilient crops and availability of wide array of genomic resources in pigeonpea, it is imperative to conduct more heterosis studies targeting climate-smart traits.

6.5 ROLE OF GENOMIC DATABASES AND BIOINFORMATICS TOOLS FOR CLIMATE-SMART PIGEONPEA

6.5.1 Overview of Public Domain Genomics Databases Available for Pigeonpea

Sequencing efforts in various crops led to huge amounts of data that needed to be stored in the most organized and accessible manner in the databases. In the case of pigeonpea, a rough draft genome was not available up to 2011. Sequencing efforts led by two research groups resulted in two genome assemblies of a popular variety "Asha." Wherein the first group led by Varshney et al. (2012) assembled >70% of the entire 833 Mb genome with the help of Illumina sequencing platform, and another group led by Singh et al. (2012) assembled nearly 60% genome assembly of pigeonpea genome using 454 sequencing system. These reference genome assemblies have been improved by various sequencing efforts. Garg et al. (2022) a combined Hi-C sequencing technique to improve chromosome-length genome assemblies of legumes including pigeonpea. The sequencing data were used to anchor, order, orient, and correct misjoins in genome assemblies. Genomic resources developed in this study are publicly available through an online repository, "Legumepedia." Similarly, Marla et al. (2020) improved draft assembly of pigeonpea. It combined two existing assemblies, A1 and A2 by assembly reconciliation approach to fill the gaps and to give high quality and near complete genome assembly of pigeonpea. Once genome sequence information became available, sequence data has been saved in various genomic databases that are available in the public domain. National Center for Biotechnology Information (NCBI) stores genome sequence data for pigeonpea with an assembly size of 648.3 Mb. These data were submitted by Indian Agricultural Research Institute (IARI) and National Research Center for Plant Biotechnology (NRCPB) (https://www.ncbi.nlm.nih.gov/). PulseDB is another general database that contains genomic information related to legumes including pigeonpea (https://www.pulsedb.org/). Apart from these generic databases, there are some dedicated databases for genomic information in pigeonpea crop. International Initiative for Pigeonpea Genomics (IIPG) is one such database that provides genomic information, as well as SSRs, SNPs, and QTL data for pigeonpea crops. Molecular marker information is critical for successful implementation of different breeding techniques in any crop. Pigeonpea Microsatellite Database (PIPEMicroDB) is dedicated for identification of desired number of molecular markers on specific positions of chromosomes (http://webapp.cabgrid.res.in/pigeonpea/) in pigeonpea. Additionally, modulation of gene expression is being reported to be associated with various agronomic traits in several crops, whereby transcription factors (TFs) are master regulators of gene expression. With this view, various databases containing TFs information have been designed in several crops. Similarly, Pigeonpea Transcription Factors Database (PpTFDB) was designed for pigeonpea that offers public domain data on pigeonpea TFs (https://ngdc.cncb.ac.cn/databasecommons/database/id/4902; Singh et al., 2017).

6.5.2 Tools and Software Available for Genome and Trait Analysis in Pigeonpea

Apart from databases and repositories, bioinformatics tools and software's are required for analysis of genomic data. Generally, the aforementioned databases contain their dedicated tools for analysis of genomic sequences. In the case of NCBI database, we can use basic local alignment search tool (BLAST) to search for similarity of any sequence with genome assembly of pigeonpea. Moreover, it also contains information about genes and proteins that are discovered in pigeonpea. Similarly, PulseDB database contains tools that can be used to search for any information related to genomics (by finding sequence similarity using BLAST), genetics (by finding QTLs), and breeding aspects (by searching traits). Another database, PIPEMicroDB, contains automated primer design tool that can be used for location wise, as well as chromosome wise search of primers in pigeonpea. It is also equipped with "primer 3," a basic primer designing tool for designing primers for selected markers (Sarika et al., 2013). Moreover, some generic software, tools, and programs can be effectively used in pigeonpea genomic research. ProtParam online server was used to find molecular properties of TFs and Blast2GO to assign gene ontology to TFs (Singh et al., 2017). In addition, there are several other bioinformatics tools that were designed for genomic analyses in other crops that are also helpful in pigeonpea research. The quality of raw sequence data can be checked by FastQC (v0.11.8), SNPs, structural variants (SVs) and insertion deletions (InDels) annotated using SnpEFF software and primers design using NCBI primer blast tool (Singh et al., 2024c).

Besides tools and software that are used for genome sequence analysis, specific tools are required for trait mapping in pigeonpea as ultimately most studies have an objective of finding genomic region associated with a trait of interest. Some commonly available tools and software can be used in trait mapping in pigeonpea crops. QTL IciMapping software version 4.1 based on inclusive composite interval mapping method was used to map QTLs for high-selfing flower and seed quality traits in pigeonpea (Yadav et al., 2019). Similarly, marker trait associations (MTAs) can be detected using common software, such as TASSEL v. 2.1 software was used to find MTAs for resistant against FW variant 2 in pigeonpea (Patil et al., 2017). In brief, once genome sequence information is available for pigeonpea, a vast variety of software's and tools are available that can be used for genomics related research such as search of molecular markers, annotation of genes, analysis of sequence quality, detection of QTLs and many other related aspects can be explored.

6.5.3 Genomics, Big data and Artificial Intelligence (AI) to Achieving Climate-Resilient Crop

The amount of genomics, big data (phenomics and enviromics data) is gradually increasing due to a large leap in the field of genome sequencing and high-throughput phenotyping methods. Consequently, there is a need to develop techniques and models that can simultaneously process these different types of data to help in breeding climate-resilient crops. Although there are many databases that store these data and there are some techniques such as GWAS and GS that integrate genomics and phenomics data, and it can help in breeding climate-resilient crops, these techniques generally do not integrate environmental data, resulting in reduced prediction accuracy. However, integration of environmental data can be rewarding as in case of conducted studies, integration of environmental data with genomics and phenomics data resulted in increased prediction accuracy of various GS models in wheat (Montesinos-Lopez et al., 2024) and maize (Millet et al., 2019). Consequently, it is crucial to integrate environmental data into various models and techniques used in plant breeding. Additionally, when the goal is to develop climate-resilient crop varieties for present and for future, it becomes cardinal to integrate current and predicted future environmental conditions to models to breed for desired climate-resilient crop varieties. Though techniques like GWAS and GS showed increased prediction accuracy, it can be quite challenging for these moderate techniques to handle large and complex data. As a result, there is a need for techniques that can handle large amounts of genomics, phenomics and enviromics data efficiently to give desired results.

AI can be a solution for handling large amounts of data and integrating different types of data such as genomics, phenomics, and enviromics data with more efficiency. In the field of phenomics, a large amount of complex image data is taken through drone-based imaging, thermal imaging, chlorophyll florescence, and LiDAR (Light Detection and Ranging). For breeding of climate-resilient crops, the data are generally related to various traits that confer resistance to several types of stresses such as drought stress, heat stress, salinity stress, and cold stress. AI can effectively process such large amounts of image data quickly and extract meaningful biological information out of it, thereby associating such information to traits that confer stress resistance. Similarly, in the case of genomics, generally it involves genome sequence, gene expression, and RNA expression data, and in the case of enviromics, it usually involves information related to various parameters such as temperature, rainfall, soil type, and humidity. Both of these types of data can be large, and deriving meaningful conclusions is difficult. However, AI, especially machine learning and deep learning methods can handle and integrate genomics, phenomics, and enviromics data easily and learn from such data that helps in increasing efficiency of these methods in future predictions. Nevertheless, use of AI for production of climate-resilient crops on a wide scale is still uncommon in countries around the world, and most of the AI related methods are often used in major crops such as rice, wheat, and maize, and such developments are lacking in pigeonpea improvement. In future, integration of genomics, phenomics, and enviromics data and the use of AI can possibly give outstanding results for breeding climate-resilient varieties of pigeonpea.

6.5.3.1 Challenges and Limitations

Numerous factors limit genetic progress in all legume crops, including pigeonpea, which hinder their readiness for climate resilience. The genetic progress in pigeonpea breeding is hampered by a narrow genetic base (limited genetic resources), monophyletic evolution (while in cereals multi-event evolution), repeated use of similar breeding lines (mostly from primary GPs), and underutilization of wild resources (Singh et al., 2022a). The genetic variability is primarily limited within primary GPs, which shows wide phenotypic diversity but limited genetic diversity for key climate-smart traits. Many pigeonpea varieties in India are derived from landraces and released for commercial cultivation, leading to a compromise in the genetic diversity during variety development efforts (Saxena et al., 2016). To enhance climate resilience of pigeonpea cultivars, there is a need to explore the secondary and tertiary GPs for multiple stress tolerance and yield components traits. Nevertheless, challenges have remained in utilizing wild species of pigeonpea due to inadequate agronomic traits, very long crop duration, and limited available resources. These limitations, coupled with difficulties in crossing and insufficient high-throughput phenotyping facilities, restrict the exploitation of genetic diversity from secondary and tertiary GPs, which hinder the development of climate-resilient pigeonpea varieties (Bohra et al., 2010; Jha et al., 2024).

Pigeonpea climate resilience is challenged by complex G × E × M (genotype, environment, and management) interactions, specifically for abiotic stress tolerance traits, which are polygenic in nature (Jha et al., 2017). Effective breeding requires multiyear, multilocation screenings (Singh et al., 2023) and the integration of high-throughput phenotyping and genomic data to identify stable, resilient donors for crop improvement. Current gene-editing tools, like CRISPR/Cas9, have major challenges such as inefficient transformation protocol, low regeneration rates, and insufficient functional genomic information, making them unsuitable for precise trait modifying in pigeonpea. Apart from that, pigeonpea gets less attention by researchers and receives very limited research funding compared to other major crops, which slows the progress to understand the complex trait mechanism. The lack of consistent support has hindered the translation of genomic advancements into breeding pipelines, limiting the development of climate-resilient pigeonpea varieties for diverse environments. Overall, pigeonpea breeding faces genetic resource limitation, technical challenges, and insufficient attention for research funding; these all factor in the hindering of progress in developing climate-resilient cultivars. Future focus on increase in the investment in genomic tools as well as on multilocation screenings is essential to improve the multi-stress tolerance and exploit genetic diversity for climate-smart pigeonpea.

6.5.3.2 Future Prospects for Climate-smart Pigeonpea

With the help of high-quality whole genome assemblies, scientists can pinpoint the genes related to stress tolerance, thus allowing them to precisely manipulate these genes to improve crop performance in growing stress scenarios (Kumar et al., 2023). Genomics also plays a vital role in discovering genetic diversity from diverse GPs and within the pigeonpea populations, providing breeders with the opportunity to explore underutilized or wild relatives of the crop for valuable traits. This approach enables the development of hybrid varieties that harness the strengths of different genetic backgrounds, leading to improved productivity and adaptability. By accelerating the breeding process and increasing the precision of trait selection, genomics offers significant potential for developing pigeonpea varieties that are better suited to withstand changing environmental conditions. The integration of molecular markers with other advanced breeding technologies, such as CRISPR/Cas9, holds tremendous potential for accelerating the genetic improvement of pigeonpea. For example, CRISPR/Cas9 could be used to precisely edit genes associated with disease resistance or insect tolerance, complementing the work done using MAS and GS. The combination of these genomic tools with climate-smart traits will enable the development of pigeonpea varieties that are not only high-yielding but also resilient to the stresses posed by changing climate conditions. As sequencing technologies become more affordable and accessible, the genetic improvement of pigeonpea through molecular markers, MAS, and GS is expected to contribute significantly to global food security, especially in regions dependent on this important crop. Ultimately, all these tools can help to ensure food security for the millions of farmers and communities that rely on pigeonpea as a crucial source of nutrition and income, even as climate change presents new challenges.

6.6 CONCLUSION

Pigeonpea is an important food legume crop for food security, specifically in drought-prone areas worldwide. Despite the genomic advancement, many challenges like limited crossability and stagnant productivity persist in pigeonpea. Modern genomic tools and technologies, including NGS, MAS, GWAS, GS, and genome editing, are helpful in accelerating the progress of developing climate-resilient cultivars, ensuring the crop's sustainability under changing climates. Recent advancements in pigeonpea genomics have significantly enhanced breeding efforts to focus on improvement of several traits such as yield components, drought and heat tolerance, disease and pest resistance, and improve nutritional quality. An efficient integration of advanced genomic tools with high-throughput phenotyping in pigeonpea breeding holds great promise for understanding plants' responses to complex stress conditions. An enhanced understanding of plant stress response is imperative to develop sustainable agricultural practices for pigeonpea cultivation in resource-limited regions. Databases and various bioinformatics tools will also play a key role in developing climate-resilient pigeonpea cultivars. Collaborative efforts are contributing to advance genetic research in pigeonpea. By leveraging modern approaches such as GAB, gene editing, and hybrid production techniques, pigeonpea breeders can enhance crop adaptability under fluctuating climate conditions to ensure stable productivity and sustainability. Integrating climate-smart traits with advanced genomic and phenomics tools can accelerate the development of climate-resilient pigeonpea cultivars.

REFERENCES

Aggarwal A, Jeena AS, Mehra K, Kumar B, Kashyap S, Yadav DK, Maurya AK, Venkatesh SC, Singla P, Bohra A (2025). Speed-bred crops for food security and sustainable agriculture. *Planta* 262:34

Amandeep, Singh, G., Singh, I., Gill, R. K., Sharma, P., Pratap, A., Gupta, S., ... & Siddique, K. H.M. (2024). Back to wild: Designing future climate-resilient cultivars of urdbean and mungbean. In *Legume Crop Wild Relatives* (pp. 13–31). CRC Press. https://www.taylorfrancis.com/chapters/edit/10.1201/9781003434535-2/back-wild-amandeep-gurjeet-singh-inderjit-singh-gill-parul-sharma-aditya-pratap-sanjeev-gupta-harsh-nayyar-kadambot-siddique

Ariyanayagam, R. P., & Spence, J. A. (1978). A possible gene source for early, day-length neutral pigeon peas, *Cajanus cajan* (L.) Millsp. *Euphytica*, *27*, 505–509.

Bakala, H. S., Devi, J., Singh, G., & Singh, I. (2024). Drought and heat stress: insights into tolerance mechanisms and breeding strategies for pigeonpea improvement. *Planta*, *259*(5), 123.

Bakala, H. S., Mandahal, K. S., Sarao, L. K., & Srivastava, P. (2021). Breeding wheat for biotic stress resistance: Achievements, challenges and prospects. *Current Trends in Wheat Research*, *12*, 11–34.

Bakala, H. S., Singh, G., & Srivastava, P. (2020). Smart breeding for climate resilient agriculture. In *Plant Breeding-Current and Future Views*. IntechOpen. https://www.intechopen.com/chapters/74649

Basu, P. S., Singh, U. M. M. E. D., Kumar, A. N. I. L., Praharaj, C. S., &Shivran, R. K. (2016). Climate change and its mitigation strategies in pulses production. *Indian Journal of Agronomy*, *61*(4), 71–82.

Benitez-Alfonso, Y., Soanes, B. K., Zimba, S., Sinanaj, B., German, L., Sharma, V., ... & Foyer, C. H. (2023). Enhancing climate change resilience in agricultural crops. *Current Biology*, *33*(23), R1246–R1261.

Bohra, A. (2013). Emerging paradigms in genomics-based crop improvement. *The Scientific World Journal*, *2013*(1), 585467.

Bohra, A., Chand Jha, U., Godwin, I. D., & Kumar Varshney, R. (2020a). Genomic interventions for sustainable agriculture. *Plant Biotechnology Journal*, *18*(12), 2388–2405.

Bohra, A., Gandham, P., Rathore, A., Thakur, V., Saxena, R. K., Satheesh Naik, S. J., Varshney, R. K., & Singh, N. P. (2021a). Identification of microRNAs and their gene targets in cytoplasmic male sterile and fertile maintainer lines of pigeonpea. *Planta*, *253*, 59.

Bohra, A., Jha, A., Singh, I. P., Pandey, G., Pareek, S., Basu, P. S., Chaturvedi, S. K., & Singh, N. P. (2017). Novel CMS lines in pigeonpea [*Cajanus cajan* (L.) Millspaugh] derived from cytoplasmic substitutions, their effective restoration and deployment in hybrid breeding. *The Crop Journal*, *5*, 89–94.

Bohra, A., Jha, U. C., Adhimoolam, P., Bisht, D., & Singh, N. P. (2016). Cytoplasmic male sterility (CMS) in hybrid breeding in field crops. *Plant Cell Reports*, *35*, 967–993.

Bohra, A., Jha, U. C., Kavi Kishor, P. B., Pandey, S., & Singh, N. P. (2014). Genomics and molecular breeding in lesser explored pulse crops: Current trends and future opportunities. *Biotechnology Advances*, *32*, 1410–1428.

Bohra, A., Kilian, B., Sivasankar, S., Caccamo, M., Mba, C., McCouch, S. R., & Varshney, R. K. (2022a). Reap the crop wild relatives for breeding future crops. *Trends in Biotechnology*, *40*(4), 412–431.

Bohra, A., Mallikarjuna, N., Saxena, K., Upadhyaya, H. D., Vales, M. I., & Varshney, R. K. (2010). Harnessing the potential of crop wild relatives through genomics tools for pigeonpea improvement. *Journal of Plant Biology*, *37*(1), 83–98.

Bohra, A., Prasad, G., Rathore, A., Saxena, R. K., Satheesh Naik, S. J., Pareek, S., Jha, R., Pazhamala, L., Datta, D., Pandey, G., Tiwari, A., Maurya, A. K., Soren, K. R., Akram, M., Varshney, R. K., & Singh, N. P. (2021c). Global gene expression analysis of pigeonpea with male sterility conditioned by A2 cytoplasm. *The Plant Genome*, *14*, e20132.

Bohra, A., Rathore, A., Gandham, P., Saxena, R. K., Satheesh Naik, S. J., Dutta, D., Singh, I. P., Singh, F., Rathore, M., Varshney, R. K., & Singh, N. P. (2021b). Genome-wide comparative transcriptome analysis of the A4-CMS line ICPA 2043 and its maintainer ICPB 2043 during the floral bud development of pigeonpea. *Functional & Integrative Genomics*, *21*, 251–263.

Bohra, A., Saxena, K. B., Varshney, R. K., & Saxena, R. K. (2020b). Genomics-assisted breeding for pigeonpea improvement. *Theoretical and Applied Genetics*, *133*(5), 1721–1737.

Bohra, A., Saxena, R. K., Gnanesh, B. N., Saxena, K., Byregowda, M., Rathore, A., … & Varshney, R. K. (2012). An intra-specific consensus genetic map of pigeonpea [*Cajanus cajan* (L.) Millspaugh] derived from six mapping populations. *Theoretical and Applied Genetics*, *125*, 1325–1338.

Bohra, A., Singh, I. P., Yadav, A. K., Pathak, A., Soren, K. R., Chaturvedi, S. K., & Nadarajan, N. (2015). Utility of informative SSR markers in the molecular characterization of cytoplasmic genetic male sterility-based hybrid and its parents in pigeonpea. *National Academy Science Letters*, *38*, 13–19.

Bohra, A., Tiwari, A., Kaur, P., Ganie, S. A., Raza, A., Roorkiwal, M., … & Varshney, R. K. (2022b). The key to the future lies in the past: insights from grain legume domestication and improvement should inform future breeding strategies. *Plant and Cell Physiology*, *63*(11), 1554–1572.

Bohra, A., Tiwari, A., Pareek, S., Joshi, R., Satheesh Naik, S. J., Kumari, K., Verma, R. L., Parihar, A. K., Patil, P. G., & Dixit, G. P. (2025). Past and future of cytoplasmic male sterility and heterosis breeding in crop plants. *Plant Cell Rep*, *44*, 33.

Bohra, A., Tiwari, A., Satheesh Naik, S. J., Maurya, A. K., Yadav, V., Datta, D., … & Varshney, R. K. (2022c). Breeding and genomics of pigeonpea in the post-NGS era. In *Underutilised Crop Genomes* (pp. 281–300). Cham: Springer International Publishing.

Bopape, F. L., Gwata, E. T., Hassen, A. I., & Zhou, M. M. (2021). Symbiotic efficiency of pigeonpea (*Cajanus cajan* (L.) Huth) with different sources of nitrogen. *Plant Genetic Resources*, *19*(4), 312–315.

Bopape, F. L., Hassen, A. I., Chiulele, R. M., Shonhai, A., & Gwata, E. T. (2023). The Genome of a Pigeonpea Compatible Rhizobial Strain '10ap3' Appears to Lack Common Nodulation Genes. *Genes*, *14*(5), 1084.

Choudhary, S., Tak, N., Bissa, G., Chouhan, B., Choudhary, P., Sprent, J. I., … & Gehlot, H. S. (2020). The widely distributed legume tree Vachellia (Acacia) nilotica subsp. indica is nodulated by genetically diverse Ensifer strains in India. *Symbiosis*, *80*, 15–31.

FAOSTAT (2023) https://www.fao.org/faostat/en/#data/QCL

Gangashetty, P. I., Belliappa, S. H., Bomma, N., Kanuganahalli, V., Sajja, S. B., Choudhary, S., … & Pandey, M. K. (2024). Optimizing speed breeding and seed/pod chip based genotyping techniques in pigeonpea: A way forward for high throughput line development. *Plant Methods*, *20*(1), 27.

Garg, V., Dudchenko, O., Wang, J., Khan, A. W., Gupta, S., Kaur, P., … & Varshney, R. K. (2022). Chromosome-length genome assemblies of six legume species provide insights into genome organization, evolution, and agronomic traits for crop improvement. *Journal of advanced research*, *42*, 315–329.

Ghosh, R., Tarafdar, A., Kasi Rao, M., Katravath, S., & Sharma, M. (2023). Pigeonpea crop improvement: Genomics and post-genomics. In *Smart Plant Breeding for Field Crops in Post- genomics Era* (pp. 271–292). Singapore: Springer Nature Singapore.

Halladakeri, P., Gudi, S., Akhtar, S., Singh, G., Saini, D. K., Hilli, H. J., … & Mir, R. R. (2023). Meta- analysis of the quantitative trait loci associated with agronomic traits, fertility restoration, disease resistance, and seed quality traits in pigeonpea (*Cajanus cajan* L.). *The Plant Genome*, *16*(3), e20342.

Jadhav, D. R., Mallikarjuna, N., Sharma, H. C., & Saxena, K. B. (2012). Introgression of Helicoverpaarmigera resistance from Cajanus acutifolius-a wild relative from secondary gene pool of pigeon pea (*Cajanus cajan*). *Asian Journal of Agricultural Sciences*, *4*(4), 242–248.

Jaganathan, D., Ramasamy, K., Sellamuthu, G., Jayabalan, S., & Venkataraman, G. (2018). CRISPR for crop improvement: an update review. *Frontiers in Plant Science*, *9*, 985.

Jain, D., Kumari, A., Saheewala, H., Sanadhya, S., Maheshwari, D., Meena, R. H., … & Mohanty, S. R. (2020). Biochemical, functional and molecular characterization of pigeon pea rhizobia isolated from semi-arid regions of India. *Archives of Microbiology*, *202*, 1809–1816.

Jha, U. C., Bohra, A., Parida, S. K., & Jha, R. (2017). Integrated "omics" approaches to sustain global productivity of major grain legumes under heat stress. *Plant Breeding*, *136*(4), 437–459.

Jha, U. C., Nayyar, H., Sharma, K. D., von Wettberg, E. J. B., & Siddique, K. H. (Eds.). (2024). *Legume Crop Wild Relatives: Their Role in Improving Climate Resilient Legumes*. CRC Press. https://www.routledge.com/Legume-Crop-Wild-Relatives-Their-Role-in-Improving-Climate-Resilient-Legumes/ChandJha-Nayyar-DevSharma-BishopvonWettberg-HMSiddique/p/book/9781003434535

Jorrin, B., Maluk, M., Atoliya, N., Kumar, S. C., Chalasani, D., Tkacz, A., … & Poole, P. S. (2021). Genomic diversity of pigeon pea (*Cajanus cajan* L. Millsp.) endosymbionts in India and selection of potential strains for use as agricultural inoculants. *Frontiers in plant science*, *12*, 680981.

Jyoti, S. D., Singh, G., Pradhan, A. K., Tarpley, L., Septiningsih, E. M., & Talukder, S. K. (2024). Rice breeding for low input agriculture. *Frontiers in Plant Science*, *15*, 1408356.

Khan, A. W., Garg, V., Sun, S., Gupta, S., Dudchenko, O., Roorkiwal, M., Chitikineni, C., Bayer, P. E., Shi, C., Upadhyaya, H. D., Bohra, A., Bharadwaj, C., Mir, R. R., Baruch, K., Yang, B., Coyne, C. J., Bansal, K. C., Nguyen, H. T., Ronen, G., Aiden, E. L., Veneklaas, E., Siddique, K. H. M., Liu, X., Edwards, D., & Varshney, R. K. (2024). *Cicer* super-pangenome provides insights into species evolution and agronomic trait loci for crop improvement in chickpea. *Nature Genetics*, *56*, 1225–1234.

Khan, H. A., Siddique, K. H., & Colmer, T. D. (2017). Vegetative and reproductive growth of salt- stressed chickpea are carbon-limited: sucrose infusion at the reproductive stage improves salt tolerance. *Journal of Experimental Botany*, *68*(8), 2001–2011.

Khoury, C. K., Castañeda-Alvarez, N. P., Achicanoy, H. A., Sosa, C. C., Bernau, V., Kassa, M. T., … &Struik, P. C. (2015). Crop wild relatives of pigeonpea [*Cajanus cajan* (L.) Millsp.]: Distributions, ex situ conservation status, and potential genetic resources for abiotic stress tolerance. *Biological Conservation*, *184*, 259–270.

Krishnamurthy, L., Upadhyaya, H. D., Saxena, K. B., &Vadez, V. (2012). Variation for temporary waterlogging response within the mini core pigeonpea germplasm. *The Journal of Agricultural Science*, *150*(3), 357–364.

Kudapa, H., Bharti, A. K., Cannon, S. B., Farmer, A. D., Mulaosmanovic, B., Kramer, R., … & Varshney, R. K. (2012). A comprehensive transcriptome assembly of pigeonpea (*Cajanus cajan* L.) using Sanger and second-generation sequencing platforms. *Molecular Plant*, *5*(5), 1020–1028.

Kulkarni, N. K., Reddy, A. S., Kumar, P. L., Vijaynarasimha, J., Rangaswamy, K. T., Muniyappa, V., … & Reddy, D. V. R. (2003). Broad-based resistance to pigeonpea sterility mosaic disease in accessions of *Cajanus scarabaeoides* (L.) Benth. *Indian Journal of Plant Protection*, *31*(1), 6–11.

Kumar, J., Choudhary, A. K., Gupta, D. S., & Kumar, S. (2019). Towards exploitation of adaptive traits for climate-resilient smart pulses. *International journal of molecular sciences*, *20*(12), 2971.

Kumar, M., Prusty, M. R., Pandey, M. K., Singh, P. K., Bohra, A., Guo, B., & Varshney, R. K. (2023). Application of CRISPR/Cas9-mediated gene editing for abiotic stress management in crop plants. *Frontiers in plant science*, *14*, 1157678.

Kumar, V., Khan, A. W., Saxena, R. K., Garg, V., & Varshney, R. K. (2016). First-generation HapMap in *Cajanus* spp. reveals untapped variations in parental lines of mapping populations. *Plant Biotechnology Journal*, *14*(8), 1673–1681.

Maesen, L. V. D. (1990). Pigeonpea: origin, history, evolution, and taxonomy.

Malhotra, E. V., Jain, R., Tyagi, S., Raman, K. V., Bansal, S., Aminedi, R., &Pattanayak, D. (2022). Comparative analysis of herbivory responsive miRNAs to delineate pod borer (Helicoverpa armigera) resistance mechanisms in *Cajanus cajan* and its wild relative *Cajanus scarabaeoides*. *Plant Cell Reports*, *41*(4), 1147–1161.

Mallikarjuna, N., Jadhav, D., Reddy, M. V., & Dutta-Tawar, U. (2005). Introgression of Phytophthora blight disease resistance from *Cajanus platycarpus* into short duration pigeonpea [*Cajanus cajan* (L.) Millsp.]. *Indian Journal of Genetics and Plant Breeding*, *65*(04), 261–263.

Mallikarjuna, N., Jadhav, D., & Reddy, P. (2006). Introgression of Cajanus platycarpus genome into cultivated pigeonpea, *C. cajan*. *Euphytica*, *149*, 161–167.

Mallikarjuna, N., & Moss, J. P. (1995). Production of hybrids between Cajanus platycarpus and Cajanus cajan. *Euphytica*, *83*, 43–46.

Mallikarjuna, N. M., Senthilvel Senapathy, S. S., Jadhav, D. R., Kulbhushan Saxena, K. S., Sharma, H. C., Upadhyaya, H. D., … & Rajeev Varshney, R. V. (2011). Progress in the utilization of *Cajanus platycarpus* (Benth.) Maesen in pigeonpea improvement.

Mapfumo, P., Adjei-Nsiah, S., Mtambanengwe, F., Chikowo, R., & Giller, K. E. (2013). Participatory action research (PAR) as an entry point for supporting climate change adaptation by smallholder farmers in Africa. *Environmental Development*, *5*, 6–22.

Marla, S. S., Mishra, P., Maurya, R., Singh, M., Wankhede, D. P., Kumar, A., … & Kumar, R. (2020). Refinement of draft genome assemblies of Pigeonpea (*Cajanus cajan*). *Frontiers in Genetics*, *11*, 607432.

Mediga, K. R., Sunkad, G., Kulkarni, S., Chandran, U. S., Ghosh, R., Kshirsagar, D., … & Sharma, M. (2023). Multi-environment testing based G× E interactions reveal stable host-plant resistance against sterility mosaic disease in pigeonpea. *Agronomy*, *13*(12), 2859.

Megha, Singh N., Sharma, M., Langyan, S., & Kumar Singh, N. (2024). Genome wide association study of antioxidant activity in pigeonpea germplasm. *Discover Food*, *4*(1), 82.

Millet, E. J., Kruijer, W., Coupel-Ledru, A., Alvarez Prado, S., Cabrera-Bosquet, L., Lacube, S., … & Tardieu, F. (2019). Genomic prediction of maize yield across European environmental conditions. *Nature genetics*, *51*(6), 952–956.

Mir, R. R., Kudapa, H., Srikanth, S., Saxena, R. K., Sharma, A., Azam, S., … & Varshney, R. K. (2014). Candidate gene analysis for determinacy in pigeonpea (*Cajanus* spp.). *Theoretical and Applied Genetics*, *127*, 2663–2678.

Mohanty, P., Allakhverdiev, S. I., & Murata, N. (2007). Application of low temperatures during photoinhibition allows characterization of individual steps in photodamage and the repair of photosystem II. *Photosynthesis research*, *94*, 217–224.

Møller, I. M., Jensen, P. E., & Hansson, A. (2007). Oxidative modifications to cellular components in plants. *Annual Review of Plant Biology*, *58*(1), 459–481.

Montesinos-López, O. A., Herr, A. W., Crossa, J., Montesinos-López, A., & Carter, A. H. (2024). Enhancing winter wheat prediction with genomics, phenomics and environmental data. *BMC Genomics*, *25*(1), 544.

Nayyar, H., Chander, K., Kumar, S., & Bains, T. (2005). Glycine betaine mitigates cold stress damage in chickpea. *Agronomy for Sustainable Development*, *25*(3), 381–388.

Njaci, I., Ngugi-Dawit, A., Oduor, R. O., Kago, L., Williams, B., Hoang, L. T. M., … & Ghimire, S. R. (2020). Comparative analysis delineates the transcriptional resistance mechanisms for pod borer resistance in the pigeonpea wild relative *Cajanus scarabaeoides* (L.) thouars. *International Journal of Molecular Sciences*, *22*(1), 309.

Obala, J., Saxena, R. K., Singh, V. K., Kale, S. M., Garg, V., Kumar, C. S., … & Varshney, R. K. (2020). Seed protein content and its relationships with agronomic traits in pigeonpea is controlled by both main and epistatic effects QTLs. *Scientific Reports*, *10*(1), 214.

Oldroyd, G. E., Murray, J. D., Poole, P. S., & Downie, J. A. (2011). The rules of engagement in the legume-rhizobial symbiosis. *Annual Review of Genetics*, *45*(1), 119–144.

Ormeño-Orrillo, E., Menna, P., Almeida, L. G. P., Ollero, F. J., Nicolás, M. F., Pains Rodrigues, E., … & Martínez-Romero, E. (2012). Genomic basis of broad host range and environmental adaptability of *Rhizobium tropici* CIAT 899 and Rhizobium sp. PRF 81 which are used in inoculants for common bean (*Phaseolus vulgaris* L.). *BMC Genomics*, *13*, 1–26.

Pang, J., Turner, N. C., Khan, T., Du, Y. L., Xiong, J. L., Colmer, T. D., … & Siddique, K. H. (2017). Response of chickpea (*Cicer arietinum* L.) to terminal drought: leaf stomatal conductance, pod abscisic acid concentration, and seed set. *Journal of Experimental Botany*, *68*(8), 1973–1985.

Patil, P. G., Dubey, J., Bohra, A., Mishra, R. K., Saabale, P. R., Das, A., … & Singh, N. P. (2017). Association mapping to discover significant marker-trait associations for resistance against Fusarium wilt variant 2 in pigeonpea [*Cajanus cajan* (L.) Millspaugh] using SSR markers. *Journal of Applied Genetics*, *58*, 307–319.

Pazhamala, L., Saxena, R. K., Singh, V. K., Sameerkumar, C. V., Kumar, V., Sinha, P., … & Varshney, R. K. (2015). Genomics-assisted breeding for boosting crop improvement in pigeonpea (*Cajanus cajan*). *Frontiers in Plant Science*, *6*, 50.

Pisias, M. T., Bakala, H. S., McAlvay, A. C., Mabry, M. E., Birchler, J. A., Yang, B., & Pires, J. C. (2022). Prospects of feral crop de novo redomestication. *Plant and Cell Physiology*, *63*(11), 1641–1653.

Prince, S. J., Murphy, M., Mutava, R. N., Durnell, L. A., Valliyodan, B., Shannon, J. G., & Nguyen, H. T. (2017). Root xylem plasticity to improve water use and yield in water-stressed soybean. *Journal of Experimental Botany*, *68*(8), 2027–2036.

Priyanka, B., Sekhar, K., Reddy, V. D., & Rao, K. V. (2010). Expression of pigeonpea hybrid proline rich protein encoding gene (CcHyPRP) in yeast and Arabidopsis affords multiple abiotic stress tolerance. *Plant Biotechnology Journal*, *8*(1), 76–87.

Quan, W., Liu, X., Wang, H., & Chan, Z. (2016). Comparative physiological and transcriptional analyses of two contrasting drought tolerant alfalfa varieties. *Frontiers in Plant Science*, *6*, 1256.

Ramakrishna, G., Kaur, P., Singh, A., Yadav, S. S., Sharma, S., Singh, N. K., & Gaikwad, K. (2021). Comparative transcriptome analyses revealed different heat stress responses in pigeonpea (*Cajanus cajan*) and its crop wild relatives. *Plant Cell Reports*, *40*, 881–898.

Reddy, L. J., Saxena, K. B., Jain, K. C., Umaid Singh, U. S., Green, J. M., Sharma, D., … & Nene, Y. L. (1997). Registration of high-protein pigeonpea elite germplasm ICPL 87162.

Saeed, F., Chaudhry, U. K., Raza, A., Charagh, S., Bakhsh, A., Bohra, A., Ali, S., Chitikineni, A., Saeed, Y., Visser, R. G. F., Siddique, K. H. M., & Varshney, R. K. (2023). Developing future heat-resilient vegetable crops. *Functional & Integrative Genomics*, *23*, 47.

Saini, P., Aziz, D., Rafiq, S., Saini, P., Puthiamadom, N., Rohela, G. K., … & Singh, G. (2024) Utilizing advanced omics approaches for extricating the hidden genetic variability from wild repository for reviving genetic gains in cereals with major emphasis on rice. In *Omics and System Biology Approaches for Delivering Better Cereals* (pp. 350–400). CRC Press. https://www.taylorfrancis.com/chapters/edit/10.1201/9781032693385-13/utilizing-advanced-omics-approaches-extricating-hidden-genetic-variability-wild-repository-reviving-genetic-gains-cereals-major-emphasis-rice-pawan-saini-danishta-aziz-summira-rafiq-pooja-saini-neeraja-puthiamadom-gulab-khan-rohela-charan-singh-arya-nair-visakh-gayathri-shiv-shankar-singh-gurjeet-singh

Sandhu, J. S., Gupta, S. K., & Khanna, V. (2009). Introgression of yield and component traits through wide hybridization in pigeonpea. *The Journal of Crop Improvement*, *36*, 50–54.

Sarika, Arora V., Iquebal, M. A., Rai, A., & Kumar, D. (2013). PIPEMicroDB: microsatellite database and primer generation tool for pigeonpea genome. *Database*, *2013*, bas054.

Satheesh Naik, S. J., Padmaja, G., Tiwari, A., Bandi, S., Parihar, A. K., Bohra, A., … & Sharma, S. (2024). Genomics-aided breeding strategies for biotic stress in pigeonpea. In *Genomics-aided Breeding Strategies for Biotic Stress in Grain Legumes* (pp. 4984). Singapore: Springer Nature Singapore.

Saxena, K., Bohra, A., Choudhary, A. K., Sultana, R., Sharma, M., Pazhamala, L. T., & Saxena, R. K. (2021a). The alternative breeding approaches for improving yield gains and stress response in pigeonpea (*Cajanus cajan*). *Plant Breeding*, *140*(1), 74–86.

Saxena, K., Choudhary, A. K., Srivastava, R. K., Bohra, A., Saxena, R. K., & Varshney, R. K. (2019). Origin of early maturing pigeonpea germplasm and its impact on adaptation and cropping systems. *Plant Breeding*, *138*(3), 243–251.

Saxena, K., Saxena, R. K., & Varshney, R. K. (2017). Use of immature seed germination and single seed descent for rapid genetic gains in pigeonpea. *Plant Breeding*, *136*(6), 954–957.

Saxena, K. B., Ariyanayagam, R. P., & Reddy, L. J. (1992). Genetics of a high-selfing trait in pigeonpea. *Euphytica*, *59*, 125–127.

Saxena, K. B., Choudhary, A. K., Saxena, R. K., & Chauhan, Y. S. (2020). Can pigeonpea hybrids negotiate stresses better than inbred cultivars? *Breeding Science*, *70*(4), 423–429.

Saxena, K. B., Faris, D. G., Singh, U., & Kumar, R. V. (1987). Relationship between seed size and protein content in newly developed high protein lines of pigeonpea. *Plant Foods for Human Nutrition*, *36*, 335–340.

Saxena, K. B., Kumar, R. V., Tikle, A. N., Saxena, M. K., Gautam, V. S., Rao, S. K., ... & Varshney, R. K. (2013). ICPH 2671–the world's first commercial food legume hybrid. *Plant Breeding*, *132*(5), 479–485.

Saxena, K. B., Reddy, M. V., Bhagwat, V. R., & Sharma, S. B. (1996). Preliminary studies on the incidence of major diseases and insects in *Cajanus platycarpus* germplasm at ICRISAT Asia Center.

Saxena, K. B., Saxena, R. K., Sharma, S., Sameer Kumar, C. V., Sultana, R., Von Wettberg, E. B., & Varshney, R. K. (2018a). The drivers and methodologies for exploiting wild Cajanus genome in pigeonpea breeding. *Euphytica*, *214*, 1–20.

Saxena, K. B., Sharma, D., & Vales, M. I. (2018b). Development and commercialization of CMS pigeonpea hybrids. *Plant Breeding Reviews*, *41*, 103–167.

Saxena, K. B., Sultana, R., Bhatnagar-Mathur, P., Saxena, R. K., Chauhan, Y. S., Kumar, R. V., ... & Tikle, A. N. (2016). Accomplishments and challenges of pigeonpea breeding research in India. *Indian Journal of Genetics and Plant Breeding*, *76*(04), 467–482.

Saxena, R. K., Edwards, D., & Varshney, R. K. (2014a). Structural variations in plant genomes. *Briefings in Functional Genomics*, *13*(4), 296–307.

Saxena, R. K., Hake, A., Bohra, A., Khan, A. W., Hingane, A., Sultana, R., ... & Varshney, R. K. (2021b). A diagnostic marker kit for Fusarium wilt and sterility mosaic diseases resistance in pigeonpea. *Theoretical and Applied Genetics*, *134*, 367–379.

Saxena, R. K., Jiang, Y., Khan, A. W., Zhao, Y., Kumar Singh, V., Bohra, A., ... & Varshney, R. K. (2021c). Characterization of heterosis and genomic prediction-based establishment of heterotic patterns for developing better hybrids in pigeonpea. *The Plant Genome*, *14*(3), e20125.

Saxena, R. K., Von Wettberg, E., Upadhyaya, H. D., Sanchez, V., Songok, S., Saxena, K., ... & Varshney, R. K. (2014b). Genetic diversity and demographic history of Cajanus spp. illustrated from genome-wide SNPs. *PLoS One*, *9*(2), e88568.

Sharma, H. C., Pampapathy, G., & Reddy, L. J. (2003). Wild relatives of pigeonpea as a source of resistance to the pod fly (Melanagromyzaobtusa Malloch) and pod wasp (Tanaostigmodescajaninae La Salle). *Genetic Resources and Crop Evolution*, *50*, 817–824.

Sharma, M., Ghosh, R., Telangre, R., Rathore, A., Saifulla, M., Mahalinga, D. M., ... & Jain, Y. K. (2016). Environmental influences on pigeonpea-*Fusarium udum* interactions and stability of genotypes to Fusarium wilt. *Frontiers in Plant Science*, *7*, 253.

Sharma, P., Singh, I., Sirari, A., Khosla, G., Singh, G., Ludhar, N. K., & Singh, S. (2019). Inheritance and molecular mapping of restorer of fertility (Rf) gene in A2 hybrid system in pigeonpea (*Cajanus cajan*). *Plant Breeding*, *138*(6), 741–747.

Sharma, S., Jaba, J., Rao, P. J., Prasad, S., Gopal, N. T. V. V., Sharma, H. C., & Kilian, B. (2022). Reaping the potential of wild cajanus species through pre-breeding for improving resistance to pod borer, *Helicoverpa armigera*, in cultivated pigeonpea (*Cajanus cajan* (L.) millsp.). *Biology*, *11*(4), 485.

Singh, A., Ramakrishna, G., Singh, N. K., Abdin, M. Z., & Gaikwad, K. (2024c). Genomic insight into variations associated with flowering-time and early-maturity in pigeonpea mutant TAT-10 and its wild type parent T21. *International Journal of Biological Macromolecules*, *257*, 128559.

Singh, A., Sharma, A. K., Singh, N. K., & Sharma, T. R. (2017). PpTFDB: A pigeonpea transcription factor database for exploring functional genomics in legumes. *PLoS One*, *12*(6), e0179736.

Singh, B., Mishra, S., Bohra, A., Joshi, R., & Siddique, K. H. M. (2018). Crop phenomics for abiotic stress tolerance in crop plants. In *Biochemical, Physiological and Molecular Avenues for Combating Abiotic Stress in Plants* (pp. 277–296). Academic Press, Elsevier.

Singh, G., Gudi, S., Amandeep Upadhyay, P., Shekhawat, P. K., Nayak, G., … & Ayoubi, H. (2022a). Unlocking the hidden variation from wild repository for accelerating genetic gain in legumes. *Frontiers in plant science*, *13*, 1035878.

Singh, G., Khanna, R., Kaur, R., Kaur, K., Kaur, R., Sharma, N., & Mangat, G. S. (2023). Performance under multi-environment trial for quantitative traits of rice (*Oryza sativa* L.) genotypes in North- West India (Punjab). *Ecological Genetics and Genomics*, *28*, 100190.

Singh, G., Kumar, P., Mandal, P., Aziz, D., Rafiq, S., Saini, P., … & Saini, D. K. (2024b). Plant metabolomic profiling to better understand biotic and abiotic stress resistance in cereals. In *Omics and System Biology Approaches for Delivering Better Cereals* (pp. 268–292). CRC Press. https://www.taylorfrancis.com/chapters/edit/10.1201/9781032693385-10/plant-metabolomic-profiling-better-understand-biotic-abiotic-stress-resistance-cereals-gurjeet-singh-pradeep-kumar-parul-mandal-danishta-aziz-summira-rafiq-pawan-saini-santosh-gudi-amandeep-dinesh-kumar-saini

Singh, G., Singh, I., Sharma, P., Gupta, M., & Singh, S. (2021). Devising selection criteria based on variability and association studies in segregating populations derived from an interspecific cross between *Cajanus scarabaeoides* × *C. cajan*. *Legume Research-An International Journal*, *44*(6), 615–620.

Singh, G., Singh, I., Taggar, G. K., Rani, U., Sharma, P., Gupta, M., & Singh, S. (2020). Introgression of productivity enhancing traits, resistance to pod borer and Phytopthora stem blight from *Cajanus scarabaeoides* to cultivated pigeonpea. *Physiology and Molecular Biology of Plants*, *26*, 1399–1410.

Singh, I. P., Bohra, A., Naik, S. S., & Parihar, A. K. (2024a). Pigeonpea hybrid breeding in India. *Journal of Food Legumes*, *37*(1), 1–10.

Singh, N. K., Gupta, D. K., Jayaswal, P. K., Mahato, A. K., Dutta, S., Singh, S., … & Sharma, T. R. (2012). The first draft of the pigeonpea genome sequence. *Journal of Plant Biochemistry and Biotechnology*, *21*, 98–112.

Singh, V., Sinha, P., Obala, J., Khan, A. W., Chitikineni, A., Saxena, R. K., & Varshney, R. K. (2022b). QTL-seq for the identification of candidate genes for days to flowering and leaf shape in pigeonpea. *Heredity*, *128*(6), 411–419.

Sinha, P., Singh, V. K., Saxena, R. K., Khan, A. W., Abbai, R., Chitikineni, A., … & Varshney, R. K. (2020). Superior haplotypes for haplotype-based breeding for drought tolerance in pigeonpea (*Cajanus cajan* L.). *Plant Biotechnology Journal*, *18*(12), 2482–2490.

Srivastava, N., Vadez, V., Upadhyaya, H. D., & Saxena, K. B. (2006). Screening for intra and inter specific variability for salinity tolerance in pigeonpea (*Cajanus cajan*) and its related wild species. *Journal of SAT Agricultural Research*, *2*(1), 1–12.

Subbarao, G. V., Johansen, C., Rao, J. K., & Jana, M. K. (1990). Salinity tolerance in F1 hybrids of pigeonpea and a tolerant wild relative. *Crop Science*, *30*(4), 785–788.

Sujana, G., Sharma, H. C., & Rao, D. M. (2008). Antixenosis and antibiosis components of resistance to pod borer *Helicoverpa armigera* in wild relatives of pigeonpea. *International Journal of Tropical Insect Science*, *28*, 191–200.

Tamirisa, S., Vudem, D. R., &Khareedu, V. R. (2014). Overexpression of pigeonpea stress-induced cold and drought regulatory gene (CcCDR) confers drought, salt, and cold tolerance in Arabidopsis. *Journal of Experimental Botany*, *65*(17), 4769–4781.

Tuteja, R. E. E. T. U., Saxena, R. K., Davila, J. A. I. M. E., Shah, T. R. U. S. H. A. R., Chen, W. E. N. B. I. N., Xiao, Y. L., … & Varshney, R. K. (2013). Cytoplasmic male sterility-associated chimeric open reading frames identified by mitochondrial genome sequencing of four *Cajanus* genotypes. *DNA Research*, *20*(5), 485–495.

Varshney, R. K., Chen, W., Li, Y., Bharti, A. K., Saxena, R. K., Schlueter, J. A., … & Jackson, S. A. (2012). Draft genome sequence of pigeonpea (*Cajanus cajan*), an orphan legume crop of resource-poor farmers. *Nature biotechnology*, *30*(1), 83.

Varshney, R. K., Penmetsa, R. V., Dutta, S., Kulwal, P. L., Saxena, R. K., Datta, S., … & Cook, D. R. (2010). Pigeonpea genomics initiative (PGI): an international effort to improve crop productivity of pigeonpea (*Cajanus cajan* L.). *Molecular Breeding*, *26*, 393–408.

Varshney, R. K., Thudi, M., Nayak, S. N., Gaur, P. M., Kashiwagi, J., Krishnamurthy, L., … & Viswanatha, K. P. (2014). Genetic dissection of drought tolerance in chickpea (*Cicer arietinum* L.). *Theoretical and Applied Genetics*, *127*(2), 445–462.

Vishal, A., Bantewad, S., Kota, S., Mishra, S. P., Hingane, A., & Jaba, J. (2023). Antixenosis& antibiosis mechanisms of resistance to the pod borer, *Helicoverpa armigera* in pigeonpea cultigens and hybrids. *International Journal of Tropical Insect Science*, *43*(2), 665–675.

Wang, Q., Liu, J., & Zhu, H. (2018). Genetic and molecular mechanisms underlying symbiotic specificity in legume-rhizobium interactions. *Frontiers in Plant Science*, *9*, 313.

Yadav, M., Vashisht, B. B., Jalota, S. K., Jyolsna, T., Singh, S. P., Kumar, A., … & Singh, G. (2024). Improving Water Efficiencies in Rural Agriculture for Sustainability of Water Resources: A Review. *Water Resources Management*, 1–22.

Yadav, P., Saxena, K. B., Hingane, A., Kumar, C. S., Kandalkar, V. S., Varshney, R. K., & Saxena, R. K. (2019). An "Axiom Cajanus SNP Array" based high density genetic map and QTL mapping for high-selfing flower and seed quality traits in pigeonpea. *BMC Genomics*, *20*, 1–10.

Yu, H., Lin, T., Meng, X., Du, H., Zhang, J., Liu, G., … & Li, J. (2021). A route to de novo domestication of wild allotetraploid rice. *Cell*, *184*(5), 1156–1170.

Zhang, L., Zhao, H. K., Dong, Q. L., Zhang, Y. Y., Wang, Y. M., Li, H. Y., … & Dong, Y. S. (2015). Genome-wide analysis and expression profiling under heat and drought treatments of HSP70 gene family in soybean (*Glycine max* L.). *Frontiers in Plant Science*, *6*, 773.

Zhao, Q., Feng, Q., Lu, H., Li, Y., Wang, A., Tian, Q., … & Huang, X. (2018). Pan-genome analysis highlights the extent of genomic variation in cultivated and wild rice. *Nature Genetics*, *50*(2), 278–284.

7 Enhancing Abiotic Stress Response in Pigeonpea

A Phenomics Perspective

Rakesh Choudhary, Ruchi Asati, Kadam Kiran, Dhanapati Keerthana, Subhash Chand, Aalok Shiv, A. K. Roy, Ashok Kumar Parihar, and S. K. Chaturvedi

7.1 INTRODUCTION

Pigeonpea (*Cajanus cajan*) is a nutrient-dense food grain legume grown in regions with tropical and subtropical climates. It is a diploid species (2n = 2X = 22) with a genome size of 833.07 Mbp (Varshney et al. 2012). This crop is believed to have originated in Africa and was adopted in India. Pigeonpea is commonly known as *arhar*, *tur*, or red gram. The grains contain protein (21–28%), carbohydrates, vitamins, fats, and minerals (Singh et al. 2020). It is a multipurpose legume crop where grain is used as food and livestock feed, and leftover parts are used as fodder and domestic fuelwood, making it a sustainable crop for small and marginal farmers (Saxena et al. 2019, 2021). Pigeonpea is cultivated on 6.03 million hectares (mha) of land around the world with an annual production of 5.33 million metric tons (m t) and an average productivity of 883 kg ha^{-1}. Globally, India contributes the maximum in area (81%) and production (79%), but in productivity, India is at almost half (861 kg ha^{-1}) of the country with the highest productivity, i.e., Malawi (1698 kg ha^{-1}; FAOSTAT 2022). Indian farmers produce most of the world's crop, and most of this is dried and split to be cooked as tur dal, a widely eaten source of protein, eaten with rice and roti (made out of mainly cereal flour).

In India, it is the second most important pulse crop after chickpea. It was cultivated on about 4.13 mha during 2023–2024, compared to 4.07 mha during 2022–2023. The major pigeonpea growing states are Karnataka (1.72 mha), followed by Maharashtra (1.34 mha), Telangana (0.31 mha), and Uttar Pradesh (0.29 mha; Agricultural Statistics at a Glance 2023). The first advance foodgrains' production estimates of the Department of Agriculture and Farmers Welfare, Government of India, suggested pigeonpea production in 2024–2025 is to be 3.50 mt.

Pigeonpea is an annual/perennial legume crop consumed on a large scale in South Asia and included in daily diets as a major source of protein in different parts of the

DOI: 10.1201/9781003603092-7

world. Pigeonpea seeds are dehulled, split, and used as *dhal*, and also immature green grains are consumed as vegetables in India. Pigeonpea stubble/stocks are used to cook food and make huts. Certain unique features of pulses, viz., their N fixing ability through symbiotic association with rhizobium and a deep root system enabling them to draw moisture from deeper soil. The major environmental stresses that the pigeonpea faces during its life cycle include extremes in moisture (drought and waterlogging), high temperature (heat and cold), photoperiod, and mineral-related stresses (salinity/acidity; Singh 2023) and crop responses in various ways to reduce the impact of abiotic stress (Figure 7.1). To tackle these stresses tolerant genotype has different mechanisms, which are controlled by different genes and QTLs (Table 7.1).

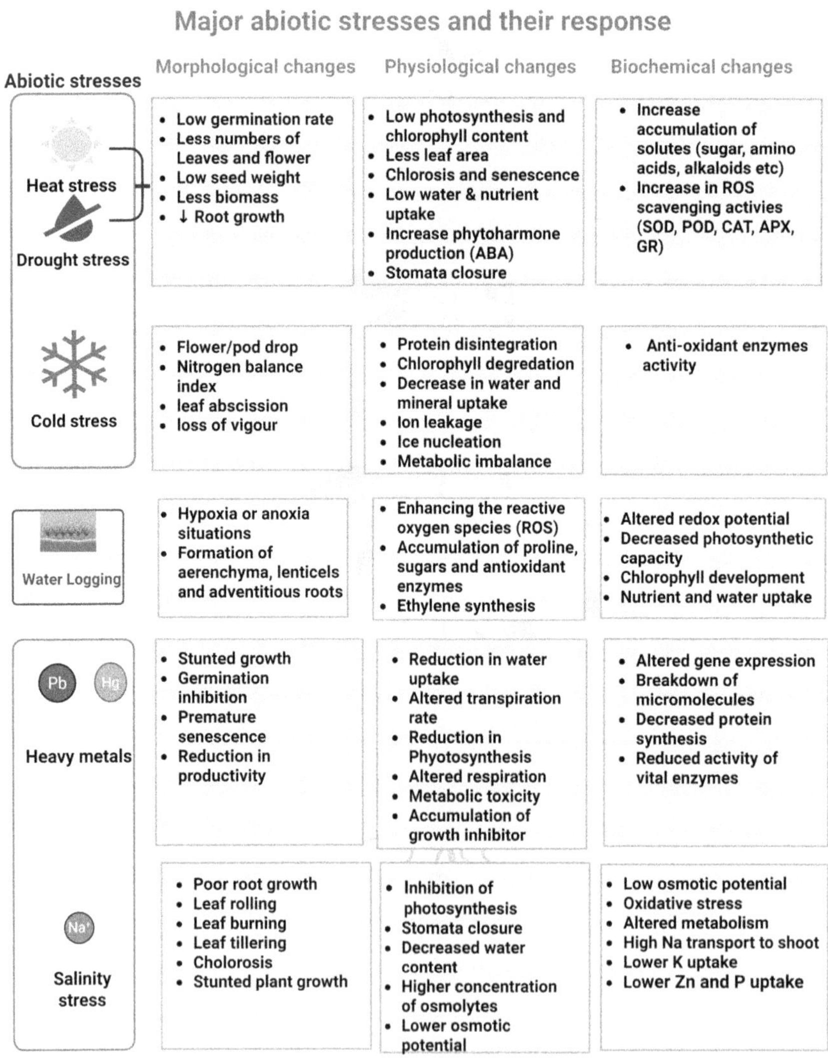

FIGURE 7.1 Major abiotic stresses of pigeonpea and their responses at different stages.

TABLE 7.1
Abiotic Stress and Tolerance Mechanisms in Pigeonpea

Abiotic Stress		Tolerance Mechanism	Genotypes	Genes/QTLs	Reference
Moisture	**Drought**	High RWC, MSI, pods per plant, high HI	LRG 30, ICPL 85063, ICPL 332, ICC 7571, ICPL 227, ICP 8744	• *CcHyPRP* • *CcCYP* • *CcCDR* • *DLP* • 51 drought responsive genes (*C. cajan_26230*, *C. cajan_39705*, *C. cajan_09181* etc) (putative candidate genes)	Priyanka et al. (2010), Sekhar et al. (2010), Tamirisa et al. (2014), Deeplanaik et al. (2013), Sinha et al. (2016)
	Water logging	Development of lenticels, aerenchyma cells, adventitious roots, and survival of the seed under submergence	ICP 5028, ICPH 2431, ICPL 87119, ICPH 2740, ICPL 149, ICPL 20241, and MAL15	–	Hingane et al. (2015)
Temperature	**Heat**	Flowering and pod setting under low temperature	ICP 4555, WRP1, JKM7, ICP8700, MAL13, NDA1, BSMR 736, ICP 995	• *CcCYP* • *qMT4.1,qMT5.1a and qMT10.1* • *CcLG03*	Basu et al. (2016), Kumawat et al. (2012), Singh et al. (2022)
	Cold	Flowering and pod setting under low temperature	IPA 7-2, Bahar, and MAL 19, ICPL 87119 ICPH 2431, ICPH 2740, ICPH 2671, ICPH 4187, MAL 9, LRG 30, Maruti, ICPL 20128, ICPL 332, ICPL 20237, ICPL 20238, Asha and MAL 15	• *CcCDR* • CcCKS	Sultana et al. (2013), Tamirisa et al. (2014), Tamirisa et al. (2017)
Salinity		Reduced translocation of Na and Cl from root to shoot	C11, ICPL 227, WRP1, GS1 and TS3, UPAS 120 and ICPL 151 ICPL 227 and Hy3C	*CcCYP* *CcCDR*	Priyanka et al. (2010), Tamirisa et al. (2014), Pazhamala et al. (2015), Subbarao and Johansen (1991)
Aluminum toxicity		Aluminum exclusion, reduced root injury, lower Al translocation in roots, and high root regrowth	IPA 7-10 and T 7, 67 B and GT 101E IPA 7-10, T 7 (late), and 67B(early)	*CcSTOP1* and *CcMATE1*	Choudhary et al. (2011a), Choudhary and Singh (2011), Daspute et al. (2018)
Heavy metal stress		High antioxidant activity, high root-to-shoot ratio		*CcMT1*	Sekhar et al. (2010)

These abiotic stresses are responsible for low productivity and lack of stability in the pigeonpea yields over the last five decades. The notable effects of environmental changes are expressed by drought and heat stress, which affect overall crop performance and ultimately lead to crop loss (Anjum et al. 2017). Climate change severely affects crop performance by different means, like disrupting crop phenology, moisture availability, rapid pathogen spreading, host crop susceptibility, soil fertility, and erosion (Mousavi-Derazmahalleh et al. 2019). The abiotic stresses disturb osmotic adjustment and produce toxic reactive oxygen species (ROS), which disturb the cellular homeostasis of the cell, resulting in significant yield losses. In pigeonpea, low temperatures (<10°C) lead to converting intracellular water into ice and consequently shrinking cells, which causes severe flower drop and wilting, resulting in yield losses and death of plants. The temperature extremes (too low/high) prevail in the northern and northeastern areas of India during the reproductive stage, primarily affecting the production rate (Roy et al., 2021). Waterlogging affects plant growth by reducing the oxygen diffusion rate between the soil and atmosphere and by changing the physical and chemical properties of the soil. Soil salinity affects pigeonpea plants through osmotic stress and interference with the uptake of mineral nutrients. Aluminum toxicity also reduces the nutrient uptake efficiency of this crop. Plants have evolved complex signaling pathways, including receptors, secondary messengers, phytohormones, and signal transducers, to detect different stresses and to adapt to changing environmental conditions. These inherent processes promote stress signal transduction and the activation of stress-responsive gene expression to maintain plant growth and productivity (Yokotani et al. 2013). Biochemically, legumes adapt to environmental stress in a variety of ways, including an abscisic acid (ABA) response, auxin signaling, osmo protection, and antioxidant defense, as well as an extensive range of signaling molecules and metabolic pathways known to be involved in drought and heat stress adaptation (Shinozaki and Yamaguchi-Shinozaki 2007). Empirical evidence is well established that one abiotic stress is often linked with several other stresses, making it difficult to know the exact cause of the crop failure.

Recent advances in automated high-throughput and imaging technologies provide high-resolution images and phenomics data for plants (Bohra et al. 2021; Singh et al. 2018). Plant phenotyping is generally known as the comprehensive assessment of multiple traits related to growth and development, physiology, plant architecture, biotic and abiotic resistance/tolerance mechanisms, yield, and the elementary measurement of distinct quantitative parameters that form the foundation for complex trait assessment (Li et al. 2014). Breeding programs are designed to select the best genotypes from a large population based on phenotypic characteristics throughout the crop cycle (Sandhu et al. 2021). Phenotyping becomes highly challenging when huge genotypes, various environments, and replicative trials are employed. Traditional phenotyping is costly, laborious, and destructive, and could decrease the significance or precision of the results. In the new era, stress phenotyping has several issues resolved by the arrival of automated, high-throughput phenotyping (HTP) devices combined with artificial intelligence (Figure 7.2). For the selection and development of climate-resilient pigeonpea

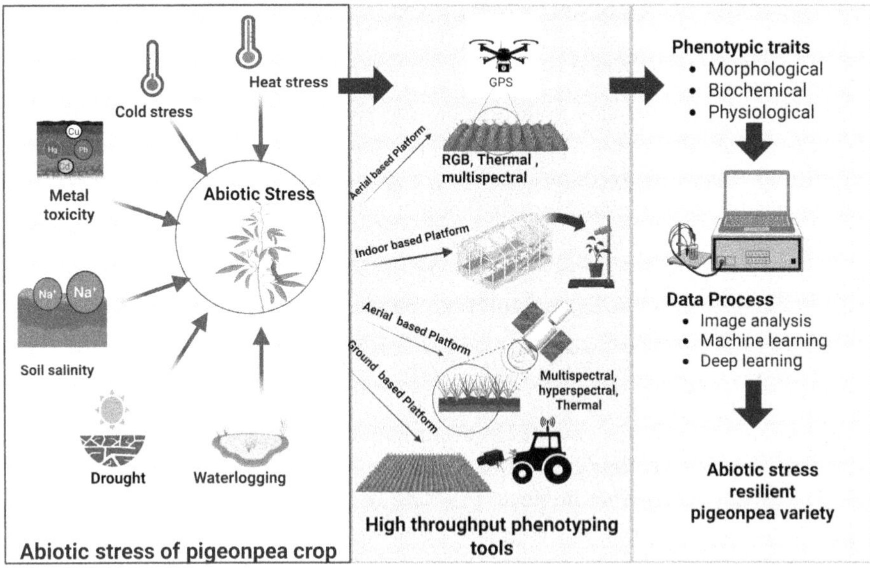

FIGURE 7.2 Role of phenomics tools in understanding the mechanisms of abiotic stress tolerance in pigeonpea.

varieties, a stepwise selection procedure in breeding programs is required. In-depth knowledge of the plant stress response is of utmost importance. In conventional breeding, phenotyping offers reliable selection and identification of improved progenies for predicted traits, but the information has not been adequately exploited for understanding the complex characteristics of multiple genes, owing to a lack of crop phenotypic data and its interaction with the environment. Irrespective of the stepwise selection procedure in breeding programs, HTP is the key to ensuring the quantity and quality of data. HTP is crucial for gaining insights into perplexed interactions among genotype, environment, and management practices (G × E × M). Thus, integration of advanced technologies in breeding programs is critical for developing pigeon pea varieties that can withstand the challenges posed by climate change (Jin et al. 2020). Phenomics tools are crucial in studying plant responses to abiotic stress by enabling high-throughput, nondestructive measurement of traits like growth rate, water use efficiency, and photosynthetic activity. Technologies such as thermal imaging, hyperspectral sensors, and chlorophyll fluorescence provide detailed insights into how plants cope with drought, salinity, and extreme temperatures. These tools allow researchers to identify stress-tolerant genotypes more efficiently, accelerating breeding programs to improve crop resilience. Phenomics data help to link phenotyping to different omics techniques, and one can fill the gap between gene response and complex targeted traits for crop improvement programs. In this chapter, we have covered various phenomics tools, which help to understand abiotic stress tolerance mechanisms in the pigeonpea and support its improvement program.

7.2 ABIOTIC STRESS AND TOLERANCE MECHANISM

7.2.1 Temperature Stress

Temperature stress includes high temperature stress or heat stress (HS) and cold stress, and high temperature also leads to drought stress (DS). DS and HS exhibit detrimental effects on pigeonpea, resulting in permanent harm to plant development and hindering normal plant growth. These stresses induce morphological, physiological, and oxidative disruption in pigeonpea, resulting in an overall reduction in productivity (Choudhary et al. 2022). In the northern and northeastern areas of India, too high or too low temperatures during the critical growth stages affect plant growth and yield.

7.2.1.1 Heat Stress

Plant response changes in case of HS according to the degree of temperature, plant type, and duration. Under extremely high temperatures, cellular damage or even cell death can occur within minutes, resulting in a complete breakdown of cellular organization. HS impacts various plant processes, including germination, growth, development, reproduction, and yield (Sher et al. 2024). It affects the stability of proteins, membranes, RNA molecules, and cytoskeletal structures differently, disrupts enzymatic reaction efficiency, and causes significant physiological challenges and metabolic imbalances in cells.

HS significantly disrupts morpho-physiological traits in common bean, including phenology, biomass partitioning, plant-water relationships, and shoot growth and elongation. HS diminishes plant biomass accumulation, leading to decreased yield, especially in tropical and subtropical regions (Ramakrishna et al. 2021). Previous studies indicate that pigeonpea production could drop by 20.8% for an increase of maximum temperature by every degree Celsius if the maximum temperature is raised throughout the crop season. HS results in the dissociation of important protein complexes and the generation of ROS (Becklin et al. 2016). In response to HS, plants upregulate the genes that code for signaling molecules and molecular chaperones, controlling a series of activities (Mishra et al. 2017).

7.2.1.2 Cold Stress

In northern India, pigeonpea is affected by low temperatures in December and January. When the minimum temperature falls below 5°C, this affects a plant's ability to grow, survive, and reproduce (Araújo et al. 2015). At freezing temperatures, intracellular water condenses into ice, resulting in wilting and poor plant health. In order to verify and acquire valid genetic information, a large number of pigeonpea genotypes must be screened for low-temperature tolerance under controlled temperature circumstances (Alagupalamuthirsolai and Vijayabharathi 2020).

7.2.1.3 Drought Stress

Being a rainfed crop grown in the *kharif* season, pigeonpea faces early and terminal DS. DS negatively affects root and shoot growth, leading to diminished overall plant growth and development. However, drought-tolerant lines exhibit significantly deeper rooting than sensitive ones in grain legumes. During the blooming and pod-filling stages, DS accelerates the senescence and shedding of older basal leaves, as observed in cowpea and pigeonpea. This increase in leaf senescence is linked to reduced nitrogen uptake, which lowers chlorophyll production and radiation efficiency. In cowpea, the combined effects of drought and monocarpy result in a similar acropetal pattern of leaf senescence, further intensifying the process under drought conditions (Khatun et al. 2021).

DS results in an elevated concentration of ROS, including highly damaging molecules like hydroxyl radicals and singlet oxygen. These ROS induce cell damage by degrading membranes and proteins, oxidizing lipids, and fragmenting DNA, ultimately leading to cell death. Legumes, like most crops, have evolved various mechanisms to mitigate the impact of drought. Their initial response to DS begins with the reception of a signal mediated by membrane receptors (McDonald and Paulsen 1997). Relative water content (RWC) of leaves and dehydration tolerance are crucial for drought tolerance. Varieties like Bahar, BSMR 853, and ICP 84031 show better osmotic adjustment under drought conditions (Basu et al. 2016).

7.2.1.4 Impact of Temperature and Drought Stress

Predominantly results in poor germination and impedes the establishment of seedlings. Additionally, HS and DS stress hinder cell elongation and mitosis, which restricts water flow from the xylem to the surrounding cells and results in inadequate growth. DS primarily limits leaf expansion due to reduced turgor pressure and a slow rate of photosynthesis. At the same time, signs of HS include growth inhibition, twig and leaf burning, leaf senescence and chlorosis, and other symptoms that reduce the total leaf area. Due to a lack of photosynthesis, both HS and DS impair embryo development of legume seeds, resulting in ovarian abortion, decreased pollen tube growth, and sterile pollen grains (Liu et al. 2019).

7.2.1.4.1 Morpho-physiological Responses

Stomatal control is the first physiological drought response, which lowers transpiration rates, photosynthesis, and carbon absorption. Through stomatal closure, ABA, a growth inhibitor, helps counteract the negative consequences of DS. In order to control leaf water and nutritional status under moisture scarcity, ABA travels through the transpiration stream and prevents leaf expansion and stomatal opening. Solute accumulation is crucial for osmotic adjustment and osmo protection during drought and HS. Solutes like nitrogen contain compounds like proline or other amino acids or hydroxyl compounds like sucrose, polyhydric alcohols, and oligosaccharides.

7.2.1.4.2 Biochemical Responses

Plant growth hormones (PGHs) such as Gibberellins, cytokinins, auxin, abscisic acid, and ethylene are crucial for a plant's ability to withstand DS and HS. These substances are considered defensive mechanisms to keep antioxidant levels high under stressful situations. By facilitating growth, development, and nutrient allocation, they aid plants in adapting to a range of environments. By generating enzymatic and non-enzymatic antioxidants like catalase, tocopherol, and ascorbate, which support ROS defense mechanisms in plant cells, the antioxidant defense system controls oxidative damage and encourages normal cellular activity. Antioxidants such as glutathione, ascorbic acid, and proline scavenge ROS to protect plants from oxidative damage and boost legume resistance to HS.

7.2.1.4.3 High-Throughput Techniques in Controlling Temperature Stress

HTP has emerged as a transformative approach, enabling rapid, accurate, and automated data collection to accelerate the development of drought-adapted cultivars. By efficiently evaluating numerous plants or genotypes, HTP significantly reduces the time needed for data acquisition and analysis. Additionally, it facilitates the screening of traits at earlier stages, eliminating the necessity of waiting till the plant matures (Bhat et al. 2022).

Advancements in imaging technologies have further enhanced the capability to monitor plant physiological responses to DS nondestructively, offering early indicators to identify drought-resistant cultivars. Image-based measurement of chlorophyll fluorescence has provided critical insights into photosynthetic efficiency in chickpeas under water-deficit conditions, aiding the identification of genotypes with enhanced stress tolerance.

7.2.2 Waterlogging

Waterlogging affects pigeonpea production where annual rainfall is 600–1500 mm (Hingane et al. 2015). Pigeonpea is grown in the semiarid tropics and low-lying areas of north-east India. Waterlogging often leads to failure of the pigeonpea crop in low-lying locations. Pigeonpea has shown some resilience against waterlogging, while the majority of early-maturing types are vulnerable. Electrolyte leakage from oxygen shortage exposes the cell to the external environment, which can lead to lipid and nucleic acid peroxidation and, eventually, death. Restoring the oxygen supply to stressed tissue through a variety of ways can be a key characteristic for waterlogging tolerance. In waterlogged plants, the development of lenticels, adventitious roots, and aerenchyma aids in oxygen restoration. Along with these, various types of biochemical changes also occur for waterlogging tolerance in pigeonpea plants, including enzymatic activity in glycolysis and fermentation and participation of antioxidants. Waterlogging causes hypoxia or anoxia in the soil, which limits plant growth and development. In order to maintain growth under waterlogging, the study found that genotype ICPL-84023 had significant nitrogen absorption and aerenchyma formation (Bansal and Srivastava 2017). Waterlogging can alter a number of variables, including total reducing sugars, superoxide dismutase, membrane stability index, number of pods per plant, pod dry weight, and seed production (Yohan et al. 2017).

7.2.2.1 Impact of Waterlogging

During waterlogging, growth, and other developmental processes, such as seed germination, vegetative growth, and subsequent reproductive growth, are impeded by the suppression of aerobic respiration (Pan et al. 2021). Moreover, ethylene and carbon dioxide levels in the root area rise sharply in wet soils, changing the functions of the soil microbiome and causing severe denitrification and the buildup of ammonium and polyphenolic chemicals (Arduini et al. 2019). Additionally, it either reduces the availability of nutrients like sulfur and nitrogen or transforms them into a form that plants are unable to absorb. The most frequent consequences of waterlogging stress on plants are ethylene buildup and oxygen shortage, which can limit root permeability, along with the development and ultimately cause cell death.

However, plants respond to waterlogging stress by controlling their morphological, biochemical, and molecular characteristics as part of their complex defensive system. Significant yield losses are also caused by waterlogging, which increases the susceptibility of pigeonpea plants to fungal diseases, including Phytophthora blight and Fusarium wilt. At physiological and biochemical levels, plants produce numerous molecules, such as osmolytes, calcium, ROS, hormones, and antioxidants, that confer waterlogging tolerance.

7.2.2.2 High-Throughput Techniques in Controlling Waterlogging Stress

HTP tools incorporate advanced technologies like robotics, computerized data analysis, spectroscopy, and high-throughput imaging. While substantial advancements have been achieved in developing methods to assess phenotypic traits linked to abiotic stress tolerance or sensitivity; however, standardized protocols for waterlogging stress phenotyping are still lacking. Currently, three tomographic techniques, X-ray computed tomography (CT), magnetic resonance imaging (MRI), and positron emission tomography (PET), are being employed for 3D root phenotyping in soils (Atkinson et al. 2019).

Additionally, a systems biology approach known as PANOMICS is essential for understanding the flow of biological information underlying complex traits. This method integrates diverse omics data, predicts cellular processes, and enables new modeling approaches. Pan-omics holds the potential to provide fresh insights into unraveling the intricate mechanisms of waterlogging signaling and signal transduction.

7.2.3 Salinity Stress

The growth response of leguminous plants to salinity varies considerably among species. Pigeonpea is a subtropical crop, and little is known about its salinity tolerance. There is not much investigation about the physiological effects of salt stress in this crop. Plant species and variants within a species differ in their level of salt tolerance, which is generally changeable. Numerous physiological, molecular, and biochemical pathways control salt tolerance, a multigenic trait. It plays a role in controlling the buildup of different solutes/osmolytes, polyamines, ROS, and proteins (Kumar et al. 2015). One suitable strategy for reducing yield losses under salinity stress circumstances may be the creation of saline-tolerant cultivars. Pigeonpea cultivars have been shown to exhibit genotype-dependent salt responses, accumulating osmolytes such as proline, glycine, betaine, and soluble sugars in response to salt stress.

7.2.3.1 Impact of Salinity Stress

The nature and magnitude of salt stress affect the growth of both halophytes and glycophytes. There is a wide range of variability among glycophytes, from species that are extremely sensitive to those that are tolerant. Because they lack adequate tolerance mechanisms, leguminous plants may be more vulnerable to salt stress and cannot thrive in saline soils. Salinity has a negative impact on reproductive growth, resulting in a marked decrease in flower, pod, seed, and seed weight output. The yield is severely harmed because pulses are more susceptible to saline soil than cereals and oilseeds. They build up too much salt, which accelerates the anthocyanin coloration in leaves and stems, hence decreasing germination and seedling colonization (Kumar et al. 2017).

Excess Na^+ buildup during salt stress disrupts protein and nucleic acid structures by interfering with hydrogen bonding and polar interactions. Accordingly, stressed pigeonpea plants have a much-reduced total soluble protein content (Singh et al. 2019). As salinity rises, pigeonpea succulence and moisture content sharply decline, suggesting a loss of turgor. It reduces the effective number and weight of pods, promotes floral shedding, and ultimately reduces the number of seeds and production (Tayyab et al., 2016).

At various levels, the growth and reproduction of plants are negatively impacted, ranging from morphology to genetics, which ultimately makes it difficult for them to survive in saline environments. Tayyab et al. (2023) investigated the effects of salinity on pigeonpea plants and discovered that the plants could withstand sea salt salinity of up to 3.5 Ece dS/m. Leaf osmotic changes at low salinity may be the cause of low moisture content and succulence, as well as increased accumulation of soluble carbohydrates and proteins. When choosing a genotype for breeding in salt tolerance, one should look for features that contribute to salinity tolerance rather than overall performance. Wild relatives of pigeonpea from secondary and tertiary gene pools (*C. scarabaeoides*, *C. albicans*, and *C. platycarpus*) have demonstrated resistance to salinity stress through characteristics like maintaining optimal transpiration rates under salinity and retaining Na+ and Cl− in roots with limited translocation to shoots. To develop tolerant varieties, it is essential to comprehend the genetic foundation of salt tolerance. To regulate the plant mechanism for producing salt-tolerant genotypes using traditional and molecular techniques, it is important to identify morphological and physiological features and determine several genetic parameters. The primary method for maintaining plant cell hydration in the face of drought and salt stress is osmotic adjustment. Osmotic pressure in the cytoplasm must be maintained because the buildup of osmo-protectants enables the first cell tolerance response, which provides adaptability, enhances plant growth, and modulates the osmotic and antioxidant defense of pigeonpea under salt stress (Garg and Noor 2009; Hanafy et al. 2013).

7.2.3.2 High-Throughput Techniques in Controlling Salinity Stress

HTP in plants allows precise evaluation of salt stress effects across various physiological levels, including photosynthesis, transpiration, ionic balance, plant senescence, yield, and other traits associated with salinity tolerance. Recent advancements in plant phenomics, such as automated and digital imaging, offer opportunities to deepen our understanding of the diverse temporal responses of genotypes under salinity stress. Tools like deep learning, active vision cell, image acquisition, and other cutting-edge phenotyping technologies can be utilized to identify, quantify, and predict plant responses to salinity with greater accuracy (Jha et al. 2019).

The vast potential of phenomics techniques for understanding salinity stress tolerance in grain legumes is yet to be fully explored. Advanced high-throughput methods, including inductively coupled plasma mass spectrometry (ICP-MS), Inductively Coupled plasma-atomic/optic emission spectrometry (ICP-AES/OES), X-ray fluorescence (XRF), laser ablation inductively coupled plasma mass spectrometry (LA-ICP-MS), and neutron activation analysis (NAA), have been employed to study them. Ionomics has also been utilized to investigate ion detoxification and homeostasis mechanisms in response to salinity, with application in cereals and legumes such as *Lotus criticus* and the glycophytic legume species *L. carniculatus* and *L. tenuis* (Ali et al. 2024).

7.2.4 Metal Toxicity

Heavy metal toxicity is emerging as one of the most serious concerns across the globe. The toxic effects of heavy metals on plants depend on various factors, including the type and concentration of heavy metals and plant species, plant growth stage, and exposure time. In plants, heavy metals are primarily absorbed through the roots and transported to the shoots, which can cause extensive damage to cellular structures and organelles, such as chloroplasts and mitochondria. This damage disrupts critical processes like photosynthesis and energy production, leading to oxidative stress characterized by the overproduction of ROS (Jaiswal et al. 2018).

These effects ultimately impair plant morphology, growth, and survival rates. While most plants exhibit sensitivity to heavy metals, certain species have evolved mechanisms to tolerate or even hyper-accumulate these toxic elements. Such hyperaccumulators sequester heavy metals in their tissues, often using specialized transporters and chelation systems, which not only allow them to survive in contaminated environments but also offer potential for phytoremediation applications (Parwez et al. 2022)

7.2.4.1 Impact of Metal Toxicity

Among all the heavy metals, cadmium and copper are the most concerning of all the heavy metals, although these are more mobile, non-biodegradable, and poisonous to plants. Excessive cadmium buildup in plants can result in serious phytotoxicity, as well as a host of physiological, morphological, and biochemical consequences, including crop death, excessive ROS generation, and pigment degradation (Hayat et al. 2021).

In response to heavy metal toxicity, resistant plants trigger a complex signaling network that frequently triggers a defense response. This defense response includes, among other things, the accumulation of antioxidants for detoxifying excess ROS, secondary metabolites for sequestering heavy metals to vacuoles, and compatible solutes for osmoregulation and metal transporter regulation.

7.2.4.2 Phenotyping Against Metal Toxicity

Metal toxicity poses a major challenge to sustainable agriculture, limiting crop productivity and soil health. Traditional breeding approaches often fall short in efficiently developing metal toxicity-tolerant crops due to their complex and multigenic nature (Raza et al. 2022). However, integrating omics technologies with artificial intelligence (AI) and HTP has emerged as a transformative strategy to overcome these limitations. Omics methodologies such as genomics, transcriptomics, proteomics, and metabolomics provide comprehensive insights into the molecular mechanisms that govern plant responses to metal toxicity (Ningombam et al. 2024). These approaches enable the identification of key genes, pathways, and biomarkers associated with tolerance traits. Coupled with AI, omics data can be analyzed and interpreted rapidly, uncovering novel targets for crop improvement. HTP further complements this integration by enabling large-scale, precise phenotyping assessments under controlled and field conditions, facilitating the selection of metal toxicity-tolerant genotypes with greater efficiency. HTP methods, leveraging automated imaging and sensing technologies, offer a robust platform for rapidly collecting detailed data on plant growth, development, and physiological responses. These methods provide critical insights into plant performance under metal toxicity by capturing parameters such as biomass, chlorophyll content, root architecture, and stress-induced changes in real-time (Ali et al. 2022).

7.3 USE OF CROP WILD RELATIVES (CWRs)

In pigeonpea, the primary gene pool lacks sources of tolerance or resistance, but several CWRs from the secondary and tertiary gene pools harbor desirable alleles for tolerance to different abiotic stresses (Table 7.2). However, there are some challenges, like cross-incompatibility, linkage drag, F_1 sterility, and morphological and phenological differences among wild and cultivated species (Bohra et al. 2010, 2022).

TABLE 7.2
List of Wild Relatives of Pigeonpea as a Potential Source of Abiotic Stress Tolerance

Sr. No.	Crop Wild Relatives	Stress Tolerance To	References
i.	*C. acutifolius*	Drought, heat, waterlogging	Hingane et al. (2015)
ii.	*C. albicans*	Salt	Choudhary et al. (2011b)
iii.	*C. cinereus*	Drought, heat	Khoury et al. (2015)
iv.	*C. lanceolatus*	Drought	
v.	*C. latisepalus*	Drought	
vi.	*C. lineatus*	Drought	
vii.	*C. cajanifolius*	Heat	
viii.	*C. platycarpus*	Heat, salt tolerance, aluminum toxicity	Choudhary et al. (2011a)
ix.	*C. sericeus*	Drought, salt tolerance, waterlogging	Choudhary et al. (2011b)
x.	*C. scarabaeoides*	Drought, salt tolerance, aluminum toxicity	Upadhyaya (2006) and Choudhary et al. (2011a)
xi.	*C. lineatus*	Waterlogging	Khoury et al. (2015)

7.4 CONVENTIONAL AND NONCONVENTIONAL PHENOTYPING TOOLS

7.4.1 Conventional Phenotyping

Plant phenotyping is a comprehensive evaluation of crop characteristics like growth, development, adaptation, quality, yield, resistance, tolerance, and architecture. Conventional phenotyping methods rely on visual observations and manual measurements of traits. These methods are labor intensive, time-consuming, low throughput, less precise, and typically damaging to plants (Chen et al. 2014). Phenotyping drought-related traits, like root architecture, requires a destructive harvest of plants, so it is impossible to repeat the measurements on the same plant. Additionally, measuring parameters like plant parts' length and width, the number of hairs, branches, seeds, seedlings, pods, or flowers, and leaf area takes much time and effort.

Phenotyping helps to evaluate and quantify the phenotypic variation among different genotypes and to characterize traits associated with tolerance. Various traits, such as plant height, leaf area, biomass, root characteristics, stress-related physiological and biochemical markers, and tolerant genotypes, can be assessed, and tolerant genotypes can be identified. It also helps to understand the genetic basis and heritability of tolerance traits.

7.4.2 Nonconventional Phenotypic Tools

Nonconventional phenotypic tools include AI, sensors for phenotyping different breeding phases, and data collection to trait modeling to increase the rate of genetic gain. HTP is the process of phenotyping plants using a variety of tools, including remote sensing, aerial phenotyping, and ground-based imaging. In recent years, a number of plant phenotyping platforms have been developed (Figure 7.3), including aerial (manned and unmanned aerial vehicle) and ground-based platforms.

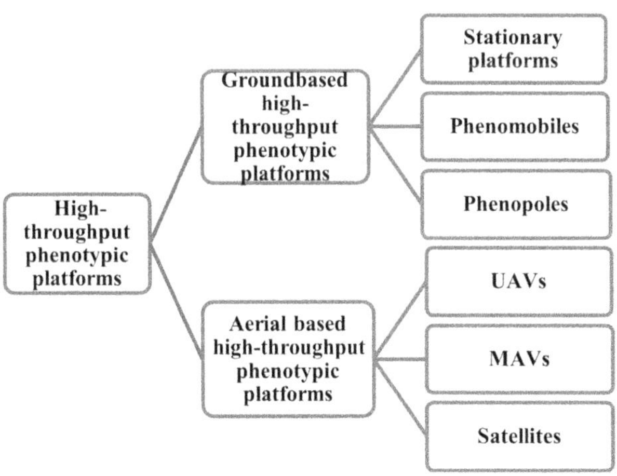

FIGURE 7.3 Classification of high high-throughput phenotypic platform.

7.4.2.1 Ground-Based Platforms of High-Throughput Phenotyping

Ground-based platforms have been developed for field applications and controlled indoor environments (laboratories and greenhouses). Based on throughput and human participation, these phenotypes can also be classified as robotic, vehicle-carried, and manually operated. Ground-based systems may collect high-resolution data for assessing phenotypic features at finer levels (e.g., the plant and organ level) and with more viewing angles than aerial systems. They also often have a larger payload to carry heavier sensors and equipment. Furthermore, ground platforms usually offer higher-quality data than aerial systems by controlling the data collection environment (such as light conditions) with well-designed enclosures. Additionally, they are less impacted by wind, allowing them to maintain a straight path and conduct more even longitudinal scanning.

7.4.2.2 Aerial-Based Platforms of High-Throughput Phenotyping

In recent years, crop phenotyping in the field has made extensive use of aerial phenotyping platforms, mostly consisting of satellites and unmanned aerial vehicles (UAVs) (Jang et al. 2020). For example, multispectral, hyperspectral, or thermal cameras can be carried by UAVs to observe crop health, indicate stress, and forecast yields across large fields. Aerial platforms are very efficient and adaptable, although they are frequently less spatially resolved than ground-based techniques, and environmental factors like cloud cover can impact them. Large-scale crop breeding often uses airborne phenotyping platforms as opposed to ground-based ones. Additionally, the primary purpose of manned aircraft photos is to assess biomass, plant height, nitrogen content, chlorophyll content, canopy temperature, and structure, and hundreds of field plots (Sankaran et al. 2015). Large-scale field studies, varietal growth regulations, and genetic variations may be conducted using satellite imagery (Lobell et al. 2019). However, they cannot be used for crop experiment monitoring because of their low image resolution (Matese et al. 2015). Multispectral (MS) cameras installed on UAVs can be used to track changes in vegetation indices (VIs) over time and space. Using the photosynthetic reaction to incident light, VIs are spectral reflectance calculations that aid in determining the presence and condition of vegetation (Steven et al. 2024). For example, chlorophyll absorbs red light; healthy plants have low reflectance in the red band and high reflectivity in the infrared band. According to Khan et al. (2018), this can change in a stressed or diseased plant that may have less chlorophyll pigment. The accuracy of VIs obtained from hyperspectral or aerially acquired MS images to measure crop health, moisture, and nutrient content has been confirmed by several studies (Cuaran and Leon 2021; Goswami et al. 2021; Shi et al. 2016). Leaf area index, biomass, and chlorophyll content are also being estimated more often using VIs produced from MS sensors (Gano et al. 2021; Li et al. 2018; Potgieter et al. 2017; Shafian et al. 2018). Yu et al. (2016) studied yield estimations by using high-resolution MS imaging data gathered from a UAV-based high-throughput platform during a full soybean growing season.

7.4.2.3 Image-Based Technologies Used to Analyze Abiotic Stress

Phenomics provides more exact and reliable data, allowing researchers to better understand the impact of abiotic stress and its processes across different plant growth stages. Plant phenotyping characteristics can be visualized using spectral (reflectance, transmittance, and emission of electromagnetic radiation) and chemical fingerprinting. The bulk of the traits needed for plant breeding may be examined using X-ray, ultraviolet (UV), visible, infrared (IR), near-infrared (NIR), far-infrared (FIR), and microwave spectral fingerprints collected via radio metric sensors and image cameras. The visible range sensors evaluate variables such as early vitality, growth rate, biomass, phenology, and plant pigments (lycopene, anthocyanin, carotenoids, and chlorophyll). Hyperspectral sensors in the UV, NIR, and FIR ranges are used to determine chemical composition (chlorophyll, flavonoids, sugar, proteins, seed quality, etc.), water status, plant tissue temperature, abiotic stress response, transpiration, photosynthetic efficiency, radiation use efficiency, and other factors. Waterlogging and automated weighing and watering stations will evaluate the weight of containers before and after watering to simulate a drought. Therefore, important physiological features determining yield and stress tolerance may be rapidly assessed for a diverse variety of plants at predetermined development intervals (GroBkinsky et al. 2015). Image-based technologies used for HTP of plants under stress are mentioned in Table 7.3.

7.4.2.3.1 Visible Light Imaging

Visible light sensors offer red, green, and blue (RGB) reflected values after detecting light with wavelengths between 400 and 700 nm. HTP frequently uses visible light imaging due to its affordability, ease of use, and accessibility. Plant biomass (Kim et al. 2018), root architecture (Ye et al. 2018), plant growth rate (Gang et al. 2022), germination rate, and yield can all be precisely measured using high-resolution RGB images.

7.4.2.3.2 Fluorescence Imaging

It is possible to learn about plant metabolism by artificially stimulating the photosystems of plants and observing the corresponding reactions. When shorter wavelength radiations are absorbed, light is released, which is known as fluorescence. A portion of the absorbed light is re-emitted when blue actinic light strikes a chloroplast because the fluorescing component of plants is the chlorophyll complex. The amount of light that is re-emitted is determined by the way the plant can metabolize the light that is gathered. Actinic light absorption by the plant is shown by the fluorescent portion of the light that is re-emitted. Incorporating actinic light with short-term blue pulse saturation allows for the estimation of physiological factors such as non-photochemical quenching and plant photosynthetic efficiency. Fluorescence imaging records these indications of fluorescence.

7.4.2.3.3 Thermal Imaging

Thermal imaging is used to visualize infrared radiations, which show an object as a fluorescing component when the temperature of blue actinic light is uniform over its surface. Thermal cameras have a spectral range of 3 to 14 µm, with the most often utilized wavelengths being 3–5 µm and 7–14 µm. Abiotic stressors like heat, salt, and drought alter the stomata of plant leaves, which in turn alter transpiration and, ultimately, the plant's temperature. Accordingly, plant temperature variations may be used via thermal imaging to track stress levels (Patel et al. 2001; Munns et al. 2010).

TABLE 7.3
Image-Based Technologies Used for HTP

Image-based Technologies	Platform	Feature	Benefits	Constraints	References
RGB	Laboratory, fixed	Shoot biomass, leaf/shoot area, identify aging through color analysis and growth rate	Environmental control is possible, relatively affordable	Only provides morphological information	Ge et al. (2016), Neilson et al. (2015), Campbell et al. (2020)
	Field, fixed	Distinguish ground cover and plants	Quickly obtain information on breeding rows for large populations	Image blur and area mismatch errors occur due to changes in light conditions	Becker and Schmidhalter (2017)
	Field, aerial	Leaf area index (LAI), canopy length, wilting characteristics, and yield estimation			Bhandari et al. (2021), Francesconi et al. (2021)
	Field, vehicle	NDVI, yield estimation			Bai et al. (2016)
	Field, portable	Root biomass, canopy color, and canopy length			Nehe et al. (2021), Shabannejad et al. (2020)
NIR	Laboratory, fixed	Water content, canopy spectrum reflectance	Low water reflectance, simple use	Limited use with a narrow reflectance range	Chen et al. (2014), Das et al. (2021)
	Field, portable	Canopy reflectance			El-Hendawy et al. (2017)
Thermal	Field, aerial, vehicle	Canopy temperature	Able to analyze increases in plant temperature	Difficulty in distinguishing plants	Rischbeck et al. (2016), Sankaran et al. (2018), Tattaris et al. (2016), Romano et al. (2016)
Hyperspectral	Laboratory, fixed	Leaf water content, canopy spectrum	Precise data detection is possible using multiple bands	Difficulty in selecting and calibrating specific bands	Rischbeck et al. (2016), Ge et al. (2016)
	Field, aerial	Plant height, flowering date, yield estimation, NDVI			Sankaran et al. (2018), Tattaris et al. (2016), Romano et al. (2016), Condorelli et al. (2018), Elsayed et al. (2015)
	Field, vehicle	Canopy reflectance, solar radiation			Rischbeck et al. (2016), Elsayed et al. (2015)
Fluorescence	Laboratory, fixed	Chlorophyll fluorescence, gas exchange, transpiration, and stomatal conductance	Automatic and rapid measurement of photosynthetic rate	Difficult to use in the field requires dark adaptation	Bürling et al. (2013), Daszkowska-Golec et al. (2019), Pshenichnikova et al. (2019), Rane et al. (2019), Mikołajczak et al. (2020), Tschiersch et al. (2017), McAusland et al. (2019)
	Field, portable	Chlorophyll fluorescence			Begovic et al. (2020), Fernández-Calleja et al. (2020), Hao et al. (2012)

When using thermal imaging to monitor a drought-resistant genotype of maize, it was found that the average temperature difference between the various genotypes under water stress was greater than 2°C (Romano et al. 2011; Zia et al. 2013). This effectively demonstrated that drought-resistant genotypes are better suited to drought with lower temperatures.

7.4.2.3.4 Imaging Spectroscopy

In phenomics, imaging spectroscopy is used to remotely sense vegetation. Plant imaging spectroscopy uses the solar energy that plants reflect. Single leaves or canopies in the visible spectral region (400–700 nm) generate a low degree of reflection.

7.5 ADVANCED PHENOMICS TOOLS FOR ASSESSING ABIOTIC STRESS

The morphological, physiological, and biochemical properties of crops are commonly measured using distant and proximal sensing techniques. These properties may then be combined to provide quantitative traits that are helpful for selection in breeding programs. It has been shown that HTP is appropriate for screening out both temporary and prolonged abiotic stresses. Recently, HTP of root system architecture has become the focus in major legumes (Ye et al. 2018). In legumes such as pigeonpea, root system architecture (RSA) directly influences the amount of moisture extracted from surrounding soil layers, which in turn influences evapotranspiration, canopy characteristics, and plant health via controlling the interactions between plants and water and nutrients. Key phenotypic imaging technologies include chlorophyll florescence imaging, visible infrared imaging, hyperspectral imaging (HSI), and multispectral imaging (MSI). HSI can be used to examine drought tolerance and nutrient deficit in pigeonpea (Singh et al. 2019). Multispectral imaging can assist in identifying the properties of agricultural plants. The wavelength of light determines the color of the object in the scene. The suggested approach employs multispectral imaging to determine pigeonpea crop quality characteristics, such as chlorophyll content in leaves and water stress (Bagherian et al., 2023). Infrared/thermal imaging and fluorescence imaging are common phenotyping methods for canopy temperature and physiological characteristics, respectively.

A range of features, including canopy traits, temperature, leaf characteristics, biomass, Normalized Difference Vegetation Index (NDVI), and even the ability to analyze physiological responses to DS without causing damage, are provided by imaging systems in various crops, which range from basic RGB systems to multispectral and hyperspectral systems (Asaari et al., 2022). In water-stress investigations, NIR devices allow for automatic watering systems and monitoring of the water content of soil and tissue (Pieruschka and Schurr 2019). The best phenotyping method for measuring diverse stresses, such as salinity stress in various crops, cereals, and pulses, has recently been thought to be thermal imaging using LemnaTec's Scanalyzer 3D (Dissanayake et al. 2020; Pineda et al. 2020). Accordingly, RSA field phenotyping with HTP has been created for a number of crops, such as wheat, rice, and maize (Ye et al. 2018). In order to understand how plants respond to abiotic stressors and non-stress phenotyping, research on the HTP system also starts with certain legumes (Huang et al. 2022; Lazarević et al. 2022). Their growing field-level use in legumes makes them more significant for learning about DS and HS tolerance through RSA. For field-based root phenomics, HTP imaging analysis—such as digital imaging of root features (DIRT) has proven to be a successful method (Das et al. 2015).

This information from HTP will help model the above and below-ground component qualities in pigeonpea, which will aid in the selection of superior donors from large populations and the assessment of elite lines and their performance in target environments through prediction-based breeding. To assist this legume to develop tolerance to these important pressures, phenomic research has the potential to greatly advance our understanding of the quantitative genetics of such complex characteristics in pigeonpea, utilizing trait modeling and HTP data. Currently, several HTP technologies are employed to evaluate features related to both biotic and abiotic stress tolerance in crops, including pigeonpea. These systems allow for the automated and effective assessment of plant responses to stress. To determine tolerance to several abiotic stresses, for example, the "GROWSCREEN FLUORO" platform is used to phenotype leaf development and chlorophyll fluorescence, while the "PHENOPSIS" platform is utilized to phenotype plant responses to soil water stress. "HyperART" is utilized for nondestructive measurement of leaf attributes, including chlorophyll content and disease severity, while the "LemnaTec 3D Scanalyzer system" enables non-invasive screening of salt tolerance qualities. "PhenoBox" is used to identify how plants react to salt stress, while "PHENOVISION" is used to identify how plants react to DS. "PhénoField" is a platform used for characterizing various abiotic stresses under field conditions, and the "PlantScreen™ Robotic XYZ System" analyzes traits associated with drought tolerance.

7.6 SPECIALIZED HIGH-THROUGHPUT PHENOTYPING SYSTEMS TO STUDY DIFFERENT STRESSES

In addition, several HTP systems have been developed for specific traits in major crops like rice, wheat, maize, and sorghum. For root and shoot traits, platforms like "RootReader3D" (Clark et al. 2011) and "GROWSCREEN-Rhizo" (Nagel et al. 2012) have been applied. For aerial monitoring of crop traits, the "Zeppelin NT aircraft" has been employed (Liebisch et al. 2015). Other notable platforms include "Phenocart" for in-field phenotyping (Crain et al. 2018), "Phenovator" for rapid screening (Flood et al. 2016), and "PHENOARCH" for greenhouse-based assessments (Brichet et al. 2017). The "Field Scanalyzer" platform enables field-scale monitoring (Virlet et al. 2016), while "CropQuant" and "MVS-Pheno" provide advanced imaging for analyzing crop traits under stress conditions (Zhou et al. 2017; Wu et al. 2020). Different types of deep learning methods have also been proposed; these methods will help to rapidly detect stomata and provide an automated high-throughput analysis platform. For stomata analysis, deep learning is generally categorized into three models: object detection, the Semantic method (Zhao et al., 2019; Minaee et al., 2021), and the instance segmentation method (Hafiz and Bhat, 2020). Stomatal data are crucial for supporting the analysis of photosynthesis, which is significantly influenced by abiotic stress. These platforms hold significant potential for HTP of traits related to stress tolerance in a variety of crops, offering valuable insights for breeding programs aimed at improving stress resilience.

7.7 CONCLUSION

Global warming is causing dramatic changes in weather patterns and is detrimental to crop production. Extreme temperatures, moisture, photoperiod, and mineral-related stressors are the most important stresses that occur during the life cycle. By altering the crop's physiology and morphology, certain stresses reduce crop growth. HTP has proven to be a successful approach for analyzing agronomic traits and gaining insight into environmental factors for plant breeding. HTP has allowed plant breeders to screen superior germplasm with higher biotic and abiotic stress resilience. By thoroughly integrating phenotypes, genotypes, and other omics data sources, it would enable precision crop breeding by revealing the genetic regulatory networks and mechanisms underlying stress-related traits. To reduce the redundancy of manual phenotyping, many sensor-based, automated/semi-automated phenotyping facilities are being used. By integrating data from advanced imaging and sensor technologies, researchers can identify pigeonpea genotypes that exhibit superior stress tolerance. This accelerates the selection of resilient varieties, improves breeding efficiency, and helps create pigeonpea cultivars that can thrive in challenging environments.

REFERENCES

Alagupalamuthirsolai, M., & Vijayabharathi, A. (2020). Genetic diversity analysis of pigeonpea (*Cajanus cajan* [L.] Millsp.) under cold stress conditions. *Journal of Pharmacognosy and Phytochemistry*, *9*(3), 2048–2053.

Ali, A., Altaf, M. T., Nadeem, M. A., Karaköy, T., Shah, A. N., Azeem, H., … Chung, Y. S. (2022). Recent advancement in omics approaches to enhance abiotic stress tolerance in legumes. *Frontiers in Plant Science*, *13*, 952759.

Ali, F., Zhao, Y., Ali, A., Waseem, M., Arif, M. A., Shah, O. U., … Wang, Z. (2024). Omics-driven strategies for developing saline-smart lentils: A comprehensive review. *International Journal of Molecular Sciences*, *25*(21), 11360.

Anjum, S. A., Ashraf, U., Zohaib, A., Tanveer, M., Naeem, M., Ali, I., … Nazir, U. (2017). Growth and development responses of crop plants under drought stress: A review.

Araújo, S. S., Beebe, S., Crespi, M., Delbreil, B., González, E. M., Gruber, V., … Patto, M. C. V. (2015). Abiotic stress responses in legumes: Strategies used to cope with environmental challenges. *Critical Reviews in Plant Sciences*, *34*(1–3), 237–280. https://doi.org/10.1080/07352689.2014.898450

Arduini, I., Baldanzi, M., & Pampana, S. (2019). Reduced growth and nitrogen uptake during waterlogging at tillering permanently affect yield components in late sown oats. *Frontiers in Plant Science*, *10*, 1087. https://doi.org/10.3389/fpls.2019.01087

Asaari, M. S. M., Mertens, S., Verbraeken, L., Dhondt, S., Inzé, D., Bikram, K., & Scheunders, P. (2022). Non-destructive analysis of plant physiological traits using hyperspectral imaging. *Computers and Electronics in Agriculture*, *195*, 106806.

Atkinson, J. A., Pound, M. P., Bennett, M. J., & Wells, D. M. (2019). Uncovering the hidden half of plants using new advances in root phenotyping. *Current Opinion in Biotechnology*, *55*, 1–8.

Bagherian, K., Bidese-Puhl, R., Bao, Y., Zhang, Q., Sanz-Saez, A., Dang, P. M., … Chen, C. (2023). Phenotyping agronomic and physiological traits in peanut under mid-season drought stress using UAV-based hyperspectral imaging and machine learning. *The Plant Phenome Journal*, *6*(1), e20081.

Bai, G., Ge, Y., Hussain, W., Baenziger, P. S., & Graef, G. (2016). A multi-sensor system for high-throughput field phenotyping in soybean and wheat breeding. *Computers and Electronics in Agriculture*, *128*, 181–192.

Bansal, R., & Srivastava, J. P. (2017). Effect of waterlogging on root anatomy and nitrogen distribution in pigeonpea (*Cajanus cajan* [L.] Millsp.). *Indian Journal of Plant Physiology*, *22*(1), 130–134.

Basu, P. S., Singh, U. M., Kumar, A., Praharaj, C. S., & Shivran, R. K. (2016). Climate change and its mitigation strategies in pulses production. *Indian Journal of Agronomy*, *61*(4), 71–82.

Becker, E., & Schmidhalter, U. (2017). Evaluation of yield and drought using active and passive spectral sensing systems at the reproductive stage in wheat. *Frontiers in Plant Science*, *8*, 379.

Becklin, K. M., Anderson, J. T., Gerhart, L. M., Wadgymar, S. M., Wessinger, C. A., & Ward, J. K. (2016). Examining plant physiological responses to climate change through an evolutionary lens. *Plant Physiology*, *172*(2), 635–649.

Begovic, L., Galic, V., Abicic, I., Loncaric, Z., Lalic, A., & Mlinaric, S. (2020). Implications of intra-seasonal climate variations on chlorophyll a fluorescence and biomass in winter barley breeding program. *Photosynthetica*, *58*(4), 995–1008.

Bhandari, M., Baker, S., Rudd, J. C., Ibrahim, A. M., Chang, A., Xue, Q., … Auvermann, B. (2021). Assessing the effect of drought on winter wheat growth using unmanned aerial system-based phenotyping. *Remote Sensing*, *13*(6), 1144.

Bhat, K. A., Mahajan, R., Pakhtoon, M. M., Urwat, U., Bashir, Z., Shah, A. A., … Zargar, S. M. (2022). Low temperature stress tolerance: An insight into the omics approaches for legume crops. *Frontiers in Plant Science*, *13*, 888710.

Bohra, A., Mallikarjuna, N., Saxena, K., Upadhyaya, H. D., Vales, M. I., & Varshney, R. K. (2010). Harnessing the potential of crop wild relatives through genomics tools for pigeonpea improvement. *Journal of Plant Biology*, *37*(1), 83–98.

Bohra, A., Satheesh Naik, S. J., Kumari, A., Tiwari, A., & Joshi, R. (2021). Integrating phenomics with breeding for climate-smart agriculture. In *Omics technologies for sustainable agriculture and global food security* (Vol. II, pp. 1–24). Springer.

Bohra, A., Tiwari, A., Kaur, P., Ganie, S. A., Raza, A., Roorkiwal, M., … Varshney, R. K. (2022). The key to the future lies in the past: Insights from grain legume domestication and improvement should inform future breeding strategies. *Plant and Cell Physiology*, *63*(11), 1554–1572.

Brichet, N., Fournier, C., Turc, O., Strauss, O., Artzet, S., Pradal, C., … Cabrera-Bosquet, L. (2017). A robot-assisted imaging pipeline for tracking the growths of maize ear and silks in a high-throughput phenotyping platform. *Plant Methods*, *13*(1), 96.

Bürling, K., Cerovic, Z. G., Cornic, G., Ducruet, J. M., Noga, G., & Hunsche, M. (2013). Fluorescence-based sensing of drought-induced stress in the vegetative phase of four contrasting wheat genotypes. *Environmental and Experimental Botany*, *89*, 51–59.

Campbell, M. T., Grondin, A., Walia, H., & Morota, G. (2020). Leveraging genome-enabled growth models to study shoot growth responses to water deficit in rice. *Journal of Experimental Botany*, *71*(18), 5669–5679.

Chen, D., Neumann, K., Friedel, S., Kilian, B., Chen, M., Altmann, T., & Klukas, C. (2014). Dissecting the phenotypic components of crop plant growth and drought responses based on high-throughput image analysis. *The Plant Cell*, *26*(12), 4636–4655.

Choudhary, A. K., & Singh, D. (2011). Screening of pigeonpea genotypes for nutrient uptake efficiency under aluminium toxicity. *Physiology and Molecular Biology of Plants*, *17*(2), 145–152.

Choudhary, A. K., Singh, D., & Iquebal, M. A. (2011a). Selection of pigeonpea genotypes for tolerance to aluminium toxicity. *Plant Breeding*, *130*(4), 492–495.
Choudhary, A. K., Sultana, R., Pratap, A., Nadarajan, N., & Jha, U. C. (2011b). Breeding for abiotic stresses in pigeonpea. *Journal of Food Legumes*, *24*(3), 165–174.
Choudhary, J. R., Get, S., Tripathi, A., Kaldate, R., Rana, M., Mehta, S., … Wani, S. H. (2022). Breeding efforts for crop productivity in abiotic stress environment. In *Augmenting crop productivity in stress environment* (pp. 63–103). Springer Nature.
Clark, R. T., MacCurdy, R. B., Jung, J. K., Shaff, J. E., McCouch, S. R., Aneshansley, D. J., & Kochian, L. V. (2011). Three-dimensional root phenotyping with a novel imaging and software platform. *Plant Physiology*, *156*(2), 455–465.
Condorelli, G. E., Maccaferri, M., Newcomb, M., Andrade-Sanchez, P., White, J. W., French, A. N., … Tuberosa, R. (2018). Comparative aerial and ground-based high-throughput phenotyping for the genetic dissection of NDVI as a proxy for drought adaptive traits in durum wheat. *Frontiers in Plant Science*, *9*, 893.
Crain, J., Mondal, S., Rutkoski, J., Singh, R. P., & Poland, J. (2018). Combining high-throughput phenotyping and genomic information to increase prediction and selection accuracy in wheat breeding. *The Plant Genome*, *11*(1), 170043.
Cuaran, J., & Leon, J. (2021). Crop monitoring using unmanned aerial vehicles: A review. *Agricultural Reviews*, *42*(2), 121–132.
Das, A., Schneider, H., Burridge, J., Ascanio, A. K. M., Wojciechowski, T., Topp, C. N., … Bucksch, A. (2015). Digital imaging of root traits (DIRT): A high-throughput computing and collaboration platform for field-based root phenomics. *Plant Methods*, *11*(1), 51.
Das, B., Sahoo, R. N., Pargal, S., Krishna, G., Verma, R., Viswanathan, C., … Gupta, V. K. (2021). Evaluation of different water absorption bands, indices and multivariate models for water-deficit stress monitoring in rice using visible-near infrared spectroscopy. *Spectrochimica Acta Part A: Molecular and Biomolecular Spectroscopy*, *247*, 119104.
Daspute, A. A., Kobayashi, Y., Panda, S. K., Fakrudin, B., Tokizawa, M., … Koyama, H. (2018). Characterization of *CcSTOP1*; a C2H2-type transcription factor regulates Al tolerance gene in pigeonpea. *Planta*, *247*(1), 201–214.
Daszkowska-Golec, A., Collin, A., Sitko, K., Janiak, A., Kalaji, H. M., & Szarejko, I. (2019). Genetic and physiological dissection of photosynthesis in barley exposed to drought stress. *International Journal of Molecular Sciences*, *20*(24), 6341.
Deeplanaik, N., Kumaran, R. C., Venkatarangaiah, K., Shivashankar, S. K. H., Doddamani, D., & Telkar, S. (2013). Expression of drought responsive genes in pigeonpea and *in silico* comparison with soybean cDNA library. *Journal of Crop Science and Biotechnology*, *16*(4), 243–251.
Dissanayake, R., Kahrood, H. V., Dimech, A. M., Noy, D. M., Rosewarne, G. M., Smith, K. F., … Kaur, S. (2020). Development and application of image-based high-throughput phenotyping methodology for salt tolerance in lentils. *Agronomy*, *10*(12), 1992.
El-Hendawy, S. E., Hassan, W. M., Al-Suhaibani, N. A., & Schmidhalter, U. (2017). Spectral assessment of drought tolerance indices and grain yield in advanced spring wheat lines grown under full and limited water irrigation. *Agricultural Water Management*, *182*, 1–12.
Elsayed, S., Rischbeck, P., & Schmidhalter, U. (2015). Comparing the performance of active and passive reflectance sensors to assess the normalized relative canopy temperature and grain yield of drought-stressed barley cultivars. *Field Crops Research*, *177*, 148–160.
Fernandez-Calleja, M., Monteagudo, A., Casas, A. M., Boutin, C., Pin, P. A., Morales, F., & Igartua, E. (2020). Rapid on-site phenotyping via field fluorimeter detects differences in photosynthetic performance in a hybrid–parent barley germplasm set. *Sensors*, *20*(5), 1486.
Flood, P. J., Kruijer, W., Schnabel, S. K., van der Schoor, R., Jalink, H., Snel, J. F., … Aarts, M. G. (2016). Phenomics for photosynthesis, growth and reflectance in *Arabidopsis thaliana* reveals circadian and long-term fluctuations in heritability. *Plant Methods*, *12*(1), 14.

Food and Agriculture Organization of the United Nations. (2022). *FAOSTAT 2022*. http://www.fao.org/faostat/en/

Francesconi, S., Harfouche, A., Maesano, M., & Balestra, G. M. (2021). UAV-based thermal, RGB imaging and gene expression analysis allowed detection of Fusarium head blight and gave new insights into the physiological responses to the disease in durum wheat. *Frontiers in Plant Science*, *12*, 628575.

Gang, M. S., Kim, H. J., & Kim, D. W. (2022). Estimation of greenhouse lettuce growth indices based on a two-stage CNN using RGB-D images. *Sensors*, *22*(15), 5499.

Gano, B., Dembele, J. S. B., Ndour, A., Luquet, D., Beurier, G., Diouf, D., & Audebert, A. (2021). Using UAV-borne multispectral imaging for field phenotyping of shoot biomass, leaf area index and height of West African sorghum varieties under two contrasted water conditions. *Agronomy*, *11*(5), 850.

Garg, N., & Noor, Z. (2009). Genotypic differences in plant growth, osmotic and antioxidative defence of *Cajanus cajan* (L.) Millsp. modulated by salt stress. *Archives of Agronomy and Soil Science*, *55*(1), 3–33.

Ge, Y., Bai, G., Stoerger, V., & Schnable, J. C. (2016). Temporal dynamics of maize plant growth, water use, and leaf water content using automated high-throughput RGB and hyperspectral imaging. *Computers and Electronics in Agriculture*, *127*, 625–632.

Goswami, S., Choudhary, S. S., Chatterjee, C., Mailapalli, D. R., Mishra, A., & Raghuwanshi, N. S. (2021). Estimation of nitrogen status and yield of rice crop using unmanned aerial vehicle equipped with multispectral camera. *Journal of Applied Remote Sensing*, *15*(4), 042407.

Großkinsky, D. K., Svensgaard, J., Christensen, S., & Roitsch, T. (2015). Plant phenomics and the need for physiological phenotyping across scales to narrow the genotype-to-phenotype knowledge gap. *Journal of Experimental Botany*, *66*(18), 5429–5440.

Hafiz, A. M., & Bhat, G. M. (2020). A survey on instance segmentation: State of the art. *International Journal of Multimedia Information Retrieval*, *9*(3), 171–189.

Hanafy, M. S., El-Banna, A., Schumacher, H. M., Jacobsen, H. J., & Hassan, F. S. (2013). Enhanced tolerance to drought and salt stresses in transgenic faba bean (*Vicia faba* L.) plants by heterologous expression of the PR10a gene from potato. *Plant Cell Reports*, *32*(5), 663–674.

Hao, D., Chao, M., Yin, Z., & Yu, D. (2012). Genome-wide association analysis detecting significant single nucleotide polymorphisms for chlorophyll and chlorophyll fluorescence parameters in soybean (*Glycine max*) landraces. *Euphytica*, *186*(3), 919–931.

Hayat, K., Khan, A., Bibi, F., Murad, W., Fu, Y., Batiha, G. E. S., … Al-Harrasi, A. (2021). Effect of cadmium and copper exposure on growth, physiochemicals, and medicinal properties of *Cajanus cajan* L. *Metabolites*, *11*(11), 769.

Hingane, A. J., Saxena, K. B., Patil, S. B., Sultana, R., Srikanth, S., Mallikarjuna, N., … Kumar, C. V. S. (2015). Mechanism of waterlogging tolerance in pigeonpea. *Indian Journal of Genetics and Plant Breeding*, *75*(2), 208–214.

Huang, X., Zheng, S., & Zhu, N. (2022). High-throughput legume seed phenotyping using a handheld 3D laser scanner. *Remote Sensing*, *14*(2), 431.

Jaiswal, S. K., Naamala, J., & Dakora, F. D. (2018). Nature and mechanisms of aluminium toxicity, tolerance and amelioration in symbiotic legumes and rhizobia. *Biology and Fertility of Soils*, *54*(3), 309–318.

Jang, G., Kim, J., Yu, J. K., Kim, H. J., Kim, Y., Kim, D. W., … Chung, Y. S. (2020). Cost-effective unmanned aerial vehicle (UAV) platform for field plant breeding application. *Remote Sensing*, *12*(6), 998.

Jha, U. C., Bohra, A., Jha, R., & Parida, S. K. (2019). Salinity stress response and "omics" approaches for improving salinity stress tolerance in major grain legumes. *Plant Cell Reports*, *38*(3), 255–277.

Jin, X., Zarco-Tejada, P. J., Schmidhalter, U., Reynolds, M. P., Hawkesford, M. J., Varshney, R. K., … Li, S. (2020). High-throughput estimation of crop traits: A review of ground and aerial phenotyping platforms. *IEEE Geoscience and Remote Sensing Magazine*, *9*(1), 200–231.

Khan, Z., Rahimi-Eichi, V., Haefele, S., Garnett, T., & Miklavcic, S. J. (2018). Estimation of vegetation indices for high-throughput phenotyping of wheat using aerial imaging. *Plant Methods*, *14*(1), 20.

Khatun, M., Sarkar, S., Era, F. M., Islam, A. M., Anwar, M. P., Fahad, S., … Islam, A. A. (2021). Drought stress in grain legumes: Effects, tolerance mechanisms and management. *Agronomy*, *11*(12), 2374.

Khoury, C. K., Castañeda-Álvarez, N. P., Achicanoy, H. A., Sosa, C. C., Bernau, V., Kassa, M. T., … Struik, P. C. (2015). Crop wild relatives of pigeonpea (*Cajanus cajan* [L.] Millsp.): Distributions, ex situ conservation status, and potential genetic resources for abiotic stress tolerance. *Biological Conservation*, *184*, 259–270.

Kim, D. W., Yun, H. S., Jeong, S. J., Kwon, Y. S., Kim, S. G., Lee, W. S., & Kim, H. J. (2018). Modeling and testing of growth status for Chinese cabbage and white radish with UAV-based RGB imagery. *Remote Sensing*, *10*(4), 563.

Kumar, J., Singh, S., Singh, M., Srivastava, P. K., Mishra, R. K., Singh, V. P., & Prasad, S. M. (2017). Transcriptional regulation of salinity stress in plants: A short review. *Plant Gene*, *11*, 160–169.

Kumar, V., Singh, A., Mithra, S. A., Krishnamurthy, S. L., Parida, S. K., Jain, S., … Mohapatra, T. (2015). Genome-wide association mapping of salinity tolerance in rice (*Oryza sativa*). *DNA Research*, *22*(2), 133–145.

Kumawat, G., Raje, R. S., Bhutani, S., Pal, J. K., Mithra, A. S., Gaikwad, K., … Singh, N. K. (2012). Molecular mapping of QTLs for plant type and earliness traits in pigeonpea (*Cajanus cajan* L. Millsp.). *BMC Genetics*, *13*(1), 84.

Lazarević, B., Carović-Stanko, K., Živčak, M., Vodnik, D., Javornik, T., & Safner, T. (2022). Classification of high-throughput phenotyping data for differentiation among nutrient deficiencies in common bean. *Frontiers in Plant Science*, *13*, 931877.

Li, J., Shi, Y., Veeranampalayam-Sivakumar, A. N., & Schachtman, D. P. (2018). Elucidating sorghum biomass, nitrogen and chlorophyll contents with spectral and morphological traits derived from an unmanned aircraft system. *Frontiers in Plant Science*, *9*, 1406.

Li, L., Zhang, Q., & Huang, D. (2014). A review of imaging techniques for plant phenotyping. *Sensors*, *14*(11), 20078–20111.

Liebisch, F., Kirchgessner, N., Schneider, D., Walter, A., & Hund, A. (2015). Remote, aerial phenotyping of maize traits with a mobile multi-sensor approach. *Plant Methods*, *11*(1), 9.

Liu, Y., Li, J., Zhu, Y., Jones, A., Rose, R. J., & Song, Y. (2019). Heat stress in legume seed setting: Effects, causes, and future prospects. *Frontiers in Plant Science*, *10*, 938.

Lobell, D. B., Di Tommaso, S., You, C., Djima, I. Y., Burke, M., & Kilic, T. (2019). Sight for sorghums: Comparisons of satellite- and ground-based sorghum yield estimates in Mali. *Remote Sensing*, *12*(1), 100.

Matese, A., Toscano, P., Di Gennaro, S. F., Genesio, L., Vaccari, F. P., Primicerio, J., … Gioli, B. (2015). Intercomparison of UAV, aircraft and satellite remote sensing platforms for precision viticulture. *Remote Sensing*, *7*(3), 2971–2990.

McAusland, L., Atkinson, J. A., Lawson, T., & Murchie, E. H. (2019). High-throughput procedure utilising chlorophyll fluorescence imaging to phenotype dynamic photosynthesis and photoprotection in leaves under controlled gaseous conditions. *Plant Methods*, *15*(1), 109.

McDonald, G. K., & Paulsen, G. M. (1997). High temperature effects on photosynthesis and water relations of grain legumes. *Plant and Soil*, *196*(1), 47–58.

Mikołajczak, K., Ogrodowicz, P., Ćwiek-Kupczyńska, H., Weigelt-Fischer, K., Mothukuri, S. R., Junker, A., … Krajewski, P. (2020). Image phenotyping of spring barley (*Hordeum vulgare* L.) RIL population under drought. *Frontiers in Plant Science*, *11*, 743.

Minaee, S., Boykov, Y., Porikli, F., Plaza, A., Kehtarnavaz, N., & Terzopoulos, D. (2021). Image segmentation using deep learning: A survey. *IEEE Transactions on Pattern Analysis and Machine Intelligence*, *44*(7), 3523–3542.

Mishra, S., Singh, R., Kumar, R., Kalia, A., & Panigrahy, S. R. (2017). Impact of climate change on pigeon pea. *Economic Affairs*, *62*(3), 455–457.

Mousavi-Derazmahalleh, M., Bayer, P. E., Hane, J. K., Valliyodan, B., Nguyen, H. T., Nelson, M. N., … Edwards, D. (2019). Adapting legume crops to climate change using genomic approaches. *Plant, Cell & Environment*, *42*(1), 6–19.

Munns, R., James, R. A., Sirault, X. R., Furbank, R. T., & Jones, H. G. (2010). New phenotyping methods for screening wheat and barley for beneficial responses to water deficit. *Journal of Experimental Botany*, *61*(13), 3499–3507.

Nagel, K. A., Putz, A., Gilmer, F., Heinz, K., Fischbach, A., Pfeifer, J., … Schurr, U. (2012). GROWSCREEN-Rhizo: A novel phenotyping robot enabling simultaneous measurements of root and shoot growth. *Functional Plant Biology*, *39*(11), 891–904.

Nehe, A. S., Foulkes, M. J., Ozturk, I., Rasheed, A., York, L., Kefauver, S. C., … Morgounov, A. (2021). Root and canopy traits explain drought tolerance in winter wheat. *PLOS ONE*, *16*(4), e0242472.

Neilson, E. H., Edwards, A. M., Blomstedt, C. K., Berger, B., Møller, B. L., & Gleadow, R. M. (2015). High-throughput shoot imaging to examine dynamic phenotypic responses of a C4 cereal crop. *Journal of Experimental Botany*, *66*(7), 1817–1832.

Ningombam, L., Hazarika, B. N., Yumkhaibam, T., Heisnam, P., & Singh, Y. D. (2024). Heavy metal priming plant stress tolerance deciphered through physiological, biochemical, molecular and omics mechanisms. *South African Journal of Botany*, *168*, 16–25.

Pan, J., Sharif, R., Xu, X., & Chen, X. (2021). Mechanisms of waterlogging tolerance in plants. *Frontiers in Plant Science*, *11*, 627331.

Parwez, R., Naeem, M., Aftab, T., Ansari, A. A., Gill, S. S., & Gill, R. (2022). Heavy metal toxicity and tolerance in medicinal legumes. In *Hazardous and trace materials in soil and plants* (pp. 141–177). Academic Press.

Patel, N. R., Mehta, A. N., & Shekh, A. M. (2001). Canopy temperature and water stress quantification in rainfed pigeonpea. *Agricultural and Forest Meteorology*, *109*(3), 223–232.

Pazhamala, L., Saxena, R. K., Singh, V. K., Sameerkumar, C. V., Kumar, V., Sinha, P., … Varshney, R. K. (2015). Genomics-assisted breeding in pigeonpea. *Frontiers in Plant Science*, *6*, 50.

Pieruschka, R., & Schurr, U. (2019). Plant phenotyping: Past, present, and future. *Plant Phenomics*.

Pineda, M., Barón, M., & Pérez-Bueno, M. L. (2020). Thermal imaging for plant stress detection and phenotyping. *Remote Sensing*, *13*(1), 68.

Potgieter, A. B., George-Jaeggli, B., Chapman, S. C., Laws, K., Suárez Cadavid, L. A., Wixted, J., … Hammer, G. L. (2017). Multispectral UAV imaging for assessing seasonal leaf area dynamics in sorghum. *Frontiers in Plant Science*, *8*, 1532.

Priyanka, B., Sekhar, K., Reddy, V. D., & Rao, K. V. (2010). Expression of pigeonpea hybrid proline-rich protein gene confers abiotic stress tolerance. *Plant Biotechnology Journal*, *8*(1), 76–87.

Pshenichnikova, T. A., Doroshkov, A. V., Osipova, S. V., Permyakov, A. V., Permyakova, M. D., Efimov, V. M., & Afonnikov, D. A. (2019). Pubescence traits in wheat associated with photosynthesis under water stress. *Planta*, *249*(3), 839–847.

Ramakrishna, G., Kaur, P., Singh, A., Yadav, S. S., Sharma, S., Singh, N. K., & Gaikwad, K. (2021). Comparative transcriptome analysis of heat stress responses in pigeonpea. *Plant Cell Reports*, *40*(5), 881–898.

Rane, J., Sharma, D., Ekatpure, S., Aher, L., Kumar, M., Prasad, S. V. S., … Singh, N. P. (2019). Photosystem II tolerance in wheat under desiccation. *Photosynthetica*, *57*(4), 1243–1253.

Raza, A., Tabassum, J., Zahid, Z., Charagh, S., Bashir, S., Barmukh, R., … Varshney, R. K. (2022). Advances in omics approaches for improving toxic metal tolerance. *Frontiers in Plant Science*, *12*, 794373.

Rischbeck, P., Elsayed, S., Mistele, B., Barmeier, G., Heil, K., & Schmidhalter, U. (2016). Data fusion for yield prediction in drought-stressed barley. *European Journal of Agronomy*, *78*, 44–59.

Romano, G., Zia, S., Spreer, W., Sanchez, C., Cairns, J., Araus, J. L., & Müller, J. (2011). Thermography for high-throughput phenotyping of maize. *Computers and Electronics in Agriculture*, *79*(1), 67–74.

Romano, M., Chulkov, A., Sommier, A., Balageas, D., Vavilov, V., Batsale, J. C., & Pradere, C. (2016). Broadband sub-terahertz camera based on photothermal conversion. *Journal of Infrared, Millimeter, and Terahertz Waves*, *37*(5), 448–461.

Roy, A., Kolady, D., Paudel, B., Yumnam, A., Mridha, N., Chakraborty, D., & Singh, N. U. (2021). Climate change trends in North-Eastern India. *Journal of Environmental Biology*, *42*(6), 1415–1424.

Sandhu, K., Patil, S. S., Pumphrey, M., & Carter, A. (2021). Multitrait machine- and deep-learning models for genomic selection. *The Plant Genome*, *14*(3), e20119.

Sankaran, S., Khot, L. R., Espinoza, C. Z., Jarolmasjed, S., Sathuvalli, V. R., Vandemark, G. J., … Pavek, M. J. (2015). Low-altitude aerial imaging systems for crop phenotyping. *European Journal of Agronomy*, *70*, 112–123.

Sankaran, S., Zhou, J., Khot, L. R., Trapp, J. J., Mndolwa, E., & Miklas, P. N. (2018). High-throughput phenotyping in dry bean using UAV imagery. *Computers and Electronics in Agriculture*, *151*, 84–92.

Saxena, K., Bohra, A., Choudhary, A. K., Sultana, R., Sharma, M., Pazhamala, L. T., & Saxena, R. K. (2021). Alternative breeding approaches in pigeonpea. *Plant Breeding*, *140*(1), 74–86.

Saxena, K., Choudhary, A. K., Srivastava, R. K., Bohra, A., Saxena, R. K., & Varshney, R. K. (2019). Origin of early maturing pigeonpea germplasm. *Plant Breeding*, *138*(3), 243–251.

Sekhar, K., Priyanka, B., Reddy, V. D., & Rao, K. V. (2010). Overexpression of pigeonpea cyclophilin confers abiotic stress tolerance. *Plant, Cell & Environment*, *33*(8), 1324–1338.

Shabannejad, M., Bihamta, M. R., Majidi-Hervan, E., Alipour, H., & Ebrahimi, A. (2020). High-throughput image analysis improves genomic prediction accuracy. *Plant Methods*, *16*(1), 146.

Shafian, S., Rajan, N., Schnell, R., Bagavathiannan, M., Valasek, J., Shi, Y., & Olsenholler, J. (2018). UAV-based remote sensing for monitoring sorghum. *PLOS ONE*, *13*(5), e0196605.

Sher, A., Noor, M. A., Li, H. X., Nasir, B., Manzoor, M. A., Hussain, S., … Hussain, S. (2024). Heat stress effects on legumes. *Plant Stress*, *13*, 100537.

Shi, Y., Thomasson, J. A., Murray, S. C., Pugh, N. A., Rooney, W. L., Shafian, S., … Yang, C. (2016). UAVs for high-throughput phenotyping. *PLOS ONE*, *11*(7), e0159781.

Shinozaki, K., & Yamaguchi-Shinozaki, K. (2007). Gene networks involved in drought stress response. *Journal of Experimental Botany*, *58*(2), 221–227.

Singh, A., Singh, P. K., Sharma, A. K., Singh, N. K., Sonah, H., Deshmukh, R., & Sharma, T. R. (2019). WRKY gene family in pigeonpea under stress. *Plants*, *8*(7), 214.

Singh, B., Mishra, S., Bohra, A., Joshi, R., & Siddique, K. H. M. (2018). Crop phenomics for abiotic stress tolerance. In *Biochemical, physiological and molecular avenues for combating abiotic stress* (pp. 277–296). Academic Press.

Singh, G., Gudi, S., Amandeep, Upadhyay, P., Shekhawat, P. K., Nayak, G., … Ayoubi, H. (2022). Unlocking hidden variation from wild legumes. *Frontiers in Plant Science, 13*, 1035878.

Singh, N. (2023). Abiotic stresses in pigeonpea. In *Plant abiotic stress responses and tolerance mechanisms*. IntechOpen.

Singh, N., Rai, V., & Singh, N. K. (2020). Multi-omics strategies to enhance seed quality in pigeonpea. *The Nucleus, 63*(3), 249–256.

Sinha, P., Pazhamala, L. T., Singh, V. K., Saxena, R. K., Krishnamurthy, L., Azam, S., … Varshney, R. K. (2016). Drought-responsive genes in pigeonpea. *Frontiers in Plant Science, 6*, 1065.

Steven, M. D., Malthus, T. J., & Baret, F. (2024). Toward standardization of vegetation indices. In *Remote sensing handbook* (Vol. I, pp. 338–364). CRC Press.

Subbarao, G. V., & Johansen, C. (1991). Potential for genetic improvement in salinity tolerance in pigeonpea.

Sultana, R., Choudhary, A. K., Pal, A. K., Saxena, K. B., Prasad, B. D., & Singh, R. (2013). Abiotic stresses in major pulses. In *Approaches to plant stress and their management* (pp. 173–190). Springer India.

Tamirisa, S., Vudem, D. R., & Khareedu, V. R. (2014). Overexpression of pigeonpea cold and drought regulatory gene confers stress tolerance. *Journal of Experimental Botany, 65*(17), 4769–4781.

Tamirisa, S., Vudem, D. R., & Khareedu, V. R. (2017). Cyclin-dependent kinase regulatory subunit gene confers stress tolerance. *Frontiers in Plant Science, 8*, 165.

Tattaris, M., Reynolds, M. P., & Chapman, S. C. (2016). Comparison of remote sensing approaches for high-throughput phenotyping. *Frontiers in Plant Science, 7*, 1131.

Tayyab, A., Azeem, M., Qasim, M., Ahmed, N., & Ahmad, R. (2016). Salt stress responses of pigeon pea (*Cajanus cajan*) on growth, yield and some biochemical attributes. *Pakistan Journal of Botany, 48*(4), 1353–1360.

Tayyab, M., Sanaullah, M., & Wakeel, A. (2023). Sodium substitutes potassium requirements of sugar beet. *Journal of Plant Nutrition and Soil Science, 186*(4), 464–472.

Tschiersch, H., Junker, A., Meyer, R. C., & Altmann, T. (2017). Automated kinetic chlorophyll fluorescence analyses. *Plant Methods, 13*(1), 54.

Upadhyaya, H. D. (2006). Improving pigeonpea with the wild. *SATrends*, 62, 1.

Varshney, R. K., Chen, W., Li, Y., Bharti, A. K., Saxena, R. K., Schlueter, J. A., … Jackson, S. A. (2012). Draft genome sequence of pigeonpea. *Nature Biotechnology, 30*(1), 83–89.

Virlet, N., Sabermanesh, K., Sadeghi-Tehran, P., & Hawkesford, M. J. (2016). Field Scanalyzer robotic phenotyping platform. *Functional Plant Biology, 44*(1), 143–153.

Wu, S., Wen, W., Wang, Y., Fan, J., Wang, C., Gou, W., & Guo, X. (2020). MVS-Pheno phenotyping platform. *Plant Phenomics*.

Ye, H., Roorkiwal, M., Valliyodan, B., Zhou, L., Chen, P., Varshney, R. K., & Nguyen, H. T. (2018). Genetic diversity of root system architecture in legumes. *Journal of Experimental Botany, 69*(13), 3267–3277.

Yohan, Y., Umamahesh, V., Sudhakar, P., Reddiramu, Y., & Sreedhar, D. (2017). Influence of waterlogging on pigeonpea. *International Journal of Pure and Applied Bioscience, 5*(4), 1862–1868.

Yokotani, N., Sato, Y., Tanabe, S., Chujo, T., Shimizu, T., Okada, K., … Nishizawa, Y. (2013). WRKY76 roles in rice stress tolerance. *Journal of Experimental Botany, 64*(16), 5085–5097.

Yu, N., Li, L., Schmitz, N., Tian, L. F., Greenberg, J. A., & Diers, B. W. (2016). UAV-based soybean yield estimation. *Remote Sensing of Environment*, *187*, 91–101.

Zhao, Z. Q., Zheng, P., Xu, S. T., & Wu, X. (2019). Object detection with deep learning: A review. *IEEE Transactions on Neural Networks and Learning Systems*, *30*(11), 3212–3232.

Zhou, J., Reynolds, D., Cornu, T. L., Websdale, D., Orford, S., Lister, C., … Griffiths, S. (2017). CropQuant phenotyping platform. *bioRxiv*, 161547.

Zia, S., Romano, G., Spreer, W., Sanchez, C., Cairns, J., Araus, J. L., & Müller, J. (2013). Infrared thermal imaging for identifying water-stress tolerant maize. *Journal of Agronomy and Crop Science*, *199*(2), 75–84.

8 Pigeonpea Pest Dynamics in a Changing Climate Scenario

GK Sujayanand, Anup Chandra, and Asish K. Rout

8.1 INTRODUCTION

Pigeonpea, *Cajanus cajan* (L.) Millsp. is an important food legume crop cultivated in India and in many semiarid tropical countries. It is mainly cultivated in rainfed areas where assured irrigation is not available. In some pockets, it is also grown as a mixed or intercrop with sorghum or bajra. The maturity of pigeonpea varies according to the agro-ecology where it is grown as short duration, medium duration, and long duration (Bohra et al. 2020; Saxena et al. 2019, 2021). The insect-pest scenario and its dominance also vary according to the maturity groups. Further, India ranks first in the area (81%) and production (79%) of pigeonpea in the world (FAOSTAT 2022). The productivity of pigeonpea in India during 2022 was quite lower (861 kg/ha) than the global average of 883 kg/ha and the productivity of Tanzania (1127 Kg/ha) and Malawi (1698 Kg/ha; Chandra et al. 2025). The biotic constraints like insect pests and climate change pose a pressing challenge in enhancing the productivity of pigeonpea. The shifting of monsoon, increasing heat stress, and moisture-deficient soils due to climate change have profoundly impacted grain legume production in the tropics rather than temperate regions of the world. The abiotic factors directly affect the pulse crop productivity and indirectly affect the pest population dynamics. Insect-pest population dynamics focuses on the study of changes in the insect population and their different stages. Further, it attributes the various abiotic factors or biotic factors responsible for the changes in the insect population. The insect-pest population incidence and its dynamics were greatly influenced by environmental factors (rainfall, temperature, humidity, wind speed, etc.), availability of host crops, and presence or absence of natural enemies (predators, parasitoids, and pathogens). The monsoon vagaries further complicate the adversaries of climate change on host crop and pest phenology, distribution, etc. Therefore, it is a matter of urgency to have

DOI: 10.1201/9781003603092-8

a strategically planned integrated crop health management for minimizing the insect-pest damage on pulse crops.

8.2 CLIMATIC FACTORS INFLUENCE ON PEST DYNAMICS

The projected temperature rise in India is 1.7°C to 2°C by the 2030s and 3.3°C to 4.8°C by the 2080s over the preindustrial period (Chaturvedi et al. 2012). The increasing temperature and carbon dioxide will affect the insect-pest interaction with crop plants. For example, with the shrinking of winter duration, there may be an early onset of insect-pest infestation in long-duration pigeonpea. Increased temperature, UV radiation, low relative humidity, and rainfall anomalies render many of the biocontrol agents or biopesticides ineffective (Sharma et al. 2015). Moreover, temperature and rainfall affect the number of generations or seasonal abundance of insect pests as temperature influences their host availability and its abundance. Thus, the sudden increase in minimum temperature (>5°C) during the seventh to eighth standard meteorological week (SMW), increases abundance and infestation of the pod borer, *H. armigera* larvae in pigeonpea during 11–12th SMW (Vishwadhar et al. 2007; 2008). Similarly, other factors like rainfall during June to September and November play a pivotal role in forecasting the pod borer in Andhra Pradesh and Karnataka, whereas humidity and temperature during February decides the peak population of pod borer during March in Punjab (Srivastava et al. 2010). Furthermore, this group reported that pod borer is more adapted to ephemeral environments and its abundance is mainly determined by abiotic factors rather than biotic factors. Likewise, Jakhar et al. (2016) reported that maximum temperature, rainfall, and relative humidity were the major abiotic factors that were significantly correlated with *H. armigera* population at Sardarkrushinagar in Gujarat. The abiotic factor, like temperature, is reported to have a strong influence on viability and incubation time of *H. armigera* eggs (Dhillon and Sharma 2007). Most importantly, it has been observed that the developmental time of *H. armigera* decreases when there is a slight temperature increase of up to 5°C (Barteková and Praslička 2006). The pod fly, *Melanagromyza obtusa* incidence in pigeonpea mainly hinges on the prevailing temperature and humidity of 45th SMW of preceding year to the 4th SMW of the succeeding year (Singh et al. 2016). Notably, *Grapholita critica* Meyerick incidence in pigeonpea has a strong positive correlation with morning relative humidity (Ugale et al. 2021). The infestation was recorded from 36th SMW to 48th SMW with a peak infestation of 1.98 larvae per plant during 42nd SMW. Interestingly, Sharma et al. (2015) reported that *G. critica* incidence was lower in crops planted in September (1.36/plant) than in crops sown during June–July (2.03–2.79/plant).

8.3 INSECT PEST OF PIGEONPEA

Numerous insect pests attack pigeonpea and its pest succession varies according to the maturity group. The brief details of major insect pests of pigeonpea, including the stage of crop infestation and yield damage are given in Table 8.1. Besides, the maturity group-wise important insect pests of pigeonpea are elaborated in Table 8.2.

TABLE 8.1
Insect-Pest Infesting Pigeonpea and Their Damage

S.N.	Classification Based on Feeding Damage	Common Name	Scientific Name	Stage of Crop Infested by Insect Pest	Yield/Pod Damage
1	Pod borer complex	Legume pod borer	*Maruca vitrata* F.	Flowering and podding stage	19.50% to 24.50% Taggar et al. (2019)
		Pod borer	*Helicoverpa armigera* Hubner	Flowering and podding stage	34.90% Muchhadiya et al. (2024)
		Plume moth	*Exelastis atomosa* (Walsingham) *Sphenarches anisodactylus* Wlk.	Flowering and podding stage	5.60% Muchhadiya et al. (2024)
		Spiny pod borer	*Etiella zinckinella* (Treitschke)	Flowering and podding stage	1.5 larvae / plant Mandi et al. (2020)
		Blue butterfly	*Lampides boeticus* L.	Flowering and podding stage	6.38% Patra et al. (2016)
		Pod wasp	*Tanastigmodes cajaninae* La Salle	Flowering and podding stage	2.5% to 29.4% Durairaj (2010)
		Pod fly	*Melanagromyza obtusa* Malloch	Flowering and podding stage	23.64% Muchhadiya et al. (2024)
2	Sucking pest	Pod bug	*Clavigralla gibbosa* Spinola *Clavigralla scutellaris* (Westwood) *Riptortus pedestris* (Fabricius) *Nezara viridula (L.)*	Podding stage until harvesting	29.75% Muchhadiya et al. (2024)
		Cowbug	*Otinotus oneratus* W. *Oxyrachis tarandus* F.	Vegetative stage, flowering stage, and podding stage	
		Mealybug	*Coccidohystrix insolita* (Green)	Vegetative stage, flowering stage, and podding stage	
		Aphids	*Aphis craccivora* Koch	Vegetative stage, flowering stage, and podding stage	
		Eriophid mite	*Aceria cajani* Channabasavanna	Vegetative stage, flowering stage, and podding stage	

(Continued)

TABLE 8.1 (Continued)
Insect-Pest Infesting Pigeonpea and Their Damage

S.N.	Classification Based on Feeding Damage	Common Name	Scientific Name	Stage of Crop Infested by Insect Pest	Yield/Pod Damage
3	Defoliators	Leaf webber	*Grapholita critica* Meyrick	Vegetative stage, flowering stage, and podding stage	8.7 to 16.2 webs per 50 plants Vennila et al. (2019)
		Bihar hairy caterpillar	*Spilosoma obliqua* Walker	Vegetative stage, flowering stage, and podding stage	
		Hairy caterpillar	*Euproctis sp*	Vegetative stage, flowering stage	
4	Flower and bud devourer	Blister beetle	*Mylabris pustulata* Thunberg	Budding and Flowering stage	38.53% Anantharaju, and Muthiah (2008) 46.77% Singh et al. (2022)
5	Roots	Termites	*Odontotermes obesus*	Any stage during moisture stress	

TABLE 8.2
Key Insect Pests of Pigeonpea as per Its Maturity Group

S.N.	Zone	States	Long Duration (270 Days to Maturity)	Short Duration (180 Days to Maturity)
1	North East plain zone (NEPZ)	Assam, Bihar, west Bengal, Jharkhand, eastern U.P.	*M. obtusa, H. armigera, Apion sp*	*M. vitrata, H. armigera, G. critica, L. boeticus, E. atomosa, Mylabris spp*, and *C. gibbosa*
2	North West plain zone (NWPZ)	Western U.P., Punjab, Haryana, Rajasthan	*M. obtusa, H. armigera, Mylabris spp.*	*M. vitrata, H. armigera, G. critica, L. boeticus, E. atomosa, Mylabris spp*, and *C. gibbosa*
3	Central zone (CZ)	M.P., Chhattisgarh, Gujarat, parts of Rajasthan, Maharashtra,	*H. armigera, M. obtuse*	*C. gibbosa, H. armigera, L. boeticus, G. critica*
4	South zone (SZ)	Karnataka, Kerala, Andhra Pradesh, Telangana, Tamilnadu, and part of Madhya Pradesh, Odisha	*H. armigera, M. obtuse*	*M. vitrata, H. armigera, L. boeticus, G. critica, Mylabris spp, R. pedestris*, and *C. gibbosa*

Source: Srivastava and Joshi (2011).

8.4 KEY INSECT PESTS AND THEIR DAMAGE

The pigeonpea flowers and pods are the economic portion and are attacked by many insect pests. Among them, the pod borer complex comprising gram pod borer, spotted pod borer, spiny pod borer, plume moth, blue butterfly, and pod fly are the key insect pests that cause severe economic yield loss in pigeonpea. The detailed damage symptoms of these key insect pests are as follows:

Gram pod borer: The *Helicoverpa armigera* larvae infest the flowers and pods affecting the yield. The color of the larvae varies from yellow to green (Figures 8.1a, 8.1b, and 8.1c). A pod borer larva can damage up to 20 to 25 pods during its larval duration. The neonates scrap the young leaves or eat the young buds or sepals of flowers, while the third instar and above larvae feed on the seeds in developing pods by making big circular holes and entering into each locule. The larva develops into a pupa after completing 17 to 23 days. The pupa is dark brown in color, and it is formed under the soil. The adult males have greenish gray forewings, whereas in the females, these are yellowish to orange (Figure 8.2a).

FIGURE 8.1 Key insect pest of pigeonpea with damage symptom.

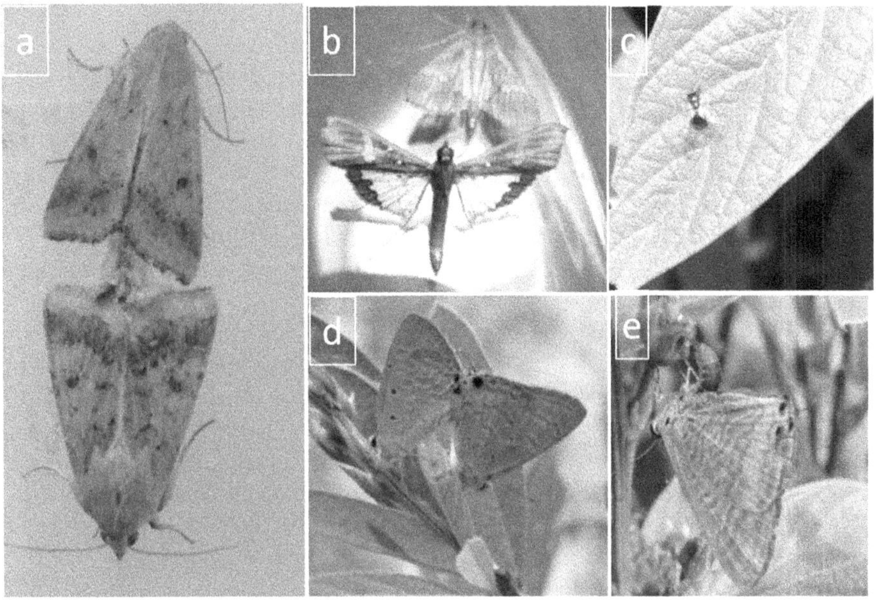

FIGURE 8.2 Adult stage of few key insect pests of pigeonpea.

Legume pod borer or spotted pod borer: The legume pod borers, *M. vitrata*, causes a typical damage symptom of webbing together of leaves, buds, flowers, pods, and shoot tips by frass. The larva damages the flowers or flower buds and or developing seeds inside the pods webbed in the frass; hence, it remains protected from the outer environment and insecticidal spray. The damage of *M. vitrata* is more pronounced in short-duration pigeonpea with a determinate growth habit and cluster-bearing types. The spotted pod borer is spread throughout pigeonpea growing regions in India. The larval stage is the damaging one, and its infestation results in drastic yield reduction by flower drop, reduced seed set, damaged pods, and secondary fungal infection in infested pods. The neonate larva is cream colored, and it feeds on flower buds (Figure 8.1d) or tender pods (Figure 8.1f), while the third instar has two dark black spots on each segment of the body except the head, which is dark black in color (Figure 8.1e). The pre-pupal stage is pale greenish in color. The pupae remain inside the larval web. An adult moth has three white spotted brown forewings and hind wings with distal brown markings (Figure 8.2b). Their wingspan is about 20 to 25 mm. The adults are nocturnal in habit and are strong fliers. They lay eggs singly near sepals or petals or flower buds or the underside of leaves. Some of the genotypes like Pusa 2001, Pusa 33, AL-15 and LRG-30, were found to be resistant based on the lowest pest resistance susceptibility rating (PRSR $\leq$ 2) recorded in three years of field screening (Sujayanand et al. 2025b). Further, the phylogenetic analysis of the *cox 1* gene of adult *M. vitrata* showed that the Indian population has shared a common ancestral origin.

Spiny pod borer: The spiny pod borer, *E. zinckenella*, primarily damages pods by webbing them together and feeding on developing seeds. Initially, the larva is pale cream in color with sparse hairs (Figure 8.1j). The mature larvae are pale green in color and turn into a dull red just a day before pupation. It generally forms a cocoon inside the soil. The adult is a slender moth that has brownish gray forewings with an orange prothorax. Forewings have a white strip along the anterior margin.

Plume moth: The plume moth species, viz., *Sphenarches anisodactylus* Walker and *Exelastis atomosa* Walsingham larvae infest developing seeds in pods. The larva is slender and pale green in color. They eat the seeds by making holes in the pods. The pupa is almost like a larva, but brown in color. The adult is grayish brown and slender in shape and the female lays green oval eggs singly on buds and pods. Four weeks are needed for completion of a generation.

Blue butterfly: Two different species infest the pigeonpea pods, viz., *Euchrysops cnejus* Fabricius (Figure 8.2d) and *Lampides boeticus* Linnaeus (Figure 8.2e). The eggs are singly laid near flower buds on sepals or petals, and it is slightly blue in color (Figure 8.2e). The larvae are greenish-blue, flat, and oval in shape (Figure 8.1h). The larvae are apodous and move like a slug. The larvae bore into green pods and damage the developing seeds. They pupate in soil or plant debris. The adults have metallic blue forewings and black spots on hind wings with a tail. The ventral side of the wings has many stripes and brown spots.

Pod fly: The adult is a small, shiny black fly (Figure 8.2c). It lays eggs inside developing pods by using its sharp ovipositor, and the neonates on hatching start feeding on the young developing green seeds by mining and making galleries in them (Figure 8.1g).

8.5 EMERGING INSECT PEST

The pod bugs are emerging as a serious problem in pigeonpea growing areas. The complex consists of three genus-infesting developing pods, viz., *Clavigralla gibbosa* (Figure 8.3d), *C. scutellaris* (Figure 8.3e), *Riptortus pedestris* (Figure 8.3c), *R. linearis*, and *Nezara viridulla* (Figure 8.3a and 8.3b). The occurrence of all three genera in Andhra Pradesh has been reported revealing prevalence of *Clavigralla* spp from 42nd to 4th SMW with a peak population of 7.6 bugs/plant during 52nd SMW. Similarly, *Riptortus spp* incidence was recorded from 31st to 5th SMW with a peak population of 2.13 bugs/plant during 52nd SMW. The detailed biology of this coreid pod bug was described by Sujayanand et al. (2025a). The green stink bug, *N. viridulla*, prevailed from 31st to 1st SMW with peak population of 0.97 bugs/plant during 43rd SMW (Udayababu et al. 2021). In the recent past, it has been noticed that the pod borer population has declined in Gujarat due to the abiotic factors like maximum temperature, rainfall, and relative humidity, whereas pod bug population *Clavigralla gibbosa* (Figure 8.3d) has increased (Jakhar et al. 2016). The mean pod bug incidence in pigeonpea has increased from 0.86 bugs/plant to 1.91 bugs/plant in a period of three years from 2011 to 2014 (Jakhar et al. 2017). Similarly, spotted pod borer, *Maruca vitrata* and pod bug, *C. gibbosa* inflicted heavy yield loss in pigeonpea (Sujithra and Chander 2014).

Sharma et al. (2015) observed that delayed sowing of pigeonpea, particularly in the month of August, changed the moderate resistance cultivar susceptible to pod borer (*H. armigera*). This group has also highlighted that the dry long spell of weather in western India favors the infestation of mealy bug, *Ceroplastodes cajani*

FIGURE 8.3 Pod bug complex of pigeonpea.

Maskell in pigeonpea. Leaf webber, *G. critica*, is also emerging as a major insect pest in many pigeonpea growing areas, including NEPZ, CZ, and SZ. During the last decade, a parasitoid *Apanteles mohandasi* Sumodan and Narendran (Braconidae: Hymenoptera) on leaf webber was first reported in Uttar Pradesh (previously reported only in Kerala; Gupta et al. 2011). Similarly, *B. relativus* was recorded for the first time as a larval parasitoid of spotted pod borer, *M. vitrata* in Karnataka apart from the common parasitoids—namely, *Phanerotoma hendecasisella* Cameron, 1905 (Braconidae: Cheloninae), and *Trathala flavoorbitalis* (Cameron, 1907; Ichneumonidae: Cremastinae; Gupta et al. 2013). Thus, the observation of new parasitoids might be due to the increasing geographic expansion of insect pests.

8.6 CHANGING PEST DYNAMICS IN PIGEONPEA

The insect-pests infestation and its developmental period are highly influenced by prevailing temperature, humidity, rainfall, and host availability. Usually, pest outbreaks occur frequently in stressed plants, as they have a weaker plant defense system compared to normal plants. For effective management of insect pests, continuous monitoring of pest incidence and its period of peak infestation is a critical indicator of whether the insect has crossed the economic threshold level (ETL) or not. Based on pest population dynamics, right management strategy can be implemented. The major insect pest population dynamics in pigeonpea has been summarized in the form of a table (Table 8.3). The gram pod borer larval population in pigeonpea was at peak in Guntur during 48th SMW (5.2/plant), and its larval population was higher than the

TABLE 8.3
Insect-Pest Dynamics in Pigeonpea

S.N.	Pest Name	Location/Year	Population Incidence: Started (SMW)	Population Incidence: Attained Peak (SMW)	Population Incidence: Last Recorded (SMW)	Maximum Pest Load Recorded	Reference
1	Pod borer, *Helicoverpa armigera* Hubner	Dharwad, Karnataka/ 2022–2023	44th	50th	2nd	3.63 larva/plant	Harshita et al. (2024)
		Guntur/2015	44th	48th	2nd	5.2 larva/plant	Sreekanth et al. (2019)
		Guntur/2015	44th	47th	4th	8 adults/trap	Sreekanth et al. (2019)
		Navsari/2014	43rd	NA	6th	15.38 larvae /plant	Jalondhara and Patel (2018)
2	Spotted pod borer	Pigeonpea/Kanpur/2014	41st	46th	47th	8 webbings/plant	Sujayanand et al. (2021)
		Pigeonpea/Kanpur/2015	41st	41st	47th	5.13 webs/plant	
		Pigeonpea/Kanpur/2016	40th	41st	47th	9.67 webs/plant	
		Pigeonpea/Kanpur/2017	41st	44th	47th	13.33 webs/plant	
		Pigeonpea/Guntur/2015	43rd	48th	1st	15.6/plant	Sreekanth et al. (2019)
		Pigeonpea/Ludhiana/2015	38th	40th	43rd	12.4 larva/plant and 24 webs/plant	Taggar et al. (2019)
		Pigeonpea/Ludhiana/2016	35th	40th	43rd	14.5 larva/plant and 24.88 webs/plant	
		Pigeonpea/Ludhiana/2017	35th	40th	44th	15.2 larva/plant and 25 webs/plant	
		Pigeonpea/ Dharwad/2022–2023	44th	50th	2nd	5.10 larva/plant	Harshita et al. (2024)
3	Pod fly	Pigeonpea/Kanpur/2013	1st	NA	NA	5 damaged pods/plant	Sujayanand et al. (2020)
		Badnapur/2014–2015	48th	51st	10th	60 larva/100 pods	Patange et al. (2017)
		Dharwad/2022–2023	48th	52nd	4th	10.8 maggots/10 pods	Harshita et al. (2024)
4	Plume moth	Pigeonpea/Varanasi/2015	5th	11th	15th	1.73 larva/plant	Rawat et al. 2017
		Pigeonpea/Varanasi/2016	5th	11th	15th	1.6 larva/plant	

5	Blue butterfly	Navsari/2014	43rd	47th	48th	1.34 larvae/plant	Jalondhara and Patel (2018)
6	Pod bugs	S.K. Nagar/*Kharif* 2011	40th	47th	51st	1.20/plant	Jakhar et al. (2017)
	C. gibbosa,	S.K. Nagar/*Kharif* 2012	39th	1st	2nd	2.31/plant	
		S.K. Nagar/*Kharif* 2013	39th	48th	2nd	2.58/plant	
		S.K. Nagar/*Kharif* 2014	42nd	50th	51st	3.98/plant	
		Paschim medinipur/*kharif* 2014 and 2015	43rd	50th	52nd	30 bugs/plant	Kar (2017)
	C. gibbosa,	Badnapur/2019	44th	51st	3rd	7.6 pod bugs/plant	Ugale et al. 2021
	C. gibbosa, *C. scutellaris Riptortus pedestris R. linearis* and *Anoplocnemis phasiana*	Dharwad/2022–2023	44th	50th	52nd	4.62 larva/plant	Harshita et al. (2024)
7	Green stink bugs *N. viridula*	Badnapur/2019	46th	52nd	3rd	6.03 pod bugs/plant	Ugale et al. 2021
8	Cow bug, *Oxyrachis tarandus*	Navsari/2014	42nd	45th	47th	2.18 bugs/plant	Jalondhara and Patel (2018)
9	*S. litura*	Guntur/2015	43rd	50	4th	78 adults/trap	Sreekanth et al. (2019)
10	Leaf webber	Pigeonpea/Pantnagar/2013	37th	41st	46th	54 larvae/100 shoots	Chakravarty and Agnihotri (2017)
		Pigeonpea/Kanpur/2013	29th	NA	NA	4 larva/plant	Sujayanand et al. (2020)
		Pigeonpea/Badnapur/2019	36th	42nd	48th	1.98 larva/plant	Ugale et al. 2021
		Pigeonpea/Banda/2021	35th	47th	49th	4.57larva/plant	Pandey et al. (2024)
11	Blister beetle	Pigeonpea/Kanpur/2013	37th	NA	NA	3 adult/plant	Sujayanand et al. (2020)
		Navsari/2014	39th	43rd	46th	3.88 beetles/plant	Jalondhara and Patel (2018)

Dharwad population (3.63/plant), whereas the adult population was on peak during 47th SMW (8/trap). Further, gram pod borer, *H. armigera*, infestation has been predicted to be higher in pigeonpea based on the study reported by Srinivasa Rao et al. (2023). They had estimated the *H. armigera* population in three different climate change periods (near, distant, and far distant) and four different resource concentration pathways (2.6, 4.5, 6.0, & 8.5) with multi-model ensemble of maximum and minimum temperature. They found that pigeonpea crop duration and *H. armigera* developmental duration reduces simultaneously, and therefore results in a larger number of annual (10–17) and seasonal (5–8) generations in a far distant period. The trend is similar in the other two periods also. The long-term data of spotted pod borer, *M. vitrata* at Kanpur (2014 to 2017) and Ludhiana (2015 to 2017), vividly portray the population buildup at both locations in the subsequent years from 8 to 13.33 webs per plant and 24 to 25 webs per plant, respectively. Further, gram pod borer, spotted pod borer, and some other lepidopteran insects were considered as migratory insect pests, and the influence of various abiotic factors with respect to climate change was vividly described by Sujayanand and Karuppaiah (2016). These data indicate the rising threat of *M. vitrata* in pigeonpea, which has to be managed through climate-smart pest management strategies. Similarly, pod bug *C. gibbosa* population had also increased in subsequent years at North Gujarat during *kharif* 2011 to 2014 from 1.2 to 3.98 per plant. Further, it is an emerging threat in pigeonpea productivity, as it can inflict a seed damage of up to 25%, which can be managed effectively by spraying homemade neem extract at the rate of 3.1 liters per hectare (Taggar et al. 2022).

8.7 INSECT-PEST MANAGEMENT UNDER CLIMATE CHANGE SCENARIO

Artificial intelligence (AI) powers monitoring of insect pests and predators for effective pest management: in the conventional monitoring of pod borer using pheromone traps, light traps should be aligned with modern technologies like sensor-based insect-pest detection for real-time monitoring of insect pests in the field. This population dynamics data can be interlinked with machine learning (ML) models using two abiotic parameters, viz., temperature and relative humidity. Most recently, in Serbia, a scikit-learn package was demonstrated for the accurate prediction of *H. armigera* (Markovic et al. 2021). Similarly, the *Helicoverpa* adult population was predicted using pheromone trap catches during 2015–2021 by deploying an artificial neural network (ANN) and ARIMA model having minimum BIC, RMSE, MAPE, MAE, and MASE values and a high R^2 value as the best fit model (Narava et al. 2022). Among the two models, they found ANN more efficient based on feed-forward neural networks for predicting adult *H. armigera* population, thereby helping the timely application of insecticides and increasing the productivity of pigeonpea. The major predators in the pigeonpea agro-ecosystem are spiders, viz., *Araenus* sp and *Clubsiona* sp. They prey on sucking insect pests like jassid, cowbugs, pod bugs, eggs of lepidopteran insects like plume moth, *H. armigera*, *M. vitrata*, blue butterfly, etc. The time series data of spiders during 2011 to 2017 were predicted with combinations of wavelet—regression and wavelet—ANN models. The accuracy of the developed wavelet-ANN was tested using root mean square error, mean absolute percentage error, and Diebold Mariano. This model will help in predicting the spider population of the seven locations studied and, thereby, in conjunction with pest defender values, it helps in reducing the insecticidal consumption, propelling the biological control along with ecological conservation (Paul et al. 2022). Thus, developments in AI can be harnessed for predicting pest and natural enemy detection and their monitoring in the pigeonpea agro-ecosystem. This will help in conserving the ecological diversity, along with sustainable production of pigeonpea by need-based spraying of biopesticides, insect growth regulators, or chemical insecticides.

Exploring wild germplasms for imparting resistant genes: The insect pest, specifically pod borer complex, can be managed by improving the host plant resistance of cultivated genotypes by crossing with wild germplasms like *C. scarabaeoides*, *C. platycarpus*, and *Cajanus acutifolius*, etc. (Datta et al. 2023; Sharma et al. 2022; Rathinam et al. 2020; Rathinam et al. 2019). The high-throughput screening method for pod borer-resistant pigeonpea was developed, and a prebreeding derivative, viz., WD 22-3-3, WDBCE 6-3-4-6, and WDBCE 6-3-4-7, was identified as pod borer resistant at ICAR-IIPR, Kanpur (Sujayanand et al., 2019, 2025c). Bommesha et al. (2012) have reported that the gram pod borer (−0.77), spotted pod borer (−0.73), and pod bugs (−0.76) shows significant negative correlation with the phenol content of immature pigeonpea pods. Whereas blue butterfly (−0.67), plume moth (−0.68*), and tur pod fly (−0.68) showed significant negative correlation with the phenol content of pods. Therefore, the phenol content of pods can be used for indirect selection of insect-pest resistant/tolerant genotypes.

Utilizing Biocontrol Agents: Due to climate change, the asynchrony between natural enemies and insect pests may happen; therefore, in order to mitigate this, nectar or pollen providing plants can be grown to enhance natural enemy activity. Romeis et al. (1999) reported that intercropping pigeonpea with sorghum increased parasitization of *H. armigera* eggs, and movement of parasitoids from sorghum to pigeonpea was also noticed (Duffield and Reddy 1997). Similarly, a reduction in *M. vitrata* larvae and damage by planting sorghum as a border crop with pigeonpea was reported at ICAR-IIPR, Kanpur (Sujayanand et al. 2021). Further, the predatory coccinellid population was also found to be higher in pigeonpea bordered with sorghum. The use of microbial biocontrol agents with multiple insecticidal genes or consortia of microbial bio-agents would help in overcoming the insect populations, such as the pod borer. The native *Bacillus thuringiensis* strains having multiple insecticidal genes like different combinations of *cry* genes, *vip3* gene, *chi36* gene, as documented by Sujayanand et al. (2023), will help in managing a wide range of polyphagous insect pests like the gram pod borer, tobacco caterpillar, and Bihar hairy caterpillar.

Chemical insecticides: The pod borer complex can be managed by spraying Indoxacarb 14.50% SC at 400 ml/ha or Indoxacarb 15.80% EC at 333 ml/ha or flubendiamide 39.35% w/w SC at 100ml/ha. The spotted pod borer can be managed by Broflanilide 300 g/l SC at 42 ml/ha. The CIBRC (Central Insecticide Board and Registration Committee), under the Directorate of plant quarantine and storage, periodically examines the insect-pest status and new chemicals that can be recommended for the crops. In case of grain legumes, there is ample scope to add few more insecticides for managing the blister beetle, leaf webber, and pod bugs.

8.8 CONCLUSION

Pigeonpea is one of the major grain legumes cultivated in India, and it is essential for achieving and maintaining nutritional security. The insect-pest scenario has changed drastically in the last four decades (Chandra et al. 2022), wherein different agro-ecological zones have different major insects; hence, there is a need for developing agro-ecological zone-wise Integrated Pest Management (IPM) modules. The available literature vividly shows that sucking pests like pod bugs, cowbugs, and mealybugs are emerging, whereas different insects in the pod borer complex, like spotted pod borer, pod fly, blue butterfly, and plume moth are also expanding. Climate change not only affects the insect-pest incidence, abundance, fecundity, and population dynamics but it also affects the natural enemies, like predators and parasitoids, that check the population buildup during adverse climatic conditions. Hence, under prevailing monsoon vagaries and climate change, it's the need of the hour to view the tri-trophic interaction as a whole, i.e., host, insect pest, and natural enemies, in order to design new strategies for insect-pest management. Furthermore, integration of the latest technologies like AI and ML for real-time monitoring and prediction of pigeonpea insect pest and natural enemies in different ecosystems in India is necessary for understanding the macro-level variations in distribution and incidence of insect pest. This certainly would help in devising climate-resilient management strategies for controlling these insect pests, as well as conserving the native natural enemies and pollinator diversity.

REFERENCES

Anantharaju P, Muthiah AR (2008) Biochemical components in relation to pests incidence of pigeonpea spotted pod borer (*Maruca vitrata*) and blister beetle (*Mylabris* spp.). *Legume Res* 31(2): 87–93.

Barteková AS, Praslička J (2006) The effect of ambient temperature on the development of cotton bollworm (*Helicoverpa armigera* Hübner, 1808). *Plant Protect Sci* 42(4): 135.

Bohra A, Saxena KB, Varshney RK, Saxena RK (2020) Genomics assisted breeding for pigeonpea improvement. *Theor Appl Genet* 133: 1721–1737.

Bommesha B, Naik MI, Mutthuraju GP et al. (2012) Effect of organic manures on biochemical components of pigeonpea, *Cajanus cajan* (L.) Millsp. and their impact on the incidence of insect pests. *Current Biotica* 6(2): 171–180.

Cameron P (1905). On the phytophagous and parasitic Hymenoptera collected by Mr. *E. Green in Ceylon. Spolia Zeylanica* 3: 67–143.

Cameron P (1907). On the parasitic Hymenoptera collected by Major C. G. Nurse in the Bombay Presidency. *J Bombay Nat Hist Soc* 17: 584–595, 1011–1012.

Chakravarty S, Agnihotri M (2017) Seasonal incidence and natural parasitization of leaf webber, *Pammene critica* (Meyrick) (Lepidoptera: Tortricidae) on pigeonpea. *Environ Ecol* 35(1): 128–133.

Chandra A, Gupta DS, Kumar Y, Tiwari TN, Patel DP, Katiyar PK, Sah U, Dixit GP (2025) Outreach programmes of IIPR, ICAR-Indian Institute of Pulses Research.

Chandra A, Sujayanand GK, Revanasidda, Bandi SM, Angami T, Kanwat M (2022) Insect pollinators and hybrid seed production: relevance to climate change and sustainability. In: Bohra A, Parihar AK, Naik SJS, Chandra A (eds) *Plant Male Sterility Systems for Accelerating Crop Improvement*. Springer, Singapore.

Chaturvedi RK, Jaideep J, Mathangi J, Bala G, Ravindranath NH (2012) Multi-model climate change projections for India under representative concentration pathways. *Curr Sci* 103: 791–802.

Datta D, Satheesh Naik SJS, Sujayanand GK, Ghritlahre S, Kumari N (2023) Pre-breeding for incorporation of biotic and abiotic stress tolerance in pigeonpea. In: *International Conference on Pulses: Smart Crops for Agricultural Sustainability and Nutritional Security, 10–12 February 2023*, NASC, New Delhi, India. (Eds Rathore M, Akram M, Nath CP, Lamichaney A, Shanmugavadivel PS, Basavaraja T, Revansidda, Rani R). p. 181.

Dhillon MK, Sharma HC (2007) Effect of storage temperature and duration on viability of eggs of *Helicoverpa armigera* (Lepidoptera: Noctuidae). *Bull Entomol Res* 97(1): 55–59.

Duffield SJ, Reddy YVR (1997) Distribution and increment of predators of *Helicoverpa armigera* in intercropped sorghum and short duration pigeonpea. *Crop Res* 14: 315–335.

Durairaj C (2010) Pod wasp, *Tanaostigmodes cajaninae* La Salle (Hymenoptera: Tanaostigmatidae): an emerging pest of pigeonpea. *Karnataka J Agric Sci* 23(1): 151–152.

FAOSTAT. (2022) https://www.fao.org/faostat/en/#data (last access: 21 January 2025).

Gupta A, Ghosh A, Baby NL, Jalali SK (2011) Morphological and molecular characterization of *Apanteles mohandasi* Sumodan & Narendran (Hymenoptera: Braconidae), a solitary endoparasitoid of *Pammene critica* Meyrick (Lepidoptera: Tortricidae), with notes on biology from India. *Entomol News* 122(4): 354–365.

Gupta A, Sujayanand GK, Bakthavatsalam N (2013) Record of three larval parasitoids (Hymenoptera: Ichneumonoidea) of *Maruca vitrata* (Fabricius) (Lepidoptera: Crambidae) from southern India. *J Biol Cont* 27(1): 53–55.

Harshita H, Gundannavar K, Kambrekar D, Yenagi BS (2024) Population dynamics of pod pest complex of pigeonpea. *J Farm Sci* 37(1): 32–35.

Jakhar BL, Singh N, Tapre PV, Ravindrababu Y, Venilla S (2017) Study the influence of climate change on *Clavigralla gibbosa* (Spinola) in pigeonpea. *J Agric Ecol* 4: 37–43.

Jakhar BL, Singh N, Venilla S et al. (2016) Influence of climate change on *Helicoverpa armigera* (Hubner) in pigeonpea. *J Agric Ecol* 2: 25–31.

Jalondhara RM, Patel DR (2018) Population dynamics of insect pests of pigeonpea in south Gujarat. *Indian J Entomol* 80(2): 207–210.

Kar A (2017) Seasonal incidence on population dynamics of pigeonpea pod bug, *Clavigralla gibbosa* (Spinola). *J Entomol Res* 41(4): 409–412.

Mandi N, Nayak BS, Nayak A et al. (2020) Eco-friendly management of pod borer complex in pigeonpea. *Int J Chem Stud* 8(1): 1300–1304.

Marković D, Vujičić D, Tanasković S et al. (2021) Prediction of pest insect appearance using sensors and machine learning. *Sensors* 21(14): 4846.

Muchhadiya DV, Patel JJ, Patel DR, Patel RB (2024) Estimation of yield losses caused by insect pests on pigeonpea (*Cajanus cajan* (L.) Millsp.). *Int J Plant Soil Sci* 36(3): 410–414.

Narava R, DV SRK, Jaba J et al. (2022) Development of temporal model for forecasting of *Helicoverpa armigera* (Noctuidae: Lepidoptera) using ARIMA and artificial neural networks. *J Insect Sci* 22(3): 2.

Pandey A, Singh SK, Singh AK et al. (2024) Assessment of the incidence of pigeonpea leaf webber, *Grapholita critica* (Meyr.) and roller, *Caloptilia soyella* (Van.) in relation to weather variables in the Bundelkhand region of Uttar Pradesh. *Legume Res* 47(9): 1613–1619.

Patange NR, Sharma OP, Chiranjeevi B (2017) Population dynamics of *Melanagromyza obtusa* (Malloch) (Diptera: Agromyzidae) and its natural parasitization in pigeonpea. *Entomon* 42(3): 201–206.

Patra S, Firake D, Thakur NA, Roy A (2016). Insect pest complex and crop losses in pigeon pea in medium altitude hill of Meghalaya. *Bioscan* 11(1): 297–300 (Suppl. on Agronomy).

Paul RK, Vennila S, Yeasin M et al. (2022) Wavelet decomposition and machine learning technique for predicting occurrence of spiders in pigeonpea. *Agron* 12(6): 1429.

Rathinam M, Mishra P, Mahato AK et al. (2019) Comparative transcriptome analyses provide novel insights into the differential response of pigeonpea (*Cajanus cajan* L.) and its wild relative (*Cajanus platycarpus* (Benth.) Maesen) to herbivory by *Helicoverpa armigera* (Hübner). *Plant Mol Biol* 101: 163–182.

Rathinam M, Roschitzki B, Grossmann J et al. (2020) Unraveling the proteomic changes involved in the resistance response of *Cajanus platycarpus* to herbivory by *Helicoverpa armigera*. *Appl Microbiol Biotechnol* 104: 7603–7618.

Rawat RK, Keval R, Ganguly S, Chakravarty S (2017) Seasonal incidence of plume moth, *Exelastis atomosa* (Lepidoptera: Pterophoridae) on long duration pigeonpea, *Cajanus cajan*. *Flora Fauna* 23(2): 445–452.

Romeis J, Shanower TG, Zebitz CPW (1999) *Trichogramma* egg parasitism of *Helicoverpa armigera* on pigeonpea and sorghum in southern India. *Entomol Exp Appl* 90(1): 69–81.

Saxena KB, Bohra A, Choudhary AK, Sultana R, Pazhamala L, Saxena RK (2021) The alternative breeding approaches for improving yield gains and stress response in pigeonpea (*Cajanus cajan* (L.) Millsp.). *Plant Breed* 140: 76–86.

Saxena KB, Choudhary A, Srivastava R, Bohra A, Saxena R, Varshney RK (2019) Origin of early maturing pigeonpea germplasm and its impact on adaptation and cropping systems. *Plant Breed* 138: 243–251.

Sharma HC, Pathania M, War AR, Pavani T, Vashisth S (2015) Climate change effects on pest spectrum and incidence in grain legumes. In: Dixit GP, Singh J, Singh NP (eds) *Pulses: Challenges and Opportunities under Changing Climate Scenario*. Indian Soc. Pulses Res. Dev. (ISPRD), 124–137.

Sharma S, Jaba J, Rao PJ et al. (2022) Reaping the potential of wild *Cajanus* species through pre-breeding for improving resistance to pod borer, *Helicoverpa armigera*, in cultivated pigeonpea (*Cajanus cajan* (L.) Millsp.). *Biol* 11(4): 485.

Singh G, Singh R, Singla A (2022) Estimation of crop losses in *Cajanus cajan* (L.) Millsp. and *Vigna radiata* (L.) Wilczek caused by blister beetle, *Mylabris pustulata* Thunberg (Coleoptera: Meloidae). *Int J Trop Insect Sci* 42(4): 3143–3150.

Singh SK, Kumar A, Agrawal R, et al. (2016) Predicting pod fly (*Melanagromyza obtusa* Malloch) incidence on long duration pigeonpea in Uttar Pradesh. *J Food Legumes* 29(1): 51–56.

Sreekanth M, Ratnam M, Ramana MV (2019) Population dynamics of pod borers on pigeonpea (*Cajanus cajan* (L.) Millsp.) in relation to abiotic factors. *Int J Bio-resour Stress Manage* 10(6): 616–620.

Srinivasa Rao M, Rama Rao CA, Raju BMK, et al. (2023) Pest scenario of *Helicoverpa armigera* (Hub.) on pigeonpea during future climate change periods under RCP-based projections in India. *Sci Rep* 13(1): 6788.

Srivastava CP, Joshi N (2011) Insect pest management in pigeon pea in Indian scenario: A critical review. *Indian J Entomol* 73(1): 63–75.

Srivastava CP, Joshi N, Trivedi TP (2010) Forecasting of *Helicoverpa armigera* populations and impact of climate change. *Indian J Agric Sci* 80(1): 3–10.

Sujayanand GK, Chandra A, Pandey S, Bhatt S (2021) Seasonal abundance of spotted pod borer, *Maruca vitrata* Fabricius in early pigeonpea [*Cajanus cajan* (L.) Millsp.] and its management through farmscaping in Uttar Pradesh. *Legume Res* 44(2): 233–239.

Sujayanand GK, Chandra A, Rout AK, Dubey P, Jagadeeswaran R (2025a) Coreidae: pod bugs emerging insect pest in legume crops and its management. *Indian Farmer* 12(9): 517–521.

Sujayanand GK, Datta D, Chandra A, Pandey S (2025b) Assessing pigeonpea genotypes for resistance against *Maruca vitrata* Fabricius and its sustainable management through biorationals. *Legume Res* 48(1): 176–182. https://doi.org/10.18805/LR-5132.

Sujayanand GK, Datta D, Katiyar A, et al. (2019) High throughput evaluation of *Cajanus scarabaeoides* pigeonpea progenies for identifying gram pod borer tolerance. *J Entomol Zool Stud* 7(2): 151–157.

Sujayanand GK, Datta D, Singh F (2020) Insect pest succession and screening for spotted pod borer tolerance in short duration pigeonpea. *J Food Leg* 33(2): 71–76.

Sujayanand GK, Dubey S, Datta D, Naik S, Tewari K, Chandra A, Devi R, Govindasamy P, Dixit GP (2025c) Identification of pod borer resistant pigeonpea: an introgression breeding with *Cajanus scarabaeoides* accompanied by integrated screening. *Arthropod Plant Interact* 19(71): https://doi.org/10.1007/s11829-025-10178-6.

Sujayanand GK, Karuppaiah V (2016) Aftermath of climate change on insect migration: a review. *Agric Rev* 37(3): 221–227.

Sujayanand GK, Pandey S, Jagadeeswaran R, Chandra A, Kumar V, Dubey S, Dubey J (2023) Characterization of entomotoxic and nematotoxic genes from indigenous *Bacillus thuringiensis* strains and their biocontrol potential. *Egyptian J Biol Pest Cont* 33(1): 76.

Sujithra M, Chander S (2014) Seasonal incidence and damage of major insect pests of pigeonpea, *Cajanus cajan* (L.). *Indian J Entomol* 76: 202–206.

Taggar GK, Singh R, Cheema HK, Singh P (2019) Relative abundance, population dynamics and damage potential of spotted pod borer, *Maruca vitrata* (Fabricius) on early pigeonpea in Punjab. *Int J Trop Insect Sci* 39: 229–234.

Taggar GK, Singh R, Chhuneja PK, et al. (2022) Insecticidal potential of home-made neem extract against pod sucking bug, *Clavigralla gibbosa* (Spinola) in pigeonpea. *Int J Trop Insect Sci* 42(2): 1913–1924.

Udayababu P, Sowjanya P, Jogarao P (2021) Seasonal dynamics of pests of pigeonpea in relation with weather in high altitude zone of Andhra Pradesh. *Acta Botanica (G)* 9: 8–20.

Ugale MV, Bantewad SD, Suradkar AL, Shaikh AB (2021) Studies on population dynamics of pigeonpea (*Cajanus cajan* L.) pod bugs, green stink bugs and leaf webber. *J Pharmacog Phytochem* 10(1): 1114–1116.

Vennila S, Paul RK, Bhat MN, et al. (2019) Approach to study of pigeonpea leaf webber [*Grapholita critica* (Meyr.)] damage dynamics and its relation to weather. *Legume Res* 42(6): 844–849.

Vishwadhar Singh SK, Kumar M, Agrawal R, Kumar A (2007) Prediction of pod borer (*Helicoverpa armigera*) infestation in short-duration pigeonpea (*Cajanus cajan*) in central Uttar Pradesh. *Indian J Agric Sci* 77(10): 701–704.

Vishwadhar Singh SK, Trivedi TP, et al. (2008) Forecasting of *Helicoverpa armigera* infestation on long duration pigeonpea in Central Uttar Pradesh. *J Food Leg* 21(3): 189–192.

9 Multi-Omics Approaches to Understand Biotic Stress Response in Pigeonpea

PS Shanmugavadivel, Arshad Husain, and M Senthil Kumar

9.1 INTRODUCTION

Pigeonpea is the second most important pulse crop of the tropics and subtropics (Bohra et al. 2020a, 2020b, Saxena et al. 2019, 2021a). Pigeonpea productivity is affected by various abiotic and biotic stresses. The major biotic stresses are caused by fungal, bacterial, viral infections, and insect pests. The disease is caused by fungal pathogens, including Fusarium wilt (*Fusarium udum*), Phytophthora blight (*Phytophthora drechsleri* sp. *cajani*), and leaf spot (*Cercospora cajani*). The major viral disease affecting pigeonpea is sterility mosaic disease (SMD), also popularly known as green plague, which is incited by the pigeonpea sterility mosaic virus (PPSMV) and spread by the eriophyid mite, *Aceria cajani*. The insect pests that are infecting the crop during field and storage such as pod borer (*Helicoverpa armigera*), pod fly (*Melanagromyza obtusa*), spotted pod borer (*Maruca vitrata*), pod wasp (*Tanaostigmodes cajaninae*), pod bugs (*Clavigralla spp*), and storage pests, such as bruchids (*Collasobruchus chinensis*), cause major losses in pigeonpea. Fusarium wilt (FW) causes yield loss ranging from 30% to 100% (Reddy et al. 1990), and wilting during the pre-podding stage causes over 50% and even up to 100% yield loss (Okiror 2002). Yield losses due to SMD vary depending on the stage of crop infection. Early infection (before 45 days) can result in yield losses ranging from 95% to 100%, while later infections can cause losses between 26% and 97% (Kumar et al. 2003). Cercospora leaf spot can cause significant yield losses, ranging from 23% to 96% under natural epiphytotic conditions (Iqbal et al. 1995). The storage pest bruchids (*Callosobruchus* sp) are the major constraints that affect the potential yield losses of up to 100%. Decades of breeding efforts have led to the development of several high-yielding pigeonpea varieties. However, pigeonpea productivity remains inadequate to meet the increasing food demands of a growing human population (Bohra et al. 2020b). This shortfall is primarily due to inherent physiological and genetic limitations, along with limited success in developing

DOI: 10.1201/9781003603092-9

race-specific resistant varieties for the diverse races of major pigeonpea pathogens. Additionally, the inability to breed varieties capable of withstanding rapidly evolving pathogens and insect pests has further hindered breeding progress (Saxena et al. 2017). In this chapter, advancements and knowledge gained about pathogens such as *Fusarium udum* and sterility mosaic disease virus, as well as the genomic regions and candidate genes related to resistance to FW, SMD, and insect pests in pigeonpea, have been discussed in detail.

9.2 PATHOGEN INFECTION AND PLANT DEFENSE MECHANISM

The occurrence and severity of plant diseases result from the interaction of three factors, namely the host plant, the pathogen, and favorable environmental conditions. Understanding both the plant host and the pathogen at the molecular level is crucial to combating the losses caused by various diseases, especially through modern breeding programs. The compatible and incompatible reactions are the outcomes of the relationship between the host plant and the pathogen interactions. A compatible reaction occurs when a pathogen successfully infects a host plant, leading to disease development by suppressing the plant's defense mechanisms. This interaction typically involves two factors: firstly, pathogen effectors that manipulate the plant's immune response or cellular processes, facilitating infection; secondly, host susceptibility genes that the pathogen exploits to promote its own growth and proliferation.

An incompatible reaction takes place when a plant successfully defends itself against a pathogen, preventing disease by the plant's effective recognition and response to the pathogen, often resulting in the activation of strong defense mechanisms, including (1) Hypersensitive Response (HR), a localized cell death at the infection site, which restricts the spread of the pathogen. (2) Pathogen Recognition: The plant detects pathogen-associated molecular patterns (PAMPs) or specific effectors through resistance (R) genes, triggering immune responses. (3) Systemic Acquired Resistance (SAR): A broad-spectrum, long-lasting defense mechanism activated throughout the plant following initial pathogen recognition. Incompatible reactions are typically associated with R gene-mediated resistance, where specific plant R proteins, such as nucleotide-binding domain and leucine-rich repeat receptors, recognize corresponding pathogen effectors, leading to a robust immune response and halting the pathogen's progress. R genes can be specific to particular races or pathotypes, making the breeding of race-specific resistance crucial to prevent the breakdown of disease tolerance in plants caused by evolving pathogens (Huang et al. 2022).

9.3 MOLECULAR BASIS OF PLANT DEFENSE SYSTEM AGAINST INVADING FUNGAL PATHOGENS

Fungi that colonize plants are detected by the plant's immune system, which further activates the host defense response. The initial plant defense responses are activated by fungal secreted effector molecules, such as the cell wall component chitin, which are referred to as PAMPs. PAMPs influence the interaction between the fungus and the plant and are recognized by membrane-localized pattern recognition receptors

(PRRs). This recognition activates the plant's first line of defense, known as PAMP-triggered immunity (PTI). Activation of PRR signaling leads to the build-up of reactive oxygen species, activation of ion channels, initiation of specific defense-related mitogen-activated protein kinase cascades, and extensive transcriptional reprogramming of the host. Together, these responses result in the production of antimicrobial and antifungal compounds, including proteinases, chitinases, and glucanases, which damage the pathogen's structures. PTI is regarded as effective against nonadapted fungi (nonhost resistance) and can influence whether a plant can serve as a host by determining resistance at the level of pathogen penetration (Lo Presti et al. 2015). In compatible interactions, fungi must evade triggering PTI or develop mechanisms to overcome or suppress it to proliferate within plants. They accomplish this by neutralizing toxic metabolites released by the plant or secreting fungal effectors—either toxic secondary metabolites or proteins—that kill host plant cells. These effectors help the fungus by protecting it, suppressing the host's immune response, or manipulating host cell physiology (Lo Presti et al. 2015).

The secreted effectors, also known as avirulence (Avr) proteins, which suppress PTI, can be detected by the plant's surveillance system, triggering a second layer of defense called effector-triggered immunity (ETI). These effectors/Avr proteins are recognized by conserved intracellular receptors known as plant resistance (R) proteins, which belong to the nucleotide-binding leucine-rich repeat (NB-LRR) class. ETI defenses often lead to a localized cell death response, known as the HR. The effectors and receptors involved in triggering ETI are highly variable and specific gene-for-gene (*R* gene and *Avr* gene) interactions: a particular *R* gene in the plant can only recognize a specific *Avr* gene product from the pathogen. If either the *R* gene in the plant or the *Avr* gene in the pathogen is absent or nonfunctional, the interaction does not occur, and the pathogen may successfully infect the plant. This R-Avr interaction drives an evolutionary arms race between plants and pathogens. Pathogens may evolve new effectors or modify existing ones to avoid detection, while plants may evolve new *R* genes to recognize the modified effectors. The variability and rapid evolution of *R* and *Avr* genes reflect this ongoing co-evolutionary battle (Lo Presti et al. 2015). To better understand ETI and PTI, readers can read the review by Lo Presti et al. (2015).

9.4 GENETICS OF DISEASE AND INSECT RESISTANCE IN PIGEONPEA

The FW disease is soil and seed borne, and hence it's difficult to manage through chemical application. Host plant resistance is the most effective strategy for the management of pigeonpea diseases, including FW. Hence, a thorough understanding and knowledge of the disease resistance at the genetic and molecular level is very important to apply modern breeding approaches for cultivar improvement. Most of the disease resistance in pigeonpea is governed by multiple genes or quantitative trait loci (QTLs), and in some instances, it's governed by a few genes. FW resistance in pigeonpea cultivars and germplasms is reported to be majorly governed by a single gene (Singh et al. 1998; Chaithanya et al. 2011), two dominant gene(s) with duplicate, inhibitory (13:3) and complementary (9:7) gene actions (Changaya et al.

2012; Joshi 1957; Pandey et al. 1996; Karimi et al. 2010; Saxena et al. 2012; Singh et al. 2016a; Mayomba 2018). However, resistance to FW is also governed by one or more recessive genes (Jain and Reddy 1995; Karimi et al. 2010; Patil et al. 2017a; Odeny et al. 2009). Many studies reported that SMD resistance is governed by one or two genes, which are mostly under the control of recessive genes with complementary and inhibitory gene actions in various genetic backgrounds (Singh et al. 1983; Sharma et al. 1984; Srinivas et al. 1997; Singh and Vishwadhar 2003; Nagaraj et al. 2004; Gnanesh et al. 2011; Ganapathy et al. 2012; Daspute et al. 2014). However, QTL analysis revealed the resistance is governed by polygenes (Gnanesh et al. 2011; Singh et al. 2017b).

High levels of resistance against pod borer *H. amigera* in wild relatives of pigeonpea have been identified by various workers. Antixenosis, antibiosis, and tolerance are the key components of resistance to *H. armigera* in pigeonpea (Kumari et al. 2006). However, there is a dearth of information on the inheritance of pod borer resistance, and the molecular marker linked with resistance in pigeonpea. Trichomes reduce insect loads and/or damage through the repellent activity of their exudates, which limit the insect contact with the plant surface. This defense mechanism has been widely utilized in several crop plants to protect against insects. Trichomes of pigeonpea wild species *Cajanus scaraboides* are associated with resistance/susceptibility to *H. armigera*. The type of trichomes (A, B, C, and D), their orientation, density, length, and the exudates on the pod wall surface play a crucial role in determining the ovipositional behavior of insects (Bernays & Champman 1994). The trichomes in the *C. scaraboides* accessions, characterized by a high density of nonglandular trichomes (types C and D) and a low density of glandular trichomes (type A), were more dominant than the trichome features found in *C. cajan*, and a single dominant gene governs antixenosis mechanism of resistance to pod borer in *C. scarabaeoides* (Verulkar et al. 1997). Accessions of wild relatives of pigeonpea, such as *C. scarabaeoides, C. sericeus*, and *C. acutifolius* are highly resistant to *H. armigera*. Accessions like ICPW 141, ICPW 278, and ICPW 280 of *C. scarabaeoides*; ICPW 214 of *Rhynchosia bracteata*; ICPW 14 of *C. albicans*; and ICPW 202 of *Flemingia stricta* demonstrated resistance to both pod fly and pod wasp (Sachan 1990).

The knowledge of inheritance and the number of genes involved in trait expression is essential for an efficient breeding program and the development of resistant cultivars. The dominant nature of inheritance makes it easier to incorporate FW resistance from resistant to susceptible cultivars following any selection method (Pastor-Corrales et al. 1994). Recessive gene-based resistance is often more stable over time, as it exerts less selective pressure on the pathogen, reducing the likelihood of resistance breakdown. In some instances, it also provides broad-spectrum resistance against multiple pathogen strains, races, or species.

9.5 GENOMIC RESOURCES OF HOST PLANT-PIGEONPEA

9.5.1 Pigeonpea Genomes: Re-sequencing Data, Transcriptome Atlas

Pigeonpea has 2n = 2x = 22 chromosomes and an estimated genome size of 858 Mbp. The genome sequencing of the pigeonpea variety Asha (ICPL87119) was completed in 2012 (Singh et al. 2012; Varshney et al. 2012). The long sequence reads obtained using 454 GS-FLX sequencing chemistry for the Asha cultivar generated 510 Mb of high-quality sequence data. This included 47,004 protein-coding genes and 12,511 transposable element genes. Of the 47,004 protein-coding genes, 1,213 were disease resistance/defense response (DR) genes (Singh et al. 2012). A total of 98 DR genes were classified into three groups: chitinases (31 genes), glucanases (56 genes), and thaumatin-like proteins (11 genes). Among the remaining 1,115 DR/R-like genes, 219 were classified into the miscellaneous category, which included genes for viral resistance, verticillium wilt resistance, bacterial blight resistance, and genes containing the LRR motif but lacking NBS, CC, or LZ motifs. Besides these, this genome facilitated the development of 437 hypervariable "Arhar" simple sequence repeat (HASSR) markers, which can be resolved on simple agarose gels for polymorphism studies and used in resource-poor laboratories for mapping studies and diversity analysis (Singh et al. 2012). The pigeonpea genome assembly, performed using the illumina next-generation sequencing platforms GA and HiSeq 2000, resulted in a genome size of 605.78 Mb with 48,680 genes. Additionally, 23,410 simple sequence repeats (SSR) markers were designed using this genome assembly (Varshney et al. 2012). The reported draft genome assemblies of *C. cajan* (Singh et al. 2012; Varshney et al. 2012) exhibit varying levels of genome completeness, along with a significant number of repeats, gaps, and segmental duplications. The refined draft genome assemblies of pigeonpea, using two assemblies reported by Singh et al. (2012) and Varshney et al. (2012), were combined using an assembly reconciliation approach, resulting in improved genome coverage of 82.4% with 51,737 genes. Among these genes, 1,301 were identified as putative resistance gene analogs, including 251 containing NBS-LRR domains and 54 NBS-encoding single-copy genes (Marla et al. 2020). Whole-genome re-sequencing of 292 *Cajanus* accessions, including 117 breeding lines, 166 landraces, 2 other varieties, and 7 accessions from three wild species (*C. cajanifolius*, *C. scarabaeoides*, and *C. platycarpus*), is available (Varshney et al. 2017). Among these, seven SMD-resistant accessions (ICP11230, ICP11238, ICP11015, ICP11096, ICP11148, ICP11059, and ICP7436) and four FW-resistant accessions (ICP14819, ICP3633, ICP8863, and ICP10240) have been re-sequenced. This will aid in identifying genes and alleles responsible for disease resistance in pigeonpea. The pigeonpea transcriptome atlas consists of 28,793 genes with differential, specific, spatiotemporal, and constitutive expression patterns, derived from 30 tissues and organs representing developmental stages from germination to senescence (Pazhamala et al. 2017). This crucial genomic information will be valuable to pigeonpea breeders and molecular biologists, aiding in the adoption of genome editing tools, transgenic methods, marker-assisted breeding, and the functional characterization and validation of alleles/genes linked to various agronomic traits and biotic stress tolerance in pigeonpea.

9.5.2 Genome-Wide Reproducible Markers, and High-Throughput Genotyping Assays

A large number of reproducible markers, such as SSRs, single nucleotide polymorphisms (SNPs), and insertion-deletions (InDels), have been developed in pigeonpea (Bohra et al. 2011; Dutta et al. 2011; Singh et al. 2012; Saxena et al. 2012; Varshney et al. 2012; Singh et al. 2017a; Bohra et al. 2017; Dutta et al. 2021). High-throughput genotyping assays for SNPs are crucial for their ultimate application in gene discovery and molecular breeding. High-density SNP arrays, illumina assays, cost-effective KASP assays, and genotyping by sequencing (GBS) have become attractive SNP genotyping tools for breeders and diagnostic purposes. GBS utilizes next-generation sequencing platforms for the simultaneous detection of polymorphisms and genotyping of individuals. However, its use in QTL mapping and breeding is often limited due to challenges in experimental procedures, data analysis, the low number of data points per line, and its incomplete coverage of all selected individuals within populations.

Saxena et al. (2018) developed the Axiom SNP array comprising 56K SNPs uniformly distributed across the genome, based on polymorphism data obtained from whole-genome re-sequencing (WGRS) of 104 pigeonpea lines. The selected SNPs are primarily located in the intergenic and intronic regions of the pigeonpea genome. Out of the 56,512 SNPs, 39,862 are found in the intergenic regions, and 7,340 are in the intronic regions. The effects of these SNPs are classified as follows: 2,903 missense mutations, 1,722 silent mutations, and 73 nonsense mutations. In contrast, Singh et al. (2020) developed a 62K SNP chip array, CcSNPnks, for the Affymetrix GeneTitan® platform based on SNP polymorphism present in single-copy genes of 45 diverse pigeonpea cultivars. This array includes 62,053 genic SNPs derived from 9,629 genes in pigeonpea, comprising single-copy genes unique to pigeonpea (4,314), single-copy genes conserved between soybean and pigeonpea (4,328), homologs of agronomically important cloned genes (156), disease resistance, DR genes (746), and multicopy genes (85). Competitive allele-specific polymerase chain reaction (KASPar) assays for 1,616 SNPs, known as PKAMs (Pigeonpea KASPar Assay Markers), have been developed based on SNPs identified from RNA-seq data of 12 pigeonpea genotypes: ICPL 87119, ICPL 87091, BSMR 736, TAT 10, ICP 7035, TTB 7, ICPL 332, ICPL 20096, ICPB 2049, ICPL 99050, ICP 28, and ICPW 94 (Saxena et al. 2012). These SNP arrays/chips have been successfully used for analyzing genetic diversity, population structure, and high-resolution QTL mapping. These high-density SNP arrays can be utilized by researchers for trait mapping through QTL mapping and association mapping approaches focused on disease resistance. They are expected to accelerate high-resolution fingerprinting and the creation of high-density linkage maps using various bi-parental and multi-parental mapping populations. Ultimately, they will contribute to marker-assisted breeding and genomic selection in pigeonpea.

9.6 GENOMIC RESOURCES OF PATHOGENS INFECTING PIGEONPEA

9.6.1 SMD Virus and Its Variants

SMD is a major biotic factor that poses a significant challenge to pigeonpea production in India. The disease is characterized by symptoms such as a bushy and pale green appearance of plants, followed by a reduction in size, an increase in the number of secondary branches, mosaic mottling of leaves, and ultimately, partial, or complete cessation of reproductive structures (Gnanesh et al. 2011). Variability in the sterility mosaic pathogen has revealed the presence of five different isolates in India, based on differences in symptom expression across various pigeonpea cultivars (Reddy et al. 1993). The isolate from Gwalior was designated as variant 1; the Badnapur and Patancheru isolates as variant 2; the Coimbatore, Kumarganj, and Pudukottai isolates as variant 3; the Bangalore and Dholi isolates as variant 4; and the Kanpur isolate as variant 5. Among these, three SMD isolates—Patancheru, Bangalore, and Coimbatore have been characterized (Reddy et al. 1993). The Bangalore isolate is considered virulent, with resistance governed by polygenic inheritance, whereas the Patancheru isolate is regarded as a mild strain, with resistance governed by digenic inheritance with dominance epistasis (Gnanesh et al. 2011).

9.6.1.1 SMV Genome

The PPSMV genome was fully sequenced using a deep sequencing approach (Elbeaino et al. 2014) and consists of five RNA segments. The largest segment, RNA-1, encodes the RNA-dependent RNA polymerase (RdRp) protein, which contains the core polymerase module with five conserved motifs (motifs A–E) of the RdRp active site. RNA-2 encodes glycoproteins (Gps), which are responsible for the formation of spike projections on the envelope membrane. RNA-3 encodes the nucleocapsid (NC) protein, which functions to encapsidate the viral genomic RNAs. RNA-4 encodes a putative movement protein (MP), which is involved in the cell-to-cell movement of the virus by modifying plasmodesmata and also acts as a gene silencing suppressor. RNA-5 encodes p5, with an unknown function in the virus life cycle. Members of the genera *Emaravirus, Tospovirus*, and *Tenuivirus* share several similarities in their genomic organization, as well as in the morphology of their particles and NC proteins (Mielke-Ehret and Mühlbach 2012). The size of the NC proteins varies among PPSMV isolates; the Patancheru and Bengaluru isolates contain a 32-kDa NC protein, while the Coimbatore isolate has a larger NC protein (35 kDa) (Jones et al. 2004; Kumar et al. 2004a). PPSMVs share similarities with other *emaraviruses* in their genomic organization and conserved 5′ and 3′-end sequences. Despite these similarities, PPSMVs exhibit overall limited sequence homology and are transmitted by eriophyid mites (Kumar et al. 2017).

Kumar et al. (2017) conducted a study comparing dsRNA profiles and small RNA (sRNA) deep sequencing from samples collected at three locations: Bengaluru, Coimbatore, and Chevella, India. Their findings revealed the presence of two types of PPSMV in Chevella, named PPSMV-I and PPSMV-II, while only PPSMV-II was detected in Bengaluru and Coimbatore. PPSMV-I contains four -ve sense single-stranded RNA segments with a 12.25 kb genome size; however, PPSMV-II contains six -ve sense single-stranded genomic RNAs with a genome of 15.13 kb. The PPSMV-I genome, consisting of four RNA segments, shows modest sequence similarity with the PPSMV-II genome, which contains six RNA segments. RNA-1 to RNA-4 of PPSMV-II exhibit high sequence homology with orthologs from fig mosaic virus (FMV), while RNA-5 and RNA-6 show modest to weak homology with orthologs from wheat mosaic virus (WoMV) and raspberry leaf blotch virus (RLBV). It is anticipated that the P5 and P6 proteins of PPSMV-II play a crucial role in the virus life cycle, performing various functions, including facilitating transmission by eriophyid mites (Elbeaino et al. 2014; Kumar et al. 2017). However, PPSMV-II contains additional non-conserved RNA segments that encode NC proteins. RNA-4 of both PPSMVs encodes the respective P4 proteins, which are believed to be involved in the cell-to-cell movement of the virus. The host range of both viruses and their vector is narrow, limited to *Nicotiana benthamiana*, a few cultivars of *Phaseolus vulgaris*, the weed host *Chrozophora rottleri*, and accessions of *C. platycarpus, C. scarabaeoides*, and *C. lanceolatus* (Sayiprathap et al. 2022).

9.6.2 Genome of Wilt-Causing Pathogen *Fusarium Udum* and Its Variants

Thirteen *F. udum* isolates from India were identified using cultural and morphological methods, along with molecular fingerprinting techniques (Dhar et al. 2012; Purohit et al. 2017). These isolates were classified into four variants and seven phylogenetic groups. Invasion of *F. udum* in healthy pigeonpea root occurs within 24–36 hours postinoculation (Purohit et al. 2017). The draft genome sequence of the pigeonpea wilt-causing pathogen *F. udum* isolate F02845 is available in the public domain (Srivastava et al. 2018). The genome assembly comprises 712 scaffolds, with a total size of 56.38 Mb, including 895,132 bp of repetitive elements, 128 long terminal repeats (LTRs), and 4,921 SSRs. The genome contains 11,829 protein-coding genes, of which 1,060 (6.55%) were identified as virulence genes, and 1,439 encode secretory proteins (Srivastava et al. 2018, 2023). Fungal effectors are classified based on their mode of deployment within the host as either extracellular effectors, which are secreted into the apoplast or xylem of the host plant, or cytoplasmic effectors, which are translocated into host cells (Huang et al. 2022). The *Secreted in Xylem (SIX)* gene family comprises effectors that encode small cysteine-rich proteins secreted by most *Fusarium oxysporum* species into the xylem of the host plant. These effector proteins play a role within the plant host by altering host cell structure, promoting infections, and suppressing plant immune responses (Rep et al. 2004). Fourteen *SIX* genes have been identified in the tomato wilt-causing pathogen *F. oxysporum f. sp. Lycopersici*. Two such *SIX* genes were identified by us in *F. udum* genome (unpublized). Moreover, these *SIX* genes are specific to isolates or races of *F. oxysporum*, and variations in the *SIX* gene sequence determine the pathogenicity of the fungus, influencing its ability to infect specific varieties of a crop. The variations in *SIX* genes can be used to differentiate the races of F. udum and will be valuable for race-specific breeding of pigeonpea in the near future.

9.7 FORWARD GENETICS AND GENOMICS-ASSISTED IDENTIFICATION OF DISEASE AND INSECT RESISTANCE LOCI IN PIGEONPEA

Forward genetics is an approach used to identify genes or genomic locations responsible for specific traits by analyzing phenotypic information of a trait obtained from a set of genotypes, along with the genetic variations observed in those selected genotypes under study (Varshney et al. 2021a, 2021b). This approach utilizes techniques such as linkage analysis, QTL mapping, association mapping (AM), and their variants (candidate gene-based association mapping, nested association mapping, genome-wide association mapping) using either bi-parental mapping populations or natural populations. These methods help to identify trait-linked markers, which ultimately facilitate the efficient selection of segregants in the breeding population by marker-assisted selection (MAS) that expedites the breeding process through indirect selection. The identification of various reproducible detection techniques for the analysis of genome polymorphisms, such as microsatellite and minisatellite repeat variations, large and small insertion-deletion (InDEL) polymorphisms, copy number variations, and SNPs, combined with the availability of low-cost genotyping technologies, has made the application of forward genetic approaches for trait mapping in pigeonpea at the genome level more accessible. The GBS approach is a cost-effective technology that simplifies the mapping of QTL through methods such as simple QTL mapping, candidate gene-based AM, and even statistically highly complicated genome-wide AM. Further, the affordability and accessibility of modern sequencing technologies are enabling the widespread use of genomic selection in legume crops, paving the way for more precise, efficient, and sustainable crop improvement for disease resistance.

9.7.1 Genomics: QTLs and Marker-Trait Association Related to Disease and Insect Resistance in Pigeonpea

Genomic regions or quantitative trait loci or qualitative trait loci associated with disease resistance can be identified using genotypic and phenotypic data of segregating populations such as F_2, doubled haploids (DH), recombinant inbred lines (RILs), near-isogenic lines (NILs), and advanced backcross lines (BC), combined with bulked segregant analysis (BSA) and QTL mapping approaches. Alternatively, natural populations can also be effectively utilized to map genomic regions conferring resistance or susceptibility through candidate gene-based or genome-wide AM approaches. Availability of a larger number of informative markers paved the way for the development of linkage maps and consensus maps, which will help in identifying QTLs for disease resistance and agronomically important traits. The attempts were made to find QTLs associated with FW, SMD resistance in pigeonpea, and a few QTL-linked markers are being employed in the pigeonpea breeding program (Table 9.1).

Kotresh et al. (2006) identified two FW susceptibility-linked RAPD markers in pigeonpea using F_2 populations derived from the crosses GS1 (FW susceptible) × ICPL87119 (FW resistant) and GS1 × ICP8863 (FW resistant) through BSA approach. The presence of two amplicons, 704 bp at the OPM03704 locus and 500 bp at the OPAC11500 locus, was associated with FW susceptibility. On the other hand, the OPGO8 RAPD marker produces FW resistance-specific amplicon in pigeonpea genotypes (Prasanthi et al. 2009). Gnanesh et al. (2011) used a QTL mapping approach to identify gene(s) or QTLs linked with SMD resistance in pigeonpea, utilizing two F_2 populations: ICP 8863 × ICPL 20097 (segregating for the Patancheru SMD isolate) and TTB 7 × ICP 7035 (segregating for both Patancheru and Bangalore SMD isolates). The study identified distinct QTLs for the Patancheru and Bangalore SMD isolates, suggesting the existence of different genes or genomic regions associated with isolate- or race-specific SMD resistance

TABLE 9.1
QTLs and MTAs Associated with FW and SMD Resistance in Pigeonpea

S.No.	Cross/Genotypes	Population	Approach	Trait	References
1	GS1 (FW susceptible) × ICPL87119 (FW resistant) and GS1 × ICP8863 (FW resistant)	F_2	BSA	FW	Kotresh et al. (2006)
2	ICP 8863 (SMD susceptible) × ICPL 20097 (SMD resistant) and TTB 7 (SMD susceptible) × ICP 7035 (SMD resistant)	F_2	QTL mapping	SMD	Gnanesh et al. (2011)
3	36 elite cultivars	Association panel	AM with SSR data	FW	Singh et al. (2013)
4	ICPL 20096 (SMD and FW tolerant) × ICPL 332 (SMD and FW susceptible)	F_7 RILs	BSA-seq	FW and SMD	Singh et al. (2016b)
	ICPL 20096, ICP 8863, ICPB 2049, ICPL 20097, ICPL 99050, ICPL 332, HPL 24, ICP 85063, ICPL 87, and ICPL 88039	Germplasm and breeding lines	WGRS		
5	ICPL 20096 (FW and SMD resistant) x ICP 332 (FW and SMD susceptible)	F_7 RILs	BSA-seq	FW and SMD	Singh et al. (2017b)
6	ICPB 2049 (FW resistant) × ICPL 99050(FW susceptible) and ICPL 20096 (FW resistant) × ICPL 332 (FW susceptible)	RILs	GBS-QTL analysis	FW and SMD	Saxena et al. (2017)
	ICPL 85063 (FW susceptible) × ICPL 87119 (FW resistant)	F_2			
7	Bahar (FW susceptible) × KPL 43 (FW resistant)	F_2	BSA with SSRs	FW	Patil et al. (2017a)
8	89 diverse pigeonpea genotypes including seven varieties, three landraces and 79 germplasm lines	Association panel	AM with SSR data	FW	Patil et al. (2017b)

in pigeonpea. Two QTLs were identified for each population in relation to the Patancheru SMD isolate: on Linkage group (LG) 9 for the ICP 8863 × ICPL 20097 population and on LG2 and LG7 for the TTB 7 × ICP 7035 population. Similarly, two QTLs were identified on LG1 and LG3 in the TTB 7 × ICP 7035 population for the Bangalore SMD isolate. However, the major QTL *qSMD4* was mapped to a 2.8 cM interval on LG7, between the SSR markers CcM1825 and CcM1895, in the

TTB 7 × ICP 7035 mapping population for the Patancheru SMD isolate, explaining 24.72% of the phenotypic variance with LOD value of 6.74. All six QTLs identified for the Bangalore SMD isolate (detected in the TTB 7 × ICP 7035 population) and the Patancheru SMD isolates (detected in both the ICP 8863 × ICPL 20097 and TTB 7 × ICP 7035 populations) inherited the "resistant" allele from the susceptible parents, ICP 8863 and TTB 7.

Singh et al. (2016b) used a combined approach of Seq-BSA (sequencing of bulks) and WGRS-based nonsynonymous SNPs (nsSNPs) with one mapping population (ICPL 20096 × ICPL 332) for identification of candidate resistance genes for FW and SMD. The parental line ICPL 20096 is resistant to both FW and SMD, and ICPL 332 is susceptible to both diseases. This study identified seven candidate SNPs present on five different chromosomes (CcLG02, CcLG07, CcLG08, CcLG10, and CcLG11). Similarly, through WGRS and haplotype analysis approach, 4 nsSNPs each for FW and SMD resistance (four on CcLG02, one on CcLG08, and three on CcLG11) showed a specific haplotype in all resistant genotypes and an alternate allele in all susceptible genotypes. These candidate nsSNPs (haplotypes) for SMD resistance were found in *C. cajan_07067*, *C. cajan_15535* (*copia-like retrotransposable*), and *C. cajan_01839* (*serine–threonine protein phosphatase*). Similarly, candidate nsSNPs (haplotypes) for FW resistance were found in *C. cajan_07078* (PHD finger protein), *C. cajan_07124 (rRNA-processing protein), C. cajan_02962 (NADH dehydrogenase)*, and *C. cajan_03203 (retrovirus-like polyprotein)*. Further, these SNPs were validated in contrasting parents for FW and SMD reactions. The causative SNPs (C to G) in *C. cajan*_01839 (*serine–threonine protein phosphatase*) associated with SMD have a deleterious effect on the susceptible genotype and are known to regulate specific signal transduction cascades. SNPs associated with the *C. cajan_03203* gene for FW resistance and the *C. cajan_01839* gene for SMD have deleterious effects on protein translation and influence conformational changes in the secondary structure of these proteins, and are implicated in various other disease resistance in other plants and model systems. The C. cajan_03203 gene exhibited a 3.45-fold down-regulation in root tissues of the FW-susceptible genotype, compared to a 0.50-fold down-regulation in the FW-resistant genotype. Similarly, the C. cajan_01839 gene showed a 2.32-fold upregulation in leaf tissues of the SMD-susceptible genotype (ICP332), in contrast to a 0.83-fold upregulation in the SMD-resistant genotype.

Singh et al. (2017b) adopted the Indel-seq approach to identify candidate genomic regions associated with FW and SMD resistance in pigeonpea. The Indel-seq approach combines WGRS with BSA and relies on InDel frequencies in extreme bulks. This study identified 16 InDels, primarily single-base pair insertion-deletion polymorphisms, affecting 26 putative candidate genes in resistant and susceptible genotypes. Notably, three InDels on CcLG02, CcLG07, and CcLG08 were significantly associated with FW resistance: (1) a "C to CA" insertion in the upstream and downstream regions of the AP-1 complex subunit sigma-2 (*C. cajan_05665*) and L-ascorbate oxidase (*C. cajan_05665*), respectively; (2) an "AT to A" deletion in the intronic region of the receptor-like protein kinase (*C. cajan_17341*); and (3) a "T to TG" deletion in the transcriptional co-repressor *SEUSS* gene. Similarly, two InDels on CcLG02 and CcLG10 were associated with SMD resistance: (1) a "CA to C" deletion affecting two genes, *C. cajan_05815* in the upstream region and *C. cajan_05816* in the downstream region, and (2) an "AT to A" deletion causing a frame-shift in an uncharacterized protein (*C.cajan_15032*).

Saxena et al. (2017) mapped the QTLs for FW resistance using two RILs (ICPB 2049 × ICPL 99050 and ICPL 20096 × ICPL 332) and one F_2 (ICPL 85063 × ICPL 87119) population following the GBS approach and identified a total of 14 QTLs with six major QTLs explaining >10% phenotypic variance. Based on comparative analysis, it was confirmed that three major QTLs/genes (*qFW11.1*, *qFW11.2* and *qFW11.3*) conferring FW resistance are present on LG11, which is common to both RIL populations. *qFW11.1* was found to be common in all three populations. Patil et al. (2017a) identified three SSR markers HASSR8, HASSR58, and HASSR121 as linked with resistance to FW race-2 following BSA approach in the F_2 mapping population derived from Bahar (FW susceptible) × KPL 43 (FW resistant). HASSR8 identified as the most closely linked (5.01 cM) with the FW resistance gene, followed by HASSR121 (10.13 cM), and HASSR58 (27.46 cM).

AM approach with Kruskal–Wallis ANOVA and simple regression statistical analysis identified six SSR markers (ASSR-1, ASSR-23, ASSR-148, ASSR-229, ASSR-363, and ASSR-366) as significantly associated with FW resistance (Singh et al. 2013). These markers were validated for their efficacy in selecting FW resistant parental lines and found ASSR-1, ASSR-23, and ASSR-148 markers as efficient for FW resistance pigeonpea breeding programs (Singh et al. 2016a). Patil et al. (2017b) identified six markers significantly associated with resistance to variant 2 of FW using an AM approach. These markers explained up to 6% of the phenotypic variance over three years of phenotypic evaluation. Among them, HASSR18 was the most stable and significant marker, accounting for 5%–6% of the phenotypic variance consistently across the years.

Saxena et al. (2021b) developed a diagnostic marker kit to identify pigeonpea lines resistant to FW and SMD. The kit was based on polymorphisms observed in two genes, *C.cajan_03691* and *C.cajan_18888*, associated with FW resistance, and four genes, *C.cajan_07858*, *C.cajan_20995*, *C.cajan_21801*, and *C.cajan_17341*, linked to SMD resistance. Few genes are located within the earlier mapped QTL intervals as reported by Singh et al. (2016b, 2017b) and Saxena et al. (2017). Variants were mined from WGRS data of 114 pigeonpea lines. The selected variants effectively differentiated between susceptible and resistant genotypes for FW and SMD. These 20 polymorphisms were converted into Kompetitive Allele-Specific PCR (KASP) genotyping assays and validated across 74 pigeonpea genotypes, including parents of segregating populations, released varieties, elite breeding materials, and 18 BC_1F_1 individual plants. Of these markers, 19 were found to be polymorphic within the selected panel, with polymorphism information content (PIC) values ranging from 0.28 to 0.37. The gene *C.cajan_03691*, located in the FW resistance genomic region on CcLG11, is predicted to encode the protein Pumilio homolog 6. This gene exhibits three synonymous mutations and one missense mutation involving a "C to A" nucleotide change. The gene *C.cajan_18888*, situated on CcLG07, encodes the protein allene oxide cyclase and harbors a synonymous mutation characterized by a "G to T" allele variation. The *C.cajan_20995* gene, located on CcLG04, plays a crucial role in imparting resistance to SMD. It encodes the Myb protein 1 and contains a missense mutation with an "A to G" nucleotide change.

Meta-analysis of 498 QTLs associated with agronomic traits, fertility restoration, disease resistance, and seed quality traits in pigeonpea led to the identification of 34 meta-QTLs (MQTLs) with an average confidence interval of 2.54 cM. Of these, 12 were high-confidence MQTLs, encompassing 2,255 gene models associated with specific super families, including: (i) glycoside hydrolase, (ii) protein kinase-like domain, (iii) leucine-rich repeat domain, (iv) glucanase domain, and (v) ribonuclease H-like. Notably, these MQTLs contain ten candidate genes linked to resistance against FW and SMD, including calmodulin-binding domain, WRKY domain, ankyrin repeat, POZ domain, transcription factor TGA-like domain, and P-loop containing nucleoside triphosphate hydrolase. The calmodulin-binding proteins and P-loop containing nucleoside triphosphate hydrolases include several kinases involved in disease resistance signaling pathways. WRKY proteins regulate the response to biotic stresses by modulating the salicylic acid, jasmonic acid, and ethylene signaling pathways (Halladakeri et al. 2023).

9.7.2 Other-omics Approaches in Identifying Candidate Genes, Pathways Associated with Disease and Insect Pest Resistance in Pigeonpea

9.7.2.1 Transcriptomic Approaches in Biotic Stress of Pigeonpea

Functional characterization of a gene, specifically its expression pattern, provides insights into tissue and stage specificity, as well as its association with responses to wounding, biotic, and abiotic stress (Kudapa et al. 2012; Bohra et al. 2021a, 2021b). The expression pattern of a gene can be analyzed using absolute or relative quantification methods, such as semi-quantitative PCR or real-time quantitative PCR. A comprehensive expression analysis of all genes in a genome can be conducted using microarray or RNA-seq/transcriptome approaches. However, with the advent of low-cost next-generation sequencing technologies and powerful statistical algorithms for RNA-seq data analysis, RNA-seq has largely surpassed microarray-based studies for expression analysis. Transcriptome analysis is crucial for understanding dynamic biological processes, such as developmental stages, responses to environmental stress, disease mechanisms, and the impact of genetic modifications. RNA-seq enables a thorough examination of gene expression across the entire genome, including the detection of low-abundance transcripts. RNA-seq also facilitates the identification of novel genes that are not annotated earlier, the detection of alternatively spliced forms, the identification of noncoding RNAs, and the detection of mutations in exons or coding sequences (CDS). This comprehensive approach provides a deeper understanding of gene regulation and cellular functions. There are reports of comprehensive transcript analysis using the RNA-seq approach for studying FW and pod borer incidence in pigeonpea (Rathinam et al. 2019; Njaci et al. 2020; Soren et al. 2021; Purohit et al. 2021; Patil et al. 2021; Ghosh et al. 2022).

RNA-seq data from a comparative analysis of Bahar (FW susceptible) and KPL-44 (FW resistant) pigeonpea cultivars at late infection stages, specifically 72 hours postinoculation (HPI) and 96 HPI, in response to the *F. udum* isolate FU12 identified potential regulatory genes involved in complex spatial-temporal genome regulation for fungal disease resistance. Genes coding for defensin-like protein, disease resistance proteins such as At4g27190, RGA1, RGA2, RGA3, RGA4, RPP13, RPM1, RPP8, and RPM1, as well as major pollen allergen, mildew-resistant protein, MLO-like protein, MLP-like protein, pro-hevein, stress-induced protein SAM, wound-induced protein WIN, S-norcoclaurine synthase, and TMV-resistant protein were among the significantly regulated under pathogen infection (Soren et al. 2021).

Transcriptome profiling of Pigeonpea FW susceptible (ICP 2376) and FW resistant (ICP 8863) genotypes challenged with *F. udum* isolate MTCC 2204 at the invasion stage of the pathogenesis process through cDNA-AFLP identified 42 differentially regulated gene fragments similar to known functions such as stress or DR, wound response, transport, signal transduction, metabolism, vascular development, transcription factors, and transcription regulation, DNA repair, circadian rhythm, and cell component (Purohit et al. 2021). Patil et al. (2021) studied ten defense responsive genes during FW infection in resistant and susceptible pigeonpea cultivars following real-time PCR analysis. The expression of pathogenesis-related protein-coding genes *PR2* and *PR10* was found to be higher in resistant cultivars (Vaishali and Maruti) than in susceptible cultivars. Similarly, *PAL2 (Phenylalanine ammonia lyase-2)* expression was higher in resistant cultivars at all stages, and it has a role in converting phenylalanine to coumaric acid. *CaPAL1* in pepper acts as a positive regulator of SA-dependent defense signaling during microbial pathogen infection, functioning through its enzymatic activity in the phenylpropanoid pathway. It also assists resistant genotypes in fortifying cell walls at the infection site by lignification and/or suberization, which helps kill the pathogen through phenols and phytoalexins, while serving as a signaling molecule. Chalcone synthase (CHS) and Isoflavone Synthase-1 (IFS1) are known to produce antifungal compounds, including isoflavonoids and phytoalexins, which help inhibit pathogen growth. Resistant cultivars exhibit high levels of *CHS8* expression in all stages after *F. udum* infection, indicating its role in inhibiting pathogen growth, thereby demonstrating its critical importance in defense mechanisms. However, *IFS1* gene expression is more pronounced in resistant cultivars at later stages of infection.

The *F3H (Flavanone 3-hydroxylase)* gene steadily increased during the initial hours of infection, then decreased in later stages of resistant genotypes; however, its expression decreased at all stages of susceptible cultivars. It involves in inducing the production of antimicrobial compounds like phenols, flavonoids, and phytoalexin. *LOX* (9(S)-Lipoxygenase) gene expression was higher in resistant cultivars during the early stages of infection, but it was higher in susceptible cultivars at later stages. The *LOX* gene plays a role in producing signaling molecules, such as jasmonic acid and its methyl ester, which boost the initial response and activate defense genes. Additionally, lipoxygenases are known to contribute to the production of fatty acid peroxides as part of the HR. *ADC* (*Arginine decarboxylase*), *SPDS* (*Spermidine synthase*), and *SPMS* (*Spermine synthase*) genes involved in polyamine metabolism were studied. The ADC and SPDS gene expression increased in early stages; however, it decreased in later stages of resistance cultivars. However, SPMS gene expression was higher in resistant cultivars than in susceptible cultivars during all stages. Polyamines (PAs) and polyamine catabolism may play multiple roles in plant resistance to pathogen infection. These include acting as compatible solutes, serving signaling functions, regulating Ca++ ion homeostasis, controlling programmed cell death and apoptosis, preserving plant cell membrane integrity, and producing efficient antioxidants and antioxidant products (Patil et al. 2021).

Transcriptome analysis of FW susceptible (ICP 2376) and FW resistant (ICP 8863) pigeonpea cultivars at 36 HPI of *F. udum* MTCC 2204 strain infection identified 14,401 upregulated, 12,257 downregulated transcripts (Ghosh et al. 2022). Among these, 7,877 and 5,469 transcripts were exclusively upregulated and downregulated, respectively, in resistant cultivars upon infection. Similarly, 9,537 and 8,251 transcripts were exclusively upregulated and downregulated, respectively, in susceptible cultivars. Notably, 826 transcripts were upregulated in susceptible cultivars but downregulated in resistant genotypes, while 913 transcripts were upregulated in resistant genotypes but downregulated in susceptible genotypes (Ghosh et al. 2022). Stress- and defense-related genes such as *peroxidase73, RPM1, RIN4, RPP8, RPS2, MLO-like protein, wound-induced protein WIN, WRKY70, WRKY33, PR1, TMV-resistant protein, WAT1, MYB46*, and *CESA* were significantly regulated upon infection with *F. udum* isolate. 41 pathogen responsive *CcNBS-LRR* genes were identified in pigeonpea, and of these, 39 and 37 genes were differentially upregulated in the resistant and susceptible cultivar against *F. udum*. Interestingly, five *NBS* genes—*CcNBS 125*, *CcNBS 244*, *CcNBS 22*, *CcNBS 136*, and *CcNBS 143*—were upregulated by 25- to 33-fold following *F. udum* infection in resistant cultivars (Bajpai et al. 2023).

Plants actively respond to insect herbivory and mechanical wounding by triggering specific biochemical changes, such as the production of phytohormones, secondary metabolites, and proteins with antimicrobial (phytoalexins), insecticidal, antioxidant properties, and other secondary metabolites. These responses occur either at the injury site or systemically in distant, unwounded tissues (Njaci et al. 2020). The wild species of pigeonpea possess traits such as pod toughness, pod wall structure, trichomes on the pod surface, and biochemical properties like anti-nutritional factors, including proteinase inhibitors, oligosaccharides, phenols, tannins, and phytic acid. These characteristics may contribute to pod borer resistance by negatively impacting insect egg-laying, larval survival, and the extent of pod damage (Sujana et al. 2008; Sharma et al. 2009). *C. platycarpus*, belonging to the tertiary gene pool and possessing all four types of trichomes (Types A, B, C, and D), is being utilized to enhance disease resistance and other agronomic traits in pigeonpea (Mallikarjuna et al. 2005). Recently, it has been studied at the molecular level to gain a better understanding of its resistance to pod borer (Rathinam et al. 2019), and the findings suggest that wild species (*C. platycarpus*) exhibits effective transcriptomic reprogramming to defend against the insect. It demonstrates a multi-faceted response, with key involvement of SAR, redox homeostasis, and the reorganization of primary metabolites, culminating in a robust defense mechanism. The genes involved in the production of cyanogenic glycosides, brassinosteroid biosynthesis, isoflavone pathway, monoterpene synthesis and lignin biosynthesis were upregulated in response to *H. armigera* infestation in *C. platycarpus* (Rathinam et al. 2019). Genes involved in direct defense encode pathogenesis-related (PR) proteins that influence larval feeding and development. Chitinase is a PR protein that is observed to be expressed in both *C. platycarpus* and *C. cajan*. However, the difference between these two species is that chitinase genes were upregulated at the later stage of infestation in cultivated pigeonpea, whereas most of the chitinase genes were upregulated since the initiation of *H. armigera* infestation in wild species. DR genes such as *CHI4, E138*, and *PER15* exhibited a consistent increase in expression levels across all time points in the wild relatives. Another defense-related bifunctional gene, *IAAS*, is expressed in both species and plays a vital role in plant carbohydrate metabolism while also exhibiting inhibitory activity against larval gut digestive enzymes (Rathinam et al. 2019). Genes involved in the activation of phytoanticipins, such as glucosidases, have been shown to be expressed at earlier time points in the wild relative. These results altogether indicate that wild relative responds quickly to the attacking herbivore (Rathinam et al. 2019).

Njaci et al. (2020) undertook transcriptome profiling of podborer susceptible (cultivated: KAT 60/8) and tolerant (wild: *C. scarabaeoides*) pigeonpea genotypes to investigate the defense mechanisms. The wild genotype exhibited higher pest-induced gene expression, including a stronger induction of phytohormone and calcium/calmodulin signaling, transcription factors, plant volatiles, and secondary metabolite genes, compared to the cultivated control. Specifically, the wild types naturally possess distinct tolerance mechanisms by modulating biosynthetic and regulatory processes associated with secondary metabolite production, such as flavonoids, terpenes, and glucosinolates. Three important enzymes coding genes in the plant hormone JA biosynthesis pathway namely, lipoxygenase, oxophytodienoate reductase, and allene oxide synthase showed enhanced expression in wild type (WT) compared to susceptible cultivars. JA plays a key role in activating a DR against biotic attacks, including chewing insects. Methyl jasmonate (MeJA) functions as a signaling molecule associated with the biosynthesis of flavonoids and helps to manage fungal disease (Du et al. 2021). Similarly, enzymes involved in the ethylene pathway, such as 2-oxoglutarate (2OGD), 1-aminocyclopropane-1-carboxylate synthase, and 1-aminocyclopropane-1-carboxylate oxidase, showed enhanced expression in the WT upon herbivory. The *Kunitz trypsin inhibitor 2 (KTI2)* and *Nerolidol synthase 1* expression was observed to be higher in WT. Trypsin inhibitors are herbivore- and wound-induced secondary metabolites that block insect digestive proteases in the larval midgut, resulting in reduced insect growth rates or increased mortality. Nerolidol synthase 1 (NSE1) is involved in the biosynthesis of monoterpenes, a group of volatile compounds that are toxic to insects. *WRKY* transcription factors are one of the largest families of regulatory proteins involved in defense responses against insect pests and pathogens. Two such *WRKY* genes, *WRKY3* and *WRKY6*, were also significantly upregulated. Based on the mapping reads obtained from RNA-seq and WGRS data of *C. scarabaeoides* on the *C. cajan* genome, a very low percentage of reads were mapped. This indicates that *C. scarabaeoides* possesses unique unigenes that elicit distinct molecular responses, activating different pathways to combat various diseases and pest infections (Njaci et al. 2020).

9.7.2.2 Proteomic Approach in Biotic Stress of Pigeonpea

Oxalic acid (OA) is a key pathogenicity factor secreted by many fungal pathogens. OA acidifies the plant tissue surrounding the infection site, causing tissue damage. The acidification by OA may further enhance the activity of many fungal cell wall–degrading enzymes that operate in the acidic range, making the plant more susceptible to pathogen invasion. Additionally, OA is a strong chelator of divalent cations, and the sequestration of calcium disrupts the cell wall reconstruction process, further weakening the cell walls (Dong et al. 2008). Removing OA released by pathogens could enhance fungal disease tolerance in plants, which can be achieved by the oxalate oxidase (OXO) enzyme that oxidizes OA into CO_2 and Hydrogen peroxide (H_2O_2). H_2O_2 plays a crucial role in the HR in plants and has a key role, such as the following: (1) Signal transduction molecule triggering the activation of various defense-related genes and pathways in adjacent cells (2) Oxidative burst: the earliest events in the HR, and thus directly killing the invading pathogens. (3) Cell wall reinforcement by lignification and oxidative cross-linking of cell wall hydroxyproline-rich proteins and other cell wall polymers, creating a physical barrier to impede pathogen spread. (4) Induction of Programmed Cell Death prevents the pathogen from spreading to other parts of the plant. (5) Production of secondary metabolites with antimicrobial properties such as phytoalexins (Levine et al. 1994; Kuźniak and Urbanek 2000). Proteomic analysis using LC-MS/MS of wilt-infected transgenic and WT pigeonpea leaves identified 2,386 proteins, with 1,048 showing significant abundance changes—738 upregulated and 310 downregulated in transgenic plants. Notably, proteins such as HMG1/2-like protein, putative nucleosome assembly protein C364.06, DEAD-box ATP-dependent RNA helicase 3, lipoxygenase 1, annexin D1, and annexin-like protein RJ4 were significantly upregulated (Karmakar et al. 2024). Metabolomic analysis of transgenic pigeonpea revealed enhanced levels of amino acids, sugars, OA, sugar alcohols, and myo-inositol. Specifically, proteins and enzymes related to pathways in sugar and starch metabolism and inositol phosphate metabolism were upregulated in transgenic pigeonpea under wilt stress (Karmakar et al. 2024).

9.7.2.3 Multi-omics Approaches (Transcriptomics and Proteomics/Metabolomics) in Biotic Stress of Pigeonpea

Tightly regulated phenylpropanoid pathway is linked with the antibiosis in the pod borer resistance. The phenylpropanoid metabolic pathway produces over 8,000 metabolites involved in plant development (Dong and Lin 2021). This pathway branches into the production of lignins and flavonoid biosynthesis, generating secondary metabolites such as flavanones, flavonols, flavanols, and anthocyanins. Among these, flavonoids are anti-digestive, anti-feedants, resulting in toxic effects on the insect's feeding, growth, development, and reproduction (Tyagi et al. 2022). The quantification of transcripts involved in the flavonoid biosynthesis pathway in *H. armigera*-challenged *C. platycarpus* accession ICPW 068 showed dynamic upregulation of genes such as *chalcone synthase, dihydroflavonol-4-reductase, flavonoid-3′5′-hydroxylase*, *flavonol synthase*, *leucoanthocyanidin reductase*, and *anthocyanidin synthase* (Tyagi et al. 2022). Targeted LC-MS analyses revealed up to a 4-fold increase in compounds such as naringenin, kaempferol, quercetin, delphinidin, cyanidin, epigallocatechin, and epicatechin-3-gallate. When *H. armiger*a larvae were fed a diet supplemented with 100 ppm of flavonoids, they showed detrimental effects on growth and prolonged the larval period, demonstrating the role of the flavonoid pathway (Tyagi et al. 2022).

Combined transcriptome and metabolomic analyses using high-performance liquid chromatography (HPLC) found upregulation of various genes involved in signal transduction and isoflavone biosynthesis pathways, along with the accumulation of flavonoids such as genistein, apigenin, vitexin, and biochanin-A upon exogenously applied jasmonic acid (JA) in fungal disease resistance in pigeonpea (Du et al. 2021). Additionally, the key genes in the flavonoid biosynthesis pathway (*CcI2′H*, *CcHIDH*, *Cc7-IOMT*) were also significantly upregulated. Overexpression of *CcbHLH35* notably induced the upregulation of *flavonoid synthase* genes and increased the accumulation of genistein, vitexin, and apigenin. This experiment highlights the role of JA in enhancing defense against fungal pathogens.

9.8 INTEGRATED APPROACH IN IDENTIFYING CAUSAL GENES/ALLELES AND PATHWAYS

In biology, the term "omics" refers to the collective set of constituents within a cell. Omics sciences share the common goal of identifying, describing, and quantifying the biomolecules and molecular processes that contribute to the structure and function of cells and tissues. To explore the potential of pigeonpea for various biotic and abiotic stress tolerance and to understand the genetic architecture behind these traits, an integrated multi-omics approach is essential. This approach should integrate phenomics, genomics, epigenomics, transcriptomics, proteomics, lipidomics, and metabolomics data to fully understand the physiological, biochemical, and molecular interactions, specifically aiming to identify critical regulators of target trait expression (Figure 9.1).

However, only a few studies have been conducted that integrate a few omics, either QTL analysis with transcriptomics, transcriptomics with proteomics/metabolomics, or proteomics with metabolomics in biotic stress of pigeonpea. The "Ppomics" database is an integrated multi-omics platform for pigeonpea, comprising phenomics, genomics, transcriptomics, and proteomics datasets. It includes data related to pigeonpea genome assembly, trait-associated genes, genotyping data, QTLs, transcriptome data, and gene functions, and can be used for the development of superior pigeonpea breeds (Singh et al. 2024). An integrated approach involving phenotypic, genomic, transcriptomic, metabolomic, and lipidomic analyses to study the physiological and molecular responses of pigeonpea roots with respect to phosphate (Pi) starvation was applied in *C. cajan* genotype CF052777 and the study revealed that low-Pi treatment increased total root length, root surface area, and fertility transition in a pigeonpea line. The Pi-starvation upregulated four starvation responsive genes namely *phosphate transporter 1* (PHT1), *phosphoethanolamine/phosphocholine phosphatase* (PECP), *fasciclin-like arabinogalactan protein* (FLA), and *glutamate decarboxylase* (GAD; Liu et al. 2022). Pazhamala et al. (2020) integrated phenomics, transcriptomics, proteomics, and metabolomics to study fertility transition in a two-line hybrid system of pigeonpea using male sterile and fertile anthers from five developmental stages and the study identified 17 repressed genes/proteins in sterile anthers that are crucial for normal cell wall morphogenesis and tapetal cell development. The same approach can be extended to identify core regulators and pathways imparting resistance against FW,

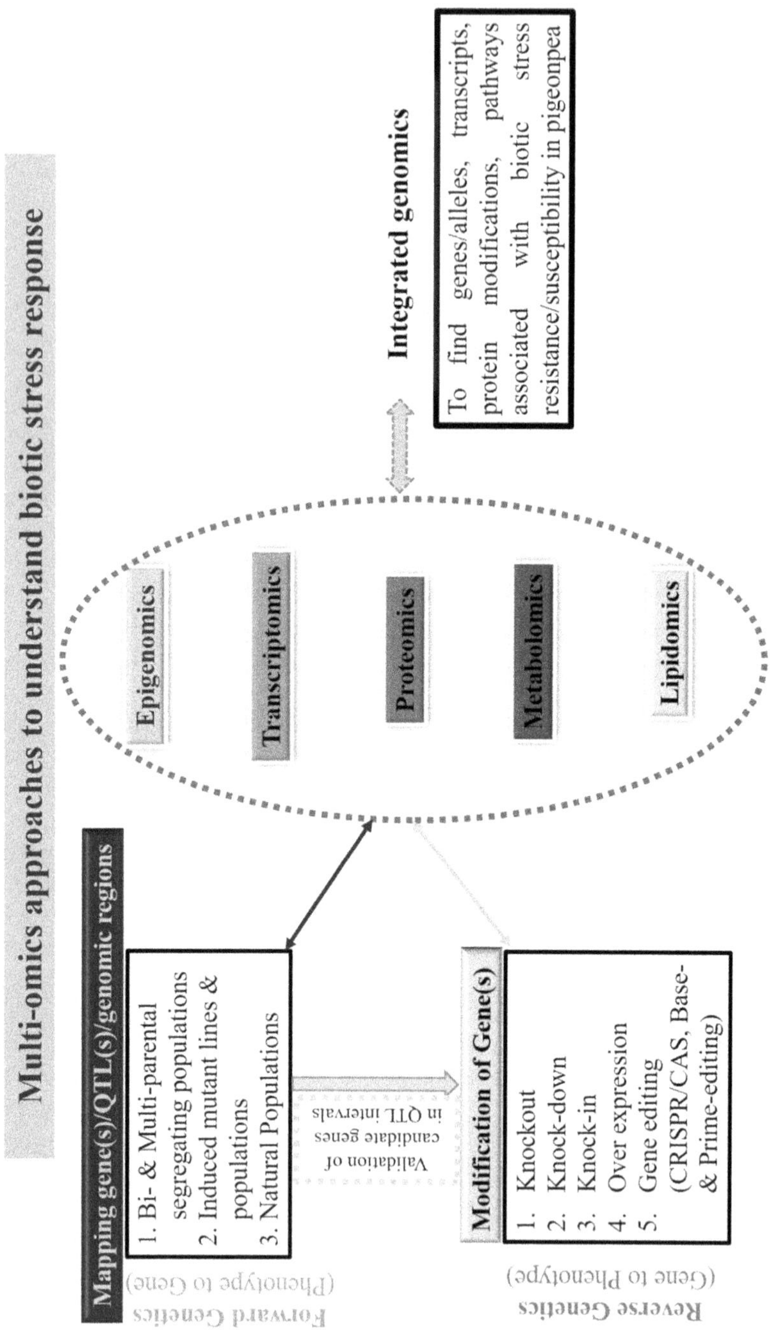

FIGURE 9.1 Integrated-Omics approach in biotic stress response of crops.

SMD and pod borer in pigeonepa. Further studies should focus on the role of microRNAs, long noncoding RNAs, and epigenetics in determining biotic stress tolerance in pigeonpea.

9.9 REVERSE GENETICS AND TRANSGENIC APPROACHES IN ACHIEVING RESISTANCE AGAINST DISEASE AND INSECTS OF PIGEONPEA

Reverse genetics is a powerful approach used in plant breeding and molecular biology to understand gene function by manipulating genes of interest. This method is particularly effective for developing resistance to diseases and insects in crops like pigeonpea, where resistance sources for pod borer are mostly available in wild species that are not easily crossable with cultivated varieties (Sharma et al. 2022). Additionally, reverse genetics is routinely used for the functional validation of candidate genes identified in QTL intervals as well as those discovered through transcriptome and proteome analyses. To perform reverse genetics, gene sequence information is essential. Researchers can induce gene silencing using technologies such as RNA interference, Antisense RNA technology, or CRISPR-Cas in pigeonpea plants to study the role of specific genes involved in disease resistance. Once resistance genes (R genes) are identified and validated, they can be used in breeding programs to improve disease resistance through marker-assisted fast breeding with gene/trait-linked markers.

Transgenic approaches with various natural and modified *Bt* genes (*cry1Ab*, *Cry1Ac*, *Cry2Aa*, *cry I E-C*, and *Cry1AcF*) are being employed globally to combat the pod borer and *S. litura* menace in pigeonpea (Surekha et al. 2005; Sharma et al. 2006; Das et al. 2016; Singh et al. 2018; Ramkumar et al. 2020; Sarkar et al. 2021). Chitinases play a vital role in fungal disease management in plants. Chitinases are enzymes that break down chitin, a major component of fungal cell walls, thereby weakening the fungal cell walls and inhibiting fungal growth and proliferation. Chitinases are either constitutively expressed or induced as part of the plant's SAR and work synergistically with other PR proteins such as glucanases to enhance the antifungal activity. Overexpression of chitinase genes in transgenic plants has been shown to enhance resistance to fungal pathogens (Kumar et al. 2004b). Pigeonpea genotypes transformed with the rice oxalate oxidase 4 (*Osoxo4*) gene significantly enhance wilt resistance by reducing or removing pathogen released OA (Karmakar et al. 2024).

9.10 GENOME EDITING FOR IMPROVING DISEASE AND INSECT TOLERANCE

Genome editing technologies offer a promising approach to improving disease and insect tolerance in pigeonpea by precise modifications to the genome (Gao 2021; Erdoğan et al. 2023; Kumar et al. 2023). Gene editing with CRISPR tools provides scientists with an option to manipulate gene expression by completely knocking out or knocking down the target genes, as well as to correct errors efficiently and precisely in specific R or S genes in plants by following base or prime editing tools (Yin and Qiu 2019). Base or prime editing tools could be an excellent solution for understanding compatible and incompatible reactions in plant-pathogen interactions.

Multiple genes can be edited simultaneously using multiplex gene editing, either for gene knockout, knockdown, or base editing (Xie and Yang 2019). The success of genome editing in enhancing disease resistance has been demonstrated in various crops (Sha et al. 2023; Mathsyaraja et al. 2024).

9.11 CONCLUSION

Identifying various resistance and susceptibility genes in pigeonpea, as well as effectors in pathogens, is crucial for understanding the mechanisms of disease resistance and susceptibility. Integrated or multi-omics approaches—sequentially applying phenomics, genomics, transcriptomics, proteomics, and metabolomics—will pave the way for deciphering genes and pathways associated with target traits. This knowledge will aid in the effective modification of plants using advanced breeding tools, such as genomic selection and gene editing technologies. Achieving broad-spectrum resistance in pigeonpea against major diseases is our ultimate goal. This can be accomplished by enhancing phenomics to independently screen plants against various races, which will help identify novel *R* genes and develop superior varieties with resistance to various diseases.

REFERENCES

Bajpai R, Sarkar A, Sarma BK (2023) Transcriptional regulation and characterization of *Fusarium udum* responsive NBS-LRR genes in pigeonpea (*Cajanus cajan* L Millsp). *J Plant Growth Regul.* https://doi.org/10.1007/s00344-023-11190-x.

Bernays EA, Champman RF (1994) *Host-plant selection by phytophagopus insects.* Chapman & Hall, New York.

Bohra A, Dubey A, Saxena RK, et al. (2011) Analysis of BAC-end sequences and development of BES-SSR markers for genetic mapping and hybrid purity assessment in pigeonpea (*Cajanus* spp). *BMC Plant Biol* 11:56.

Bohra A, Jha R, Pandey G, et al. (2017) New hypervariable SSR markers for diversity analysis, hybrid purity testing and trait mapping in pigeonpea (*Cajanus cajan* L Millspaugh). *Front Plant Sci* 8:377.

Bohra A, Jha UC, Godwin ID, Varshney RK (2020a) Genomic interventions for sustainable agriculture. *Plant Biotechnol J* 18:2388–2405. https://doi.org/10.1111/pbi.13472

Bohra A, Prasad G, Rathore A, Saxena RK, Satheesh Naik SJ, Pareek S, Jha R, Pazhamala L, Datta D, Pandey G, Tiwari A, Maurya AK, Soren KR, Akram M, Varshney RK, Singh NP (2021a) Global gene expression analysis of pigeonpea with male sterility conditioned by A2 cytoplasm. *Plant Genome* 14:e20132.

Bohra A, Rathore A, Gandham P, Saxena RK, Satheesh Naik SJ, Dutta D, Singh IP, Singh F, Rathore M, Varshney RK, Singh NP (2021b) Genome-wide comparative transcriptome analysis of the A4-CMS line ICPA 2043 and its maintainer ICPB 2043 during floral bud development of pigeonpea. *Funct Integr Genomics* 21:251–263.

Bohra A, Saxena KB, Varshney RK, Saxena RK (2020b) Genomics assisted breeding for pigeonpea improvement. *Theor Appl Genet* 133:1721–1737.

Chaithanya KB, Prasanthi L, Reddy HK, Reddy BBV (2011) Study of inheritance of fusarium wilt resistance through molecular marker analysis in pigeonpea (*Cajanus cajan* L Millsp). *Legume Res* 34:212–216.

Changaya AG, Melis R, Derera J, et al. (2012) Inheritance of resistance to fusarium wilt and yield traits in pigeonpea. *Euphytica* 186:883–896. https://doi.org/10.1007/s10681-012-0681-5

Das A, Datta S, Sujayanand GK, et al. (2016) Expression of chimeric Bt gene Cry1Aabc in transgenic pigeonpea (cv Asha) confers resistance to gram pod borer (*Helicoverpa armigera* Hubner). *Plant Cell Tiss Organ Cult* 127:705–715. https://doi.org/10.1007/s11240-016-1131-1

Daspute A, Fakrudin B, Bhairappanavar SB, et al. (2014) Inheritance of pigeonpea sterility mosaic disease resistance in pigeonpea. *Plant Pathol J* 30:188–194. https://doi.org/10.5423/PPJ.NT.10.2013.0104

Dhar V, Datta S, Chaudhary RG, et al. (2012) Pathogenic and molecular characterisations of pigeonpea wilt pathogen *Fusarium udum*. *Arch Phytopathol Plant Prot* 45:423–436. https://doi.org/10.1080/03235408.2011.587974

Dong NQ, HX Lin (2021) Contribution of phenylpropanoid metabolism to plant development and plant–environment interactions. *J Integr Plant Biol* 63:180–209. https://doi.org/10.1111/jipb.13054

Dong X, Ji R, Guo X, et al. (2008) Expressing a gene encoding wheat oxalate oxidase enhances resistance to *Sclerotinia sclerotiorum* in oilseed rape (*Brassica napus*). *Planta* 228: 331–340. https://doi.org/10.1007/s00425-008-0740-2

Du T, Fan Y, Cao H, et al. (2021) Transcriptome analysis revealed key genes involved in flavonoid metabolism in response to jasmonic acid in pigeonpea (*Cajanus cajan* L Millsp). *Plant Physiol Biochem* 168:410–422. https://doi.org/10.1016/j.plaphy.2021.10.022

Dutta S, Kumawat G, Singh BP, et al. (2021) Development of genic-SSR markers by deep transcriptome sequencing in pigeonpea (*Cajanus cajan* L Millspaugh). *BMC Plant Biol* 11:17. https://doi.org/10.1186/1471-2229-11-17

Dutta S, Kumawat G, Singh BP, Gupta DK, Singh S, Dogra V, Gaikwad K, Sharma TR, Raje RS, Bandhopadhya TK, Datta S, Singh MN, Bashasab F, Kulwal P, Wanjari KB, Varshney R, Cook DR, Singh NK. (2011) Development of genic-SSR markers by deep transcriptome sequencing in pigeonpea [Cajanus cajan (L.) Millspaugh]. *BMC Plant Biol.* 11:17. https://doi.org/10.1186/1471-2229-11-17

Elbeaino T, Digiaro M, Uppala M, Sudini H (2014) Deep sequencing of pigeonpea sterility mosaic virus discloses five RNA segments related to emaraviruses. *Virus Res* 188:27–31. https://doi.org/10.1016/j.virusres.2014.03.022

Erdoğan İ, Cevher-Keskin B, Bilir Ö, et al. (2023) Recent developments in CRISPR/Cas9 genome-editing technology related to plant disease resistance and abiotic stress tolerance. *Biology (Basel)* 12:1037. https://doi.org/10.3390/biology12071037

Ganapathy KN, Gowda BM, Ajay BC, et al. (2012) Inheritance studies of sterility mosaic disease resistance in vegetable type pigeonpea (*Cajanus cajan* L Millsp). *Aust J Crop Sci* 6:1154–1158.

Gao C (2021) Genome engineering for crop improvement and future agriculture. *Cell* 184:1621–1635.

Ghosh S, Purohit A, Hazra A, et al. (2022) Differential transcript expression profiles of susceptible and resistant pigeonpea cultivars at an early time point during *Fusarium udum* infection. *Front Genet* 13:1009127. https://doi.org/10.3389/fgene.2022.1009127

Gnanesh BN, Ganapathy KN, Ajay BC, Gowda BM (2011) Inheritance of sterility mosaic disease resistance to Bangalore and Patancheru isolates in pigeonpea (*Cajanus cajan* L Millsp). *Electron J Plant Breed* 2:218–223.

Halladakeri P, Gudi S, Akhtar S, et al. (2023) Meta-analysis of the quantitative trait loci associated with agronomic traits, fertility restoration, disease resistance, and seed quality traits in pigeonpea (*Cajanus cajan* L). *Plant Genome* 16:e20342. https://doi.org/10.1002/tpg2.20342

Huang Z, Li H, Zhou Y, et al. (2022) Predication of the effector proteins secreted by *Fusarium sacchari* using genomic analysis and heterogenous expression. *J Fungi (Basel)* 8:59. https://doi.org/10.3390/jof8010059

Iqbal SM, Ghafoor A, Basak M, Malik BA (1995) Estimation of losses in yield components of mungbean due to cercospora leaf spot. *Pak J Phytopath* 7:80–81.

Jain KC, Reddy MV (1995) Inheritance of resistance to fusarium wilt in pigeonpea (*Cajanus cajan* L Millsp). *Indian J Genet Plant Breed* 55:434–437.

Jones AT, Kumar PL, Saxena KB, et al. (2004) Sterility mosaic disease—the 'green plague' of pigeonpea: advances in understanding the etiology, transmission and control of a major virus disease. *Plant Dis* 88:436–445.

Joshi AR (1957) Genetics of resistance to disease and pests. *Ind J Genet Plant Breed* 17:305–317.

Karimi R, Owuoche JO, Silim SN (2010) Inheritance of fusarium wilt resistance in pigeonpea (*Cajanus cajan* L Millspaugh). *Indian J Genet Plant Breed* 70:271–276.

Karmakar S, Selvaraj S, Gayen D, Baig MJ (2024) A detailed proteomics and metabolomics landscape sheds light on the mechanistic insights into the resistance response of transgenic pigeon pea against wilt stress. *Plant Cell Environ.* https://doi.org/10.1111/pce.15261

Kotresh H, Fakrudin B, Punnuri SM, et al. (2006) Identification of two RAPD markers genetically linked to a recessive allele of a fusarium wilt resistance gene in pigeonpea (*Cajanus cajan* L Millsp). *Euphytica* 149:113–120. https://doi.org/10.1007/s10681-005-9059-2

Kudapa H, Bharti AK, Cannon SB, Farmer AD, Mulaosmanovic B, Kramer R, Bohra A, Weeks NT, Crow JA, Tuteja R, Shah T, Dutta S, Gupta DK, Singh A, Gaikwad K, Sharma TR, May GD, Singh NK, Varshney RK (2012) A comprehensive transcriptome assembly of pigeonpea (*Cajanus cajan* L) using Sanger and second-generation sequencing platforms. *Mol Plant* 5:1020–1028.

Kumar M, Prusty MR, Pandey MK, Singh PK, Bohra A, Guo B, Varshney RK (2023) Application of CRISPR/Cas9-mediated gene editing for abiotic stress management in crop plants. *Front Plant Sci* 14:1157678.

Kumar PL, Jones AT, Reddy DVR (2003) A novel mite-transmitted virus with a divided RNA genome closely associated with pigeonpea sterility mosaic disease. *Phytopathol* 93:71–81.

Kumar PL, Jones AT, Waliyar F (2004a) Biology, etiology and management of pigeonpea sterility mosaic disease. *Annu Rev Plant Pathol* 3:77–100.

Kumar S, Subbarao B, Hallan V (2017) Molecular characterization of emaraviruses associated with pigeonpea sterility mosaic disease. *Sci Rep* 7:11831. https://doi.org/10.1038/s41598-017-11958-8

Kumar SM, Kumar BK, Sharma KK, Devi P (2004b) Genetic transformation of pigeonpea with rice chitinase gene. *Plant Breed* 123:485–489.

Kumari AD, Sharma HC, Reddy DJ (2006) Oviposition non-preference as a component of resistance to pod borer (*Helicoverpa armigera*) in pigeonpea. *J Appl Entomol* 130:10–14.

Kuźniak E, Urbanek H (2000) The involvement of hydrogen peroxide in plant responses to stresses. *Acta Physiol Plant* 22:195–203. https://doi.org/10.1007/s11738-000-0076-4

Levine A, Tenhaken R, Dixon R, Lamb C (1994) H2O2 from the oxidative burst orchestrates the plant hypersensitive disease resistance response. *Cell* 79:583–593. https://doi.org/10.1016/0092-8674(94)90544-4

Liu C, Tai Y, Luo J, et al. (2022) Integrated multi-omics analysis provides insights into genome evolution and phosphorus deficiency adaptation in pigeonpea (*Cajanus cajan*). *Hortic Res* 9:uhac107. https://doi.org/10.1093/hr/uhac107

Lo Presti L, Lanver D, Schweizer G, et al. (2015) Fungal effectors and plant susceptibility. *Annu Rev Plant Biol* 66:513–545. https://doi.org/10.1146/annurev-arplant-043014-114623

Mallikarjuna N, Jadhav D, Reddy MV, Dutta-Tawar U (2005) Introgression of Phytophthora blight disease resistance from *Cajanus platycarpus* into short duration pigeonpea (*Cajanus cajan* L). *Indian J Genet Plant Breed* 65:261–263.

Marla SS, Mishra P, Maurya R, et al. (2020) Refinement of draft genome assemblies of pigeonpea (*Cajanus cajan*). *Front Genet* 11:607432. https://doi.org/10.3389/fgene.2020.607432

Mathsyaraja S, Lavudi S, Vutukuri PR (2024) Enhancing resistance to blast disease through CRISPR/Cas9 gene editing technology in OsHDT701 gene in RPBio-226 rice cv (*Oryza sativa* L). *J Appl Biol Biotech* 12:289–294. http://doi.org/10.7324/JABB.2024.154893

Mayomba MM (2018) Genetic inheritance of fusarium wilt resistance in pigeonpea (*Cajanus cajan* L Millsp) genotypes. *Huria J* 25:131–139.

Mielke-Ehret N, Mühlbach HP (2012) Emaravirus: a novel genus of multipartite, negative strand RNA plant viruses. *Viruses* 4:1515–1536.

Nagaraj KM, Chikkadevaiah, Kulkarni RS (2004) Inheritance of resistance to sterility mosaic virus in pigeonpea (*Cajanus cajan* L Millsp). *Indian J Genet* 64:118–120.

Njaci I, Ngugi-Dawit A, Oduor RO, et al. (2020) Comparative analysis delineates the transcriptional resistance mechanisms for pod borer resistance in the pigeonpea wild relative *Cajanus scarabaeoides* (L) Thouars. *Int J Mol Sci* 22:309. https://doi.org/10.3390/ijms22010309

Odeny DA, Githiri SM, PM Kimani (2009) Inheritance of resistance to fusarium wilt in pigeonpea (*Cajanus cajan* L Millsp). *J Anim Plant Sci* 2:89–95.

Okiror MA (2002) Genetics of resistance to *Fusarium udum* in pigeonpea (*Cajanus cajan* L Millsp). *Ind J Genet Plant Breed* 62:218–220.

Pandey RN, Pawar SE, Bhatia CR (1996) Inheritance of wilt resistance in pigeonpea. *Ind J Genet Plant Breed* 56:305–308.

Pastor-Corrales MA, Erazo OA, Estrada EI, Singh SP (1994) Inheritance of anthracnose resistance in common bean accession G 2333. *Plant Dis* 78:959–962. https://doi.org/10.1094/PD-78-0959

Patil PG, Bohra A, Dubey J, et al. (2017a) Genetic analysis and molecular resistance to race 2 of fusarium wilt in pigeonpea (*Cajanus cajan* L Millsp). *Crop Prot* 100:117–123.

Patil PG, Dubey J, Bohra A, et al. (2017b) Association mapping to discover significant marker-trait associations for resistance against fusarium wilt variant 2 in pigeonpea (*Cajanus cajan* L Millspaugh) using SSR markers. *J Appl Genet* 58:307–319. https://doi.org/10.1007/s13353-017-0400-y

Patil VL, Patel RM, Parekh VB, Pathak J, Saripalli G (2021) Differential gene expression analyses of ten defense response genes during fusarium wilt infection in resistant and susceptible pigeonpea cultivars. *Plant Stress* 2:100043.

Pazhamala LT, Chaturvedi P, Bajaj P, et al. (2020) Multiomics approach unravels fertility transition in a pigeonpea line for a two-line hybrid system. *Plant Genome* 13:e20028. https://doi.org/10.1002/tpg2.20028

Pazhamala LT, Purohit S, Saxena RK, et al. (2017) Gene expression atlas of pigeonpea and its application to gain insights into genes associated with pollen fertility implicated in seed formation. *J Exp Bot* 68:2037–2054. https://doi.org/10.1093/jxb/erx010

Prasanthi L, Reddy BBV, Rani KR, Naidu PH (2009) Molecular marker for screening fusarium wilt resistance in pigeonpea (*Cajanus cajan* L Millspaugh). *Legume Res* 32:19–24.

Purohit A, Ganguly S, Ghosh G, et al. (2017) Variability among isolates of *Fusarium udum* and the effect on progression of wilt in pigeonpea. *Eur J Plant Pathol* 149:73–87. https://doi.org/10.1007/s10658-017-1167-z

Purohit A, Ghosh S, Ganguly S, et al. (2021) Comparative transcriptomic profiling of susceptible and resistant cultivars of pigeonpea demonstrates early molecular responses during *Fusarium udum* infection. *Sci Rep* 11:22319. https://doi.org/10.1038/s41598-021-01587-7

Ramkumar N, Rathinam M, Singh S, et al. (2020) Assessment of pigeonpea (*Cajanus cajan* L) transgenics expressing Bt ICPs, Cry2Aa and Cry1AcF under nethouse containment implicated an effective control against herbivory by *Helicoverpa armigera* (Hübner). *Pest Manag Sci* 76:1902–1911. https://doi.org/10.1002/ps.5722

Rathinam M, Mishra P, Vasudevan M, Budhwar R, Mahato A, Prabha AL, Singh NK, Rao U, Sreevathsa R. (2019) Comparative transcriptome analysis of pigeonpea, (2019) Cajanus cajan (L.) and one of its wild relatives Cajanus platycarpus (Benth.) Maesen. *PLoS One*, 14(7):e0218731. https://doi.org/10.1371/journal.pone.0218731.

Reddy MV, Sharma SB, Nene YL (1990) Pigeonpea: disease management. In: Nene YL, Hall SD, Sheila VK, editors. *The pigeonpea.* CAB International; Wallingford, Oxon: pp. 303–348.

Reddy MY, Raju TN, Nene YL, et al. (1993) Variability in sterility mosaic pathogen of pigeonpea in India. *Indian Phytopathol* 46:206–212.

Rep M, van der Does HC, Meijer M, et al. (2004) A small, cysteine-rich protein secreted by *Fusarium oxysporum* during colonization of xylem vessels is required for I-3-mediated resistance in tomato. *Mol Microbiol* 53:1373–1383.

Sachan GC (1990) *Progress of host plant resistance work in chickpea and pigeonpea against Helicoverpa armigera in India. Proceedings, Host Selection Behavior of Helicoverpa armigera*, Patancheru, India, 5–7 March 1990:19–22.

Sarkar S, Roy S, Ghosh SK (2021) Development of marker-free transgenic pigeon pea (*Cajanus cajan*) expressing a pod borer insecticidal protein. *Sci Rep* 11:10543. https://doi.org/10.1038/s41598-021-90050-8

Saxena KB, Bohra A, Choudhary AK, Sultana R, Pazhamala L, Saxena RK (2021a) The alternative breeding approaches for improving yield gains and stress response in pigeonpea (*Cajanus cajan* L Millsp). *Plant Breed* 140:76–86.

Saxena KB, Choudhary A, Srivastava R, Bohra A, Saxena R, Varshney RK (2019) Origin of early maturing pigeonpea germplasm and its impact on adaptation and cropping systems. *Plant Breed* 138:243–251.

Saxena RK, Hake A, Bohra A, et al. (2021b) A diagnostic marker kit for fusarium wilt and sterility mosaic diseases resistance in pigeonpea. *Theor Appl Genet* 134:367–379. https://doi.org/10.1007/s00122-020-03702-0

Saxena RK, Rathore A, Bohra A, et al. (2018) Development and application of high-density Axiom Cajanus SNP array with 56K SNPs to understand the genome architecture of released cultivars and founder genotypes. *Plant Genome* 11. https://doi.org/10.3835/plantgenome2018.01.0005

Saxena RK, Singh VK, Kale SM, et al. (2017) Construction of genotyping-by-sequencing based high-density genetic maps and QTL mapping for fusarium wilt resistance in pigeonpea. *Sci Rep* 7:1911. https://doi.org/10.1038/s41598-017-01537-2

Saxena RK, et al. (2012) Large-scale development of cost-effective single-nucleotide polymorphism marker assays for genetic mapping in pigeonpea and comparative mapping in legumes. *DNA Res* 19:449–461. https://doi.org/10.1093/dnares/dss025

Sayiprathap BR, Patibanda AK, Prasanna Kumari V, et al. (2022) Salient findings on host range, resistance screening, and molecular studies on sterility mosaic disease of pigeonpea induced by pigeonpea sterility mosaic viruses (PPSMV-I and PPSMV-II). *Front Microbiol* 13:838047. https://doi.org/10.3389/fmicb.2022.838047

Sha G, Sun P, Kong X, et al. (2023) Genome editing of a rice CDP-DAG synthase confers multipathogen resistance. *Nature* 618:1017–1023. https://doi.org/10.1038/s41586-023-06205-2

Sharma D, Gupta SC, Rai S, Reddy MV (1984) Inheritance of resistance to sterility mosaic disease in pigeonpea. *Indian J Genet* 44:84–90.

Sharma HC, Sujana G, Rao DM (2009) Morphological and chemical components of resistance to pod borer, *Helicoverpa armigera*, in wild relatives of pigeonpea. *Arthropod Plant Interact* 3:151–161.

Sharma KK, Lavanya M, Anjaiah V (2006) Agrobacterium-mediated production of transgenic pigeonpea (*Cajanus cajan* L Millsp) expressing the synthetic Bt cry1Ab gene. *In Vitro Cell Dev Biol Plant* 42:165–173. https://doi.org/10.1079/IVP2005730

Sharma S, Jaba J, Rao PJ, et al. (2022) Reaping the potential of wild *Cajanus* species through pre-breeding for improving resistance to pod borer, *Helicoverpa armigera*, in cultivated pigeonpea (*Cajanus cajan* L Millsp). *Biol* 11:485. https://doi.org/10.3390/biology11040485

Singh AK, Rai VP, Chand R, et al. (2013) Genetic diversity studies and identification of SSR markers associated with fusarium wilt (*Fusarium udum*) resistance in cultivated pigeonpea (*Cajanus cajan*). *J Genet* 92:273–280. https://doi.org/10.1007/s12041-013-0266-7

Singh BV, Pandeya BP, Gautam PL, Beniwal SPS, Pandey MP (1983) Inheritance of resistance to sterility mosaic virus in pigeonpea. *Indian J Genet* 43:487–493.

Singh D, Sinha B, Rai VP, et al. (2016a) Genetics of fusarium wilt resistance in pigeonpea (*Cajanus cajan*) and efficacy of associated SSR markers. *Plant Pathol J* 32:95–101. https://doi.org/10.5423/PPJ.OA.09.2015.0182

Singh IP, Dhar V, Chaudhary RG (1998) Inheritance of resistance to fusarium wilt in pigeonpea. *Indian J Agric Sci* 68:729–731.

Singh IP, Vishwadhar DRP (2003) Inheritance of resistance to sterility mosaic in pigeonpea (*Cajanus cajan*). *Indian J Agric Sci* 73:414–417.

Singh N, Ujinwal M, Singh NK (2024) Ppomicsdb: a multi-omics database for genetic and molecular breeding applications in pigeonpea. *Legume Sci* 6:e220. https://doi.org/10.1002/leg3.220

Singh NK, Gupta DK, Jayaswal PK, et al. (2012) The first draft of the pigeonpea genome sequence. *J Plant Biochem Biotechnol* 21:98–112. https://doi.org/10.1007/s13562-011-0088-8

Singh S, Kumar NR, Maniraj R, et al. (2018) Expression of Cry2Aa, a *Bacillus thuringiensis* insecticidal protein in transgenic pigeonpea confers resistance to gram pod borer, *Helicoverpa armigera*. *Sci Rep* 8:8820.

Singh S, Mahato AK, Jayaswal PK, et al. (2020) A 62K genic-SNP chip array for genetic studies and breeding applications in pigeonpea (*Cajanus cajan* L Millsp). *Sci Rep* 10:4960. https://doi.org/10.1038/s41598-020-61889-0

Singh SK, Kale SM, Saxena RK, et al. (2017a) Identification and validation of insertion–deletion polymorphisms in pigeonpea. In: *InterDrought-V*, February 21–25, 2017, Hyderabad, India.

Singh VK, Khan AW, Saxena RK, et al. (2016b) Next-generation sequencing for identification of candidate genes for fusarium wilt and sterility mosaic disease in pigeonpea (*Cajanus cajan*). *Plant Biotechnol J* 14:1183–1194. https://doi.org/10.1111/pbi.12470

Singh VK, Khan AW, Saxena RK, et al. (2017b) Indel-seq: a fast-forward genetics approach for identification of trait-associated putative candidate genomic regions and its application in pigeonpea (*Cajanus cajan*). *Plant Biotechnol J* 15:906–914. https://doi.org/10.1111/pbi.12685

Soren KR, Tripathi S, Pareek S, et al. (2021) Comparative time series RNA-seq analysis of pigeonpea root tissues in response to *Fusarium udum* infection. *Front Fungal Biol* 2:664953. https://doi.org/10.3389/ffunb.2021.664953

Srinivas T, Reddy MV, Jain KC, Reddy MSS (1997) Genetics of resistance for the Patancheru isolate of sterility mosaic pathogen in pigeonpea (*Cajanus cajan* L Millsp). *Crop Improv* 24:183–188.

Srivastava AK, Kashyap PL, Chakdar H, et al. (2018) First de novo draft genome sequence of the pathogenic fungus *Fusarium udum* F02845, associated with pigeonpea (*Cajanus cajan* L Millspaugh) wilt. *Microbiol Resour Announc* 7:e01001–18. https://doi.org/10.1128/MRA.01001-18

Srivastava AK, Srivastava R, Yadav J, et al. (2023) Virulence and pathogenicity determinants in whole genome sequence of *Fusarium udum* causing wilt of pigeonpea. *Front Microbiol* 14:1066096. https://doi.org/10.3389/fmicb.2023.1066096

Sujana G, Sharma HC, Rao MD (2008) Antixenosis and antibiosis components of resistance to pod borer *Helicoverpa armigera* in wild relatives of pigeonpea. *Int J Trop Insect Sci* 28:191–200.

Surekha C, Beena MR, Arundhati A, et al. (2005) Agrobacterium-mediated genetic transformation of pigeon pea (*Cajanus cajan* L Millsp) using embryonal segments and development of transgenic plants for resistance against *Spodoptera*. *Plant Sci* 169:1074–1080.

Tyagi S, Rathinam M, Shashank PR, et al. (2022) Deciphering of pod borer (*Helicoverpa armigera* Hübner) resistance in *Cajanus platycarpus* (Benth) offers novel insights on the reprogramming and role of flavonoid biosynthesis pathway. *Toxins* 14:455. https://doi.org/10.3390/toxins14070455

Varshney R, Chen W, Li Y, et al. (2012) Draft genome sequence of pigeonpea (*Cajanus cajan*), an orphan legume crop of resource-poor farmers. *Nat Biotechnol* 30:83–89. https://doi.org/10.1038/nbt.2022

Varshney R, Saxena R, Upadhyaya H, et al. (2017) Whole-genome resequencing of 292 pigeonpea accessions identifies genomic regions associated with domestication and agronomic traits. *Nat Genet* 49:1082–1088. https://doi.org/10.1038/ng.3872

Varshney RK, Bohra A, Roorkiwal M, Barmukh R, Cowling W, Chitikineni A, Lam HM, Hickey LT, Croser J, Edwards D, Farooq M, Crossa J, Weckwerth W, Harvey M, Kumar A, Bevan MW, Siddique KHM (2021a) Rapid delivery systems for a food-secure future. *Nat Biotechnol* 39:1179–1181.

Varshney RK, Bohra A, Yu J, Graner A, Zhang Q, Sorrells ME (2021b) Designing future crops: genomics-assisted breeding comes of age. *Trends Plant Sci* 26:631–649.

Verulkar SB, Singh DP, Bhattacharya AK (1997) Inheritance of resistance to podfly and podborer in the interspecific cross of pigeonpea. *Theor Appl Genet* 95:506–508.

Xie K, Yang Y (2019) A multiplexed CRISPR/Cas9 editing system based on the endogenous tRNA processing. *Methods Mol Biol* 1917:63–73. https://doi.org/10.1007/978-1-4939-8991-1_5

Yin K, Qiu JL (2019) Genome editing for plant disease resistance: applications and perspectives. *Philos Trans R Soc Lond B Biol Sci* 374:20180322. https://doi.org/10.1098/rstb.2018.0322

10 Harnessing Speed Breeding for Pigeonpea Improvement

Ira Sharma, K Yaksha, K Dharani Sree, Shruthi H. Belliappa, Naresh Bomma, Mahesh Pujar, and Prakash Gangashetty

10.1 INTRODUCTION

Pigeonpea (*Cajanus cajan*), popularly known as red gram, is a vital legume crop prominently cultivated in tropical and subtropical regions of Asia, Africa, South America, and the Caribbean islands (Singh et al., 2012; Gangashetty et al., 2024). It holds a crucial position in sustainable agriculture, particularly in rainfed systems, due to its resilience to drought, high protein content, and ability to improve soil fertility through nitrogen fixation. Pigeonpea is listed among the most important pulses globally, alongside mung bean, urd bean, lentil, cowpea, chickpea, common bean, and dry pea, in terms of both consumption and production (Rawal & Navarro, 2019). According to the FAO, pigeonpea is cultivated over 6.09 million hectares globally, with a production of 5.01 million tons (FAO, 2022). Its adaptability to diverse agro-ecological zones and deep root architecture makes it one of the most drought-tolerant pulses, capable of osmotic adjustments superior to other legumes (Locali-Pereira et al., 2023). Additionally, its nutrient-rich grain with a significant source of vitamins (thiamine, riboflavin, B6, folate, and niacin), essential minerals (calcium, zinc, iron, phosphorus, and magnesium), coupled with a low glycemic index, elevates its demand as a "nutri-pulse" (Gangashetty et al., 2024).

Despite its nutritional and economic importance, pigeonpea production gains have stagnated due to challenges such as susceptibility to pests, diseases, and unpredictable climatic conditions. The genetic improvement of pigeonpea has been constrained by several factors, including its photoperiod sensitivity, long growth duration, and limited genetic gain (Bohra et al., 2020; Saxena et al., 2019, 2021a). Traditional breeding methods for pigeonpea production are time-consuming, often requiring 12–13 years from initial crossing to cultivar release, with only one generation advancement per year (Hickey et al., 2019; Saxena et al., 2021b; Gangashetty et al., 2024). This prolonged breeding cycle has resulted in approximately 350 improved varietal releases over the past six decades in comparison to rice, wheat, maize, posing a significant threat to the crop's potential to address global food security challenges.

DOI: 10.1201/9781003603092-10

In response to these limitations, speed breeding has emerged as a transformative innovation in plant research and breeding (Aggarwal et al., 2025). Originally developed for cereal crops like wheat, it accelerates plant growth and reproduction by optimizing environmental conditions such as light, temperature, and humidity. This technique allows up to six generations per year compared to the traditional one or two, drastically reducing the time required to develop improved cultivars. Applying speed breeding to pigeonpea represents a critical milestone in legume research, enabling scientists to overcome the crop's photoperiod sensitivity and lengthy breeding cycle.

At ICRISAT, Hyderabad, a dedicated experiment was undertaken to optimize speed breeding protocols for pigeonpea, integrating single seed descent and seed or pod chip-based genotyping. By manipulating key factors such as photoperiod, light wavelength, temperature, and crop management practices, researchers successfully advanced two to four generations per year, reducing the breeding cycle to two to four years compared to the conventional seven years (Gangashetty et al., 2024). This achievement demonstrates the potential of speed breeding to significantly expedite the development of high-yielding, stress-resistant, and nutritionally superior pigeonpea varieties.

Integration of speed breeding with advanced genomic tools, such as marker-assisted selection and genome editing, offers opportunities for targeted crop improvement. These technologies facilitate precise trait selection, and the development of varieties tailored to diverse agro-ecological conditions, changing climates, and evolving consumer preferences. The deployment of speed breeding for pigeonpea improvement marks the potential of advanced breeding technologies to address critical challenges in crop improvement, ensuring that pigeonpea continues to contribute to global nutrition, livelihood sustainability, and climate resilience.

10.2 SPEED BREEDING: CONCEPT, ORIGIN, AND EVOLUTION

Speed breeding is an advanced plant cultivation technique that involves growing plants under controlled environmental conditions, such as optimized light, temperature, and humidity, to accelerate their growth, development, and reproduction. This approach enables researchers to achieve multiple plant generations within a single year, significantly reducing the time required for breeding and genetic improvement. When combined with modern tools like marker-assisted selection and genome editing, speed breeding facilitates targeted selections and crop improvement (Hickey et al., 2019).

Speed breeding traces its roots back to the pioneering work of John Finley and Todd Herzog in the 1980s. Utilizing artificial lighting facilities, they successfully accelerated the growth of wheat crops at the U.S. Department of Agriculture's (USDA) Agricultural Research Service in Mississippi (Finley & Herzog, 1985). During the same period, NASA, in collaboration with Utah State University (USU), embarked on groundbreaking research in space breeding, aiming to address the dietary needs of astronauts. This collaboration led to the release of USU-Apogee, the first wheat variety developed using speed breeding techniques. Inspired by NASA's innovations, the University of Queensland, in 2003, embraced the challenge of further advancing this technique. They refined the process to rapidly accelerate crop generations and formally introduced the term "speed breeding" to the scientific community (Hickey et al., 2019). Since then, speed breeding has evolved as a critical tool for enhancing genetic improvement in crops, paving the way for sustainable agricultural advancements worldwide.

Among cereals, crops like wheat *(Triticum aestivum)*, barley (*Hordeum vulgare*), rice (*Oryza sativa*), sorghum (*Sorghum bicolor*), and maize (*Zea mays*) have greatly benefited from speed breeding, achieving accelerated generation cycles and enhanced breeding outcomes (Watson et al., 2018; Hickey et al., 2019). In the legumes, pigeonpea (*Cajanus cajan*), chickpea (*Cicer arietinum*), soybean (*Glycine max*), pea (*Pisum sativum*), and lentil (*Lens culinaris*) have emerged as notable examples where speed breeding has advanced genetic improvement efforts (Kumawat et al., 2022; Gangashetty et al., 2024). Oilseed crops such as rapeseed/mustard (*Brassica napus and Brassica juncea*) and sunflower (*Helianthus annuus*) have also been targeted to improve yield potential and stress resilience (Watson et al., 2018). Furthermore, vegetables like tomato (*Solanum lycopersicum*) and pepper (*Capsicum annuum*), along with tuber crops like potato (*Solanum tuberosum*), have shown promising results under optimized speed breeding conditions (Watson et al., 2018). Other crops, including cotton (*Gossypium spp.*), sugar beet (*Beta vulgaris*), and forage crops like alfalfa (*Medicago sativa*), have also utilized this approach to expedite breeding cycles and enhance desired traits (Shivakumar et al., 2023). Collectively, these studies highlight the transformative impact of speed breeding in accelerating crop improvement and addressing global agricultural challenges.

10.3 ROLE OF SPEED BREEDING IN PIGEONPEA

Traditional breeding methods for pigeonpea are constrained by its biological and environmental characteristics. The crop's extended growth duration and photoperiod dependency limit it to a single generation per year, resulting in breeding cycles that typically span 12–13 years for cultivar development. Consequently, the number of varietal releases is limited to 350 globally, a stark contrast to cereal crops, where varietal releases often exceed 3,000 in 6 decades. Furthermore, stagnant yields and the crop's susceptibility to both biotic and abiotic stresses have underscored the need for faster and more efficient breeding methods (Shivakumar et al., 2023; Gangashetty et al., 2024).

By optimizing environmental conditions, such as light, temperature, and humidity, speed breeding facilitates the development of 2–4 generations per year (Gangashetty et al., 2024). This approach significantly reduces breeding time while accelerating the incorporation of desirable traits, such as pest resistance, disease tolerance, and improved yield. The technology enabled breeders to respond swiftly to evolving pest pressures, diseases, and climate variability (Watson et al., 2018; Hickey et al., 2019).

Marker-assisted selection (MAS) and genomic selection (GS) offer precise and efficient methods for selecting complex traits, while genome editing techniques, such as CRISPR-Cas9, allows for targeted genetic modifications. The integration of these technologies with speed breeding further accelerates the development of pigeonpea varieties that are better suited to specific agro-ecological conditions and market demands (Kumawat et al., 2022; Shivakumar et al., 2023).

Globally, the application of speed breeding in pigeonpea holds significant potential to address the growing demand for pulses and enhance food security (Bohra et al. 2020).

It enabled the timely development of resilient and high-yielding varieties, ensuring sustained pulse production in the face of climate unevenness. The reduction in breeding cycle duration allows pigeonpea varieties to be adapted more rapidly to meet the global demand for pulses (Ghosh et al., 2017; Gangashetty et al., 2024).

10.4 RESEARCH EFFORTS SHORTENING THE PIGEONPEA BREEDING CYCLE

Several attempts were made to fasten the breeding cycle in the previous decade to rapidly advance the generations in pigeonpea. The different methods adopted are represented in Table 10.1. As breeding evolved, technology adoption also evolved in the past six decades.

TABLE 10.1
Methods Identified to Shorten the Breeding Cycle in Pigeonpea

S. No.	Method	Advancements	Reference
1	Mutagenic breeding	There is a very limited utilization of mutagenesis in pigeonpea. So far, only five varieties have been released through mutagenesis. High-yielding varieties, TT5 and TT6, were developed using fast neutrons. CO 3, another high-yielding variety was developed using 0.6% EMS solution. CO 5 was developed by gamma radiation. Another variety TAT10 was developed by crossing two inbred mutants, this variety has a potential to mature earlier and yield higher than the parents.	Saxena, (2008)
2	Single seed descent	An experiment was done to know the stage at which the immature seeds can germinate in the early maturing entries to advance multiple generations in a year. It was found that the seeds that develop 35 days after flower opening show 100% germination and can be used for generation advancement.	Saxena et al. (2017)
3	Genomic Assisted Breeding	Advanced Backcross-QTL method was employed in pigeonpea to transfer the novel alleles from the wild relatives to the primary gene pool. Two backcross populations were developed, i.e., ICPL 87119×ICPW 29 and ICPL 87119×ICPW 12. Among these ICPW 29 is an accession of *C. cajanifolius* species and ICPW 12 is an accession of *C. acutifolius* species.	Varshney et al. (2013)
4	Photo insensitivity	Pigeonpea being photosensitive for extended photoperiod, it was identified that three dominant genes were responsible for photosensitivity (PS_1, PS_2, and PS_3). When these occur in recessive condition, the entry is said to be photo insensitive. The advanced generation introgression lines were grouped into two sets and subjected to natural day length and artificially extended day length. ICPP 171546 and ICPP 171561 are the two introgression lines that have shown a high photo-insensitivity index.	Hussain et al. (2022) Saxena et al. (2021b)

(*Continued*)

TABLE 10.1 (Continued)
Methods Identified to Shorten the Breeding Cycle in Pigeonpea

S. No.	Method	Advancements	Reference
5	Speed breeding	An experiment was conducted to optimize speed breeding protocol in pigeonpea by modifying the photoperiod, light wavelength, temperature, and nutrient regime. A photoperiod duration of 13 h: 8 h: 13 h at vegetative: flowering: pod filling stages was found effective for reducing the breeding cycle. Broad spectrum light (5700 K LED) reduced the duration of vegetative growth and aided early pod formation. This protocol advances two to four generations per year, which otherwise takes seven years under conventional breeding. Single seed descent and seed or pod chip-based genotyping for early generation MAS, improved the precision enabling high-throughput line development.	Gangashetty et al. (2024)

10.4.1 Protocol

Speed breeding in pigeonpea targets to reduce the vegetative stage, induce early flowering and producing limited number of seeds per plant by adjusting the photoperiod, temperature, and light. The protocol was standardized by (Gangashetty et al., 2024) as in Table 10.2 and Figure 10.1. The following specifications were maintained during the optimization of a protocol.

10.4.2 Media and Planting

The bedding media for the pots consisted of red soil, sand, and vermicomposting in a 3:2:1 ratio, sterilized through pasteurization at 93°C for two hours before sowing. After cooling, the media was filled into the pots of 6 inches (6 x 4 x 6 inches), with each genotype sown in two replications with four seeds per pot, later thinned to two plants. Pot size could be further reduced to 4 inches. Manual irrigation was provided regularly until pod filling, followed by alternating irrigation until maturity. Irrigation was ceased five days prior to harvest to expedite seed ripening and drying.

10.4.3 Structure

The optimization chamber setup comprised four dark chambers, each measuring 69 × 119 × 107 cm, situated within a closed room measuring 10.05 × 13.84 m^2. These chambers were constructed using aluminum and steel alloy, featuring standing supports and pavements for pot placement. The chamber walls were constructed from 3 mm thick black opaque polycarbonate sheets to prevent light transmission, and opaque black curtains with Velcro tapes served as closing doors to ensure light proofing. To eliminate reflection from the tiled floor, a 3 mm thick black curtain was used for chamber flooring.

TABLE 10.2
Optimized Protocol for Pigeonpea Speed Breeding (Gangashetty et al., 2024)

Operations/ Activities	Sowing to Germination	Vegetative Stage	Flowering Stage	Pod Formation Stage	Maturity Stage
Pot size	4 inches	4 inches	4 inches	4 inches	4 inches
Soil mixture	3 Red soil: 2 Sand: 1 Vermi-compost	–	–	–	–
Fertilizer spray	–	Foliar spray of 1% NPK (20:20:20) after seven DAS and top up with 1% SSP after ten days of first dose followed by need based NPK spray	Need based NPK spray	Need based NPK spray	–
Micronutrient spray	–	–	Foliar spray of 1% boron spray at 30 DAS and continued based on flowering at 7 days interval	Foliar spray of 1% potassium at pod filling stage and continued based on seed set until physiological maturity	–
Irrigation	Manual/automatic irrigation once a day	Manual/ automatic irrigation once a day	Manual/automatic irrigation once a day	Manual/automatic irrigation once a day	Manual/automatic irrigation once in two days and withdrawn five days prior to harvest
Temperature	32°C–35°C Day/22°C–25°C Night	32°C–35°C Day/ 22°C–25°C Night	25°C–27°C Day/ 16°C–18°C Night	32°C–35°C Day/ 22°C–25°C Night	32°C–35°C Day/ 22°C–25°C Night
Humidity	60%–70%	60%–70%	70%–80%	60%–70%	60%–70%

Light wavelength	White: 5,700 K LED channel	White: 5,700 K LED channel	Far-Red light: 735 nm	White: 5,700 K LED channel	White: 5,700 K LED channel
Distance of light source from plant surface	50 cm	50 cm	50 cm	50 cm	50 cm
Photo period	13 h light: 11 h dark	13 h light: 11 h dark	8 h light: 16 h dark	13 h light: 11 h dark	13 h light: 11 h dark
Days of growth					
1. Extra early	1–3	4–20	21–45	41–60	61–80
2. Mid-early	1–3	4–20	21–80	81–110	111–130
3. Medium	1–3	4–20	21–90	91–120	121–150
4. Late	1–3	4–20	21–100	121–140	141–170
Critical Observations	Early seedling emergence and growth	Vigorous growth and high foliage	Early flowering	Proper pod filling and seed set	Bold and dry seed

FIGURE 10.1 Performance of pigeonpea in optimization chambers.

10.4.4 Light

Each chamber was equipped with a 600 W Heliospectra LED light featuring variable spectrum capabilities, controlled *via* HelioConnect software. The setup ensured that light treatment in one chamber did not affect others. Four Heliospectra-Elixia LED grow lights, automated by HelioConnect software, were utilized, each equipped with tunable LED channels at 450 nm, 660 nm, 735 nm, and a white 5700 K LED channel, facilitating wavelength and light intensity adjustments based on plant developmental stages. Room temperature was maintained consistently using two external air conditioners, monitored by a Tinytag Ultra-2 datalogger (model TGU-4017) to track temperature fluctuations within the optimization chambers.

10.4.5 Nutrient

In speed breeding for pigeonpea, manual foliar sprays are emphasized over basal doses of fertilizers. Initially, a 1% NPK foliar spray was applied at seven days after sowing (DAS), followed by a second spray of 1% SSP ten days later. Additional NPK sprays were given based on plant symptoms, while a 1% boron spray was applied to initiate flowering and continued until pod filling. A 1% potassium spray was administered at pod filling and continued until physiological maturity, with seed filling stages guiding the frequency. All nutrient regimes cease at physiological maturity.

10.4.6 Photoperiod and Temperature

The critical photoperiod for flowering in pigeonpea remains underexplored, referencing previous studies as a foundation (Saxena et al., 2021b). Four optimization chambers were utilized, exposing plants to 8, 9, 10, and 13 hours of light, respectively. The experiment- maintained temperatures of 25°C–27°C/16–18°C and humidity levels of 70%–80% during the flowering stage, while monitoring conditions at 32°C–35°C/22–25°C (day/night) and 60%–70% humidity during other stages of growth.

10.5 DEPLOYMENT IN LARGE-SCALE SPEED BREEDING UNITS

A large-scale speed breeding unit is a cutting-edge facility designed to optimize plant breeding through controlled environmental conditions, ensuring rapid generational advancement. One such unit for pigeonpea is housed at ICRISAT Patancheru, within a polyhouse measuring 20 meters in length and 8.4 meters in width, providing a spacious environment for large-scale breeding operations (Figures 10.2, 10.3 and 10.4). The polyhouse is covered with polycarbonate sheets on the top and sides, allowing for controlled light diffusion and temperature regulation.

For light management, artificial illumination is provided by advanced LED lights (Heliospectra, Model: Elixia), which are fully tunable for white, blue, red, and far-red light spectrums. These LEDs allow precise control of photoperiods and light intensity, facilitating optimal plant growth at different stages. The lights are Wi-Fi enabled, allowing for remote adjustments to meet the dynamic needs of the breeding process.

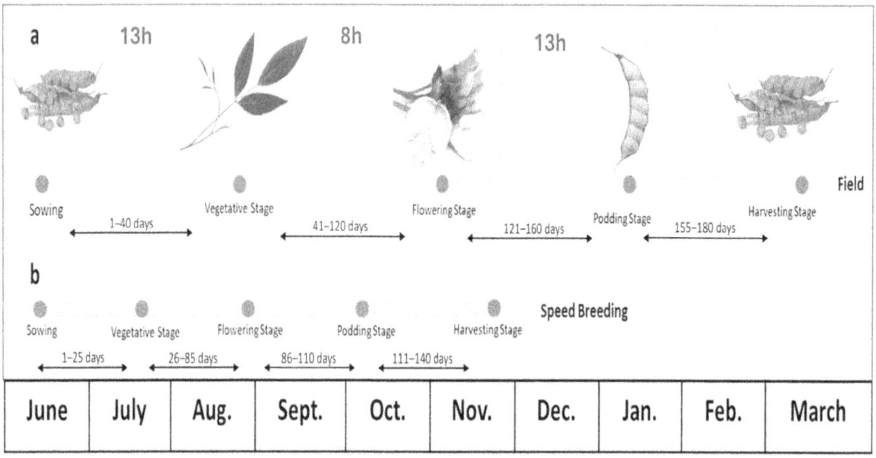

FIGURE 10.2 Duration of developmental stages. (a) Field conditions. (b) Speed breeding facility.

FIGURE 10.3 Far-red induction for early flowering in large-scale speed breeding unit.

Temperature control is achieved through an integrated cooling system, including cooling pads and axial fans, which ensure consistent temperature regulation. Circulatory fans are strategically placed to distribute cool air uniformly throughout the polyhouse, while heaters are available to raise the temperature when necessary. Humidity levels are carefully controlled using dehumidifiers, creating the ideal atmospheric conditions for plant development.

FIGURE 10.4 Large-scale speed breeding facility at ICRISAT, Hyderabad.

Nutrient delivery is optimized *via* a fertigation system, using a dosing device (BlueLab) for precise nutrient application and pH control. This system ensures that plants receive the correct nutrients at the right concentrations, maximizing their growth potential. Workbenches, equipped with irrigation systems, allow for the cultivation of seedlings in small pots or trays. These benches are movable sideways for space optimization, enhancing the unit's flexibility and efficiency. A drainage system ensures proper water management, preventing excess moisture accumulation.

In this high-tech breeding environment, all aspects of plant growth light, temperature, humidity, nutrients, and space are meticulously controlled and optimized, enabling the rapid development of pigeonpea cultivars and the successful implementation of speed breeding protocol.

10.6 IMPACT

The implementation of the optimized speed breeding protocol for pigeonpea has a transformative impact on pigeonpea improvement, enabling the advancement of two to four generations per year, as opposed to the conventional method limited to a single generation annually (Table 10.3). This acceleration in breeding cycles opens the possibility of raising pigeonpea throughout the year, eliminating seasonal constraints and allowing for continuous development. The ability to rapidly advance generations addresses the urgent need for developing resilient pigeonpea varieties capable of withstanding climate erraticism and other environmental stresses.

10.7 LIMITATIONS

Although speed breeding is an effective approach for expediting the genetic gain in various plant species, it does come with certain limitations. One of the major setbacks is the lack of advanced controlled-environment facilities that will ultimately

TABLE 10.3
Speed Breeding vs Conventional Breeding

Genotype	8 h (days)	9 h (days)	10 h (days)	13 h (days)	Control (days)	Speed Breeding Conditions			Conventional Field	No. of Days Saved		No. of Generations/ Year
						Days to Flowering (DAS)	Green Harvest Stage (DAS)	Dry Harvest Stage (DAS)	Dry Harvest Stage (DAS)	Green Harvest Stage	Dry Harvest Stage	
ICPL 11255	46	52	46	52	54	48	78	88	90	12	2	4
TS3-R	80	77	77	No flowering	No flowering	78	118	128	150	32	22	3
ICPL 87119	83	88	108	No flowering	No flowering	93	138	148	180	42	32	2
ICP 7035	104	107	104	No flowering	No flowering	105	155	165	225	70	60	2

increase establishment cost. In addition to this, sustenance of consistent supply of electricity and maintenance of an ideal temperature could be a difficult task for resource-constraint institutes for meeting the pre-requisite infrastructure (Bugbee & Koerner, 1997), and incurring costs in maintenance.

Speed breeding could result in low seed production and increased genotypic variation among plant species because of its stringent growth conditions (Gaba et al., 2021). Variation in crops may raise concerns regarding the consistence performance of crops across diverse environments, as these variations might affect the uniformity and stability of crops (Potts et al., 2023). An excessive photoperiod could lead to an increase in stress hormone levels, as well as hinder plant growth (Wanga et al., 2021).

10.8 OPPORTUNITIES IN BREEDING

10.8.1 Trait Stacking for Yield, Climate Resilience and Stress Resistance

Trait stacking can be defined as the practice of combining several desirable traits into a single crop variety, and it has gained more prominence in the era of modern agriculture (Voss-Fels et al., 2019; Potts et al., 2023). In the aforementioned context, speed breeding plays a vital role in streamlining the process. With the ability for rapid generation and to evaluate multiple generations, speed breeding aids breeders to efficiently create and analyze multi-trait combinations (Potts et al., 2023). This ultimately saves both time and resources of a crop improvement program. Therefore, breeders can develop multi-trait crop varieties with stable yield and durable resistance for biotic and abiotic stress. The accelerated trait stacking by speed breeding aids in developing crop varieties that could adapt to the demands of a dynamically changing world (Alahmad et al., 2018).

10.8.2 Creating Mapping Populations for Genomic Studies

The speed breeding approach has become an efficient technology for developing mapping populations. Traditionally, creating diverse segregating populations for genetic mapping has been an intensive and time-consuming process (Ugur & Goren, 2023). With the aid of speed breeding, researchers are now able to produce large, genetically diverse mapping populations more proficiently, facilitating essential resources for identification of genes associated to key traits (Gaba et al., 2021). This breakthrough has considerably boosted the field of MAS and quantitative trait locus analysis (Bailey, 1891; Chiurugwi et al., 2019).

10.8.3 Genomics-Assisted Speed Breeding (GASB)

Conventional selection methods like single seed descent, single pod descent, single plant selection, and clonal selection have been integrated into speed breeding during the pilot stages of trials as new age breeding advancements. These methods enables the breeder to handle populations under high density with less labor and space. These methods also expand the possibilities for maintaining genetic diversity within a crop-breeding program and extending to field environments.

Taking it a step further, the integration of fast-forward genomic breeding techniques with speed breeding could add on for further crop improvement. GASB leverages tools like MAS, marker-assisted backcrossing (MABC), GS, and genome editing to drive efficient selection and rapid generation turnover (Ćeran et al., 2024).

10.8.4 Genomic Selection with Speed Breeding

GS is an improved variant of MAS that has become a powerful tool in crop-breeding programs. It allows for more efficient selection decisions and promotes genetic advancement in targeted traits by properly arranging the training population (Saxena et al., 2021c). By combining speed breeding and GS, breeders can efficiently discover and propagate plants with desired features through early-stage trait selection, offering alternatives to conventional methods and speeding up crop development suited to environmental conditions and market demands (Sinha et al., 2023).

Each speed breeding round becomes an opportunity to apply GS, with selections focused on high genetic merit individuals. Multistage selection can be implemented where the first round selects for GEBVs across all traits of interest, followed by further selection rounds within the same cycle to refine the top-performing genotypes. With speed breeding, breeders could adapt their strategies to test and refine new traits more quickly, while GS ensures that predictions are accurate even as the breeding objectives shift. Breeders could effectively target small genetic gains in traits that would normally take longer to capture in conventional breeding programs, especially for complex traits that involve multiple genes (Jighly et al., 2019).

10.8.5 Speed Breeding with Artificial Intelligence

Integrating artificial intelligence (AI) with speed breeding offers immense potential to revolutionize the selection process in plant breeding. By incorporating AI algorithms, breeders could streamline the analysis of vast datasets produced by OMICS techniques, facilitating high-precision selection. It can also analyze complex genotypic, phenotypic, and environmental datasets. This method strengthens breeding programs by offering insights into critical agronomic traits, especially for abiotic stress tolerance. The development of next-gen AI tools that evaluate breeding values and provide nuanced analysis of genotype-environment interactions is transforming our understanding of plant resilience and adaptability (Rai, 2022).

A novel example is the AI-based physiological gravimetric system, which monitors subtle changes within the soil-plant-atmosphere continuum (SPAC), essential for assessing stress responses. Similarly, the use of UAVs and ground-based equipment integrated with machine learning in field phenomics suites provide high-throughput and precise phenotypic data collection (Singh et al. 2018, Bohra et al. 2021). This is especially valuable in controlled environments like speed breeding chambers, where accurate trait assessment can significantly enhance advancement decisions. By integrating AI within molecular techniques like MAS and GWAS, breeders could accelerate the identification of genes or QTLs tied to traits such as stress tolerance, yield, and nutritional qualities. Thus, integrating AI with speed breeding enables more informed decision-making, efficient data handling, and novel insights into plant responses, paving the way for resilient crop varieties tailored to future climate challenges.

10.9 PROSPECTS OF SPEED BREEDING IN PIGEONPEA

The future of speed breeding in pigeonpea holds promising possibilities for enhanced domestication and breeding for emerging traits like annuality, waterlogging, heat, cold and drought tolerance in terms of stresses, with a targeted focus on calcium, magnesium, and low glycemic index in grain nutrients. One significant direction involves using accelerated domestication introducing traits fitting pigeonpea as a revenue generating crop in multi- cropping systems per year. By introgressing genes from wild relatives and optimizing domestication traits, researchers could tailor pigeonpea to targeted market demands by enhancing generation turnover and trait fixation at a faster pace.

While gene editing is an advanced technique, challenges around tissue culture and laboratory containment requirements have often limited its widespread application. Modern techniques like ExpressEdit, however, could revolutionize the integration of gene editing into the speed breeding pipeline, reducing reliance on in vitro processes. This approach involves creating CRISPR-ready plants carrying the Cas9 transgene, which could then be used to rapidly produce elite inbred lines through MABC. The sgRNA subsequently directs the Cas9 enzyme to target loci within these lines, streamlining the pathway to specific genetic modifications. This combination of speed breeding and direct in planta editing could be instrumental in achieving the desired traits, like annuality, more rapidly and effectively (Hickey et al., 2019).

The use of extended photoperiods, LEDs, and other controlled-environment innovations has already expanded speed breeding's scope, reduced generation times, and enhanced phenotyping. Future advancements could include even more precise control over environmental conditions, such as breaking seed dormancy, optimizing vernalization and light quality, adjusting temperature regimes, and using elevated CO_2 levels, all customized for pigeonpea's specific phenotypic needs. Fine-tuning these parameters could amplify the efficiency and rate of trait incorporation, optimizing pigeonpea breeding to meet evolving agricultural demands (Hickey et al., 2019). With continuous refinement and the incorporation of advanced technologies, speed breeding has the potential to transform pigeonpea into a high-yielding, remunerative annual crop, supporting sustainable agriculture for the future.

REFERENCES

Aggarwal A, Jeena AS, Mehra K, Kumar B, Kashyap S, Yadav DK, Maurya AK, Venkatesh SC, Singla P, Bohra A (2025) "Speed-bred crops for food security and sustainable agriculture." *Planta* 262: 34.

Alahmad S, Dinglasan E, Leung KM, Riaz A, Derbal N, Voss-Fels KP, Able JA, Bassi FM, Christopher J, Hickey LT (2018) "Speed breeding for multiple quantitative traits in durum wheat." *Plant Methods* 14: 1–15.

Bailey LH (1891) *Some Preliminary Studies of the Influence of the Electric Arc Lamp Upon Greenhouse Plants*. Cornell University, Ithaca.

Bohra A, Satheesh Naik SJ, Kumari A, Tiwari A, Joshi R (2021) "Integrating phenomics with breeding for climate-smart agriculture." In: *Omics Technologies for Sustainable Agriculture and Global Food Security (Vol II)*, Kumar R, Shukla P, Patel HK (eds). Springer, Singapore, pp 1–24.

Bohra A, Saxena KB, Varshney RK, Saxena RK (2020) "Genomics-assisted breeding for pigeonpea improvement." *Theor Appl Genet* 133: 1721–1737.

Bugbee B, Koerner G (1997) "Yield comparisons and unique characteristics of the dwarf wheat cultivar 'USU-Apogee'." *Adv Space Res* 20(10): 1891–1894.

Ćeran M, Miladinović D, Đorđević V, Trkulja D, Radanović A, Glogovac S, Kondić-Špika A (2024) "Genomics-assisted speed breeding for crop improvement: present and future." *Front Sustain Food Syst* 8: 1–17.

Chiurugwi T, Kemp S, Powell W, Hickey LT (2019) "Speed breeding orphan crops." *Theor Appl Genet* 132: 607–616.

FAOSTAT (2022) "FAOSTAT: Food and agriculture data." https://www.fao.org/faostat/en/#data/QCL.

Finley JW, Herzog T (1985) "Accelerated growth of wheat under artificial lighting." *J Plant Physiol* 129(5): 505–510.

Gaba Y, Pareek A, Singla-Pareek SL (2021) "Raising climate-resilient crops: journey from conventional breeding to new breeding approaches." *Curr Genomics* 22(6): 450–467.

Gangashetty PI, Belliappa SH, Bomma N, Kanuganahalli V, Sajja SB, Choudhary S, Gaviyappanavar R, Bomireddy D, Anil Kumar V, Pranati J, Sharma M (2024) "Optimizing speed breeding and seed/pod chip-based genotyping techniques in pigeonpea: a way forward for high-throughput line development." *Plant Methods* 20(1): 27–39.

Ghosh G, Ganguly S, Purohit A, Chaudhuri RK, Das S, Chakraborti D (2017) Transgenic pigeonpea events expressing Cry1Ac and Cry2Aa exhibit resistance to Helicoverpa armigera. *Plant Cell Rep* 36: 1037–1051.

Hickey LT, Hafeez A, Robinson H, Jackson SA, Leal-Bertioli SC, Tester M, Gao C, Godwin ID, Hayes BJ, Wulff BB (2019) "Breeding crops to feed 10 billion." *Nat Biotechnol* 37(7): 744–754.

Hussain ME, Sharma S, Joel AJ, Kilian B (2022) "Photoperiod insensitivity in pigeonpea introgression lines derived from wild *Cajanus* species." *Agron* 12(6): 1–14.

Jighly A, Lin Z, Pembleton LW, Cogan NO, Spangenberg GC, Hayes BJ, Daetwyler HD (2019) "Boosting genetic gain in allogamous crops *via* speed breeding and genomic selection." *Front Plant Sci* 10: 1–16.

Kumawat G, Roorkiwal M, Upadhyaya HD, Khan AW, Thudi M, Varshney RK (2022) "Advances in pigeonpea genomics and breeding for accelerated crop improvement." *Theor Appl Genet* 135(3): 697–717.

Locali-Pereira AR, Boire A, Berton-Carabin C, Taboga SR, Solé-Jamault V, Nicoletti VR (2023) "Pigeon pea, an emerging source of plant-based proteins." *ACS Food Sci Technol* 3(11): 1777–1799.

Potts J, Jangra S, Michael VN, Wu X (2023) "Speed breeding for crop improvement and food security." *Crops* 3(4): 276–291.

Rai KK (2022) "Integrating speed breeding with artificial intelligence for developing climate-smart crops." *Mol Biol Rep* 49(12): 11385–11402.

Rawal, V., & Navarro, D. K. (Eds.). (2019). *The Global Economy of Pulses*. Rome: Food and Agriculture Organization of the United Nations.

Saxena K, Saxena RK, Varshney RK (2017) "Use of immature seed germination and single seed descent for rapid genetic gains in pigeonpea." *Plant Breed* 136(6): 954–957.

Saxena KB (2008) "Genetic improvement of pigeonpea—a review." *Trop Plant Biol* 1: 159–178.

Saxena KB, Bohra A, Choudhary AK, Sultana R, Pazhamala L, Saxena RK (2021a) "The alternative breeding approaches for improving yield gains and stress response in pigeonpea (*Cajanus cajan* (L.) Millsp.)." *Plant Breed* 140: 76–86.

Saxena KB, Choudhary A, Srivastava R, Bohra A, Saxena R, Varshney RK (2019) "Origin of early maturing pigeonpea germplasm and its impact on adaptation and cropping systems." *Plant Breed* 138: 243–251.

Saxena KB, Wallis ES, Chauhan YS, Byth DE (2021b) "Inheritance of photo-sensitivity in pigeonpea." *Indian J Genet Plant Breed* 81(1): 56–62.

Saxena RK, Jiang Y, Khan A, et al. (2021c) "Characterization of heterosis and genomic prediction-based establishment of heterotic pattern for developing better hybrids in pigeonpea." *Plant Genome* 14: e20125.

Shivakumar M, Patel RK, Tiwari R, Mishra DK, Singh M (2023) "Speed breeding for rapid development of crop varieties." *Agric Rev* 44(1): 1–10.

Singh B, Mishra S, Bohra A, Joshi R, Siddique KHM (2018) "Crop phenomics for abiotic stress tolerance in crop plants." In: *Biochemical, Physiological and Molecular Avenues for Combating Abiotic Stress in Plants* (Wani SH, ed). Academic Press, Elsevier, pp 277–296.

Singh NK, Gupta DK, Jayaswal PK, Mahato AK, Dutta S, Singh S, Bhutani S, Dogra V, Singh BP, Kumawat G, Pal JK (2012) "The first draft of the pigeonpea genome sequence." *J Plant Biochem Biotechnol* 21: 98–112.

Sinha D, Maurya AK, Abdi G, Majeed M, Agarwal R, Mukherjee R, Ganguly S, Aziz R, Bhatia M, Majgaonkar A, Seal S (2023) "Integrated genomic selection for accelerating breeding programs of climate-smart cereals." *Genes* 14(7): 1–37.

Ugur TAN, Goren HK (2023) "Speed breeding: an innovative approach for accelerated genetic improvement in agricultural crops and integrating with molecular-based approaches." In: *Pioneer and Contemporary Studies in Agriculture, Forest and Water Issues* (Berkan Balpetek ed.). Duvar Yayınları, Izmir, Turkey, pp 241–264.

Varshney RK, Mohan SM, Gaur PM, Gangarao NVPR, Pandey MK, Bohra A, Sawargaonkar SL, Chitikineni A, Kimurto PK, Janila P, Saxena KB (2013) "Achievements and prospects of genomics-assisted breeding in three legume crops of the semi-arid tropics." *Biotechnol Adv* 31(8): 1120–1134.

Voss-Fels KP, Cooper M, Hayes BJ (2019) "Accelerating crop genetic gains with genomic selection." *Theor Appl Genet* 132: 669–686.

Wanga MA, Shimelis H, Mashilo J, Laing MD (2021) "Opportunities and challenges of speed breeding: a review." *Plant Breed* 140(2): 185–194.

Watson A, Ghosh S, Williams MJ, Cuddy WS, Simmonds J, Rey MD, Asyraf Md Hatta, M, Hinchliffe A, Steed A, Reynolds D, Adamski NM (2018) "Speed breeding is a powerful tool to accelerate crop research and breeding." *Nat Plants* 4(1): 23–29.

11 Gene Editing for Improving Biotic Stress Resistance in Pigeonpea

Abha Tiwari, Antara Das, Alok Kumar Maurya, Ashok Kumar Parihar, and Abhishek Bohra

11.1 INTRODUCTION

The increasing global population, changing climate patterns, and diminishing agricultural resources, along with reduced crop productivity and losses after harvest, represent significant threats to worldwide food security. In addition, a change in pest-pathogen dynamics challenges global researchers to create modern crops that are both high-yielding and stress-resilient (Bakala et al., 2020; Saini et al., 2020). Legumes belong to the "Fabaceae" family, which is the third largest family of flowering plants and the second most significant after cereals in terms of agricultural crops, encompassing approximately 800 genera and 20,000 species (Smýkal & Konečná, 2014; Bohra et al., 2022a). Legumes play a vital role in human nutrition by offering affordable, nutrient-dense proteins, vitamins, and minerals, and helping to prevent chronic diseases (Arnoldi et al., 2015; Vaz Patto et al., 2015). Grain legumes account for nearly 27 percent of global food production and serve as the largest single source of vegetable protein (approximately 33%) in human diets (Young & Bharti, 2012).

Pigeonpea (*Cajanus cajan* L. Millspaugh), a pulse that was domesticated approximately 3,500 years ago from its wild relative *Cajanus cajanifolius* (Haines) in central India (Varshney et al., 2017; Bohra et al., 2022a). Over time, landraces developed at the origin site and spread in various directions for cultivation. Owing to its beneficial soil enrichment trait of fixing nitrogen, pigeonpea has adapted well to subsistence farming in the tropics and subtropical regions. However, the production of pigeonpea is not sufficient to meet the needs of the population (Bohra et al., 2020), and this problem continues to challenge pigeonpea researchers across the globe. Key factors contributing to consistently low yields of pigeonpea include the lack of resistance in high-yielding varieties, particularly to insects-pests amid changing climatic conditions, creating the biggest constraint, and relatively limiting investment in research and development.

Concerning biotic stress, pigeonpea is susceptible to over 60 pathogens, including fungi, bacteria, viruses, mycoplasma, and nematodes. Among them, sterility mosaic disease (SMD), *Phytophthora* blight (PB), and *Fusarium* wilt (FW) are prevalent across various areas. These diseases are a serious threat to the crop, but in addition,

 DOI: 10.1201/9781003603092-11

pod borer wreaks havoc on pigeonpea and chickpea crops, resulting in substantial economic losses. Pigeonpea is at least infested by 150–250 species of insect pest; infestation at post-flowering pod formation is an attraction to many pod borers. The chewing pest comprises of many pests including *Helicoverpa armigera*, *Maruca vitrata*, *Maruca testulalis*, and *Melanogromyza obtusa* causes major harm to the crop. Some other sucking pests such as *Clavigralla gibbo*sa and *Clavigralla scutellarius* are the major constraints to crop productivity in all major growing areas in India. Two of the most significant pests of pigeonpea are the lepidopteran "pod borers": *Helicoverpa armigera* and *Maruca vitrata*.

Various mitigation strategies are followed for disease management, but no disease has been completely managed and controlled. Exploitation of inherent genetic resistance mechanisms remains the most effective strategy to control plant diseases. To implement genetic resistance, one also needs to understand the interaction between the host and the pathogen because both are continuously evolving. Plants have an innate immune system that protects them against pathogen invasion. To survive, plants have evolved innate immunity as a barrier, comprising complex pathogen-interfering patterns called induced immunity, which is based on the recognition and activation of different immune receptors that sense the presence of pathogens in the plant. The plant immune system comprises a complex system of immune genes with positive or negative effects. Immune receptors (Pattern recognition receptors, PRR) and positive regulators of plant immunity contribute to the recognition of intruders and help transmit signals to activate a series of downstream events after sensing the presence of pathogens. Activation of these signaling cascades often enhances immune responses and completely evades the disease. Therefore, it is necessary to understand the complexity of plant immunity and pathogen evasion to completely inhibit infection and make plants resistant in the future. To analyze the causative genes that control the disease, it is possible to mutate them using gene-editing technologies that can further improve crop resistance.

11.2 MECHANISM OF MICROBE INTERACTION AND DISEASE TRANSMITTANCE

As mentioned in the earlier section, the primary diseases that affect pigeonpea include FW, SMD, and PB. Other diseases impact pigeonpea yields on a smaller scale, including *Alternaria* leaf spot, powdery mildew, and dry root rot, along with their respective pathogens, which can be viruses, fungi, or bacteria (Table 11.1).

To exploit the resistance traits for varietal improvement, one needs to first understand the genetic makeup of traits that contribute to resistant phenotypes. Classical genetics-based approaches, including mutation breeding, have been employed to elucidate the genetic inheritance of disease resistance in pigeonpea. For instance, Pawar and Mayee (1986) demonstrated that resistance to FW (causing yield losses from 30% to 100%) is controlled by a single dominant gene. FW, caused by *Fusarium udum*, persists in the soil and infects plants via root tips, disrupting the transport of water and minerals in the vascular system (Mishra et al., 2025a). The disease is characterized by a progressive sequence of symptoms beginning with interveinal chlorosis and reduced leaf turgidity. As the disease advances, more diagnostic signs develop, including the formation of a distinct purple band extending upward from the basal region of the stem

TABLE 11.1
Major Disease and Insect Pests of Pigonpea

Type	Disease/Pest	Causal Organism/ Vector	Yield Loss (%)	Epidemiology and Insights
Fungal	Wilt	*Fusarium udum*	30%–100% (in severe cases)	Soil-borne fungus; survives as chlamydospores; favored by high soil moisture and warm temperatures. It enters through the root system, leading to vascular discoloration and wilting symptoms.
	Phytophthora blight	*Phytophthora cajani*	20%–60%	Favored by heavy rainfall and poor drainage; spreads via water splash.
	Alternaria leaf spot	*Alternaria alternata*	10%–30%	Develops under humid conditions; secondary pathogen often infecting stressed plants.
Bacterial	Bacterial leaf spot	*Xanthomonas campestris pv. cajani*	5%–20%	Favored by warm, wet weather; spread via wind-splashed rain and contaminated seeds.
Viral	SMD	Pigeonpea sterility mosaic virus (PPSMV)	30%–95%	Transmitted by eriophyid mite (*Aceria cajani*); spread through infected alternate hosts; no seed transmission.
Nematode	Root-knot nematode	*Meloidogyne spp.*	10%–40%	Soil-borne; causes root galls; more severe under continuous cropping.
Insect pest	Pod borer	*Helicoverpa armigera*	30%–60%	Larvae feed on pods; infestation peaks during flowering to podding; favored by dry, warm climate.
	Pod fly	*Melanagromyza obtusa*	20%–40%	Larvae tunnel inside developing pods; infestation worsens under prolonged podding phase.
	Blister beetles	*Mylabris spp.*	5%–15%	Adults feed on flowers; outbreaks during flowering; more common in dry spells.
	Termites	*Odontotermes spp.*, *Microtermes spp.*	5%–20%	Attack roots and stem base; more common in sandy soils during dry periods.
	Leaf hopper/ Aphids	*Empoasca spp.*, *Aphis craccivora*	5%–25%	Sap-sucking insects; also act as vectors for viral diseases.

and longitudinal splitting of the stem tissue, revealing internal brown discoloration of the vascular bundles. Symptom expression is often irregular and may occur during both seedling and reproductive stages. Yield losses due to FW are significantly influenced by the timing of infection: up to 100% loss may occur when infection takes place prior to pod formation, approximately 67% loss if infection occurs at the pre-harvest stage, and around 30% loss has been reported at physiological maturity. In severe epidemics, total crop failure is possible. Current management strategies predominantly rely on chemical fungicides; however, their utility is constrained by limited efficacy against soil-borne inoculum and their impracticality for in-field disease suppression in established crops. Moreover, growing reports on fungicide-resistant pathogen strains underscores the urgent need to develop and implement sustainable, eco-friendly, and integrated disease management approaches. *Fusarium oxysporum* (*Fo*) reproduces asexually, resulting in minimal gene flow and a low mutation rate, leading to its classification as a pathogen with low genetic diversity. In the absence of a suitable host, *Fo* survives in the soil for extended periods as dormant chlamydospores. Upon detection of host-derived chemo-attractant signals, these spores germinate, initiating the infection cycle. The germinating hyphae grow directionally toward the host plant's roots, gaining entry either through natural openings at the intercellular junctions of cortical cells or via wounds, without the formation of specialized infection structures. Following penetration, the fungus colonizes the root cortex and breaches the endodermis to access the vascular tissue. Once within the xylem vessels, the pathogen proliferates and systemically colonizes the vascular system, thus disrupting water transport and ultimately leading to complete plant wilting. After host death, *Fo* produces abundant microconidia and macroconidia on the plant surface, which are subsequently released into the soil, serving as inoculum for future infection cycles.

In the case of SMD, resistance is contributed by four independent genetic loci—two duplicate dominant genes (*Sv1* and *Sv2*) and two duplicate recessive genes (*sv3* and *sv4*)—as reported by Saxena (2008). SMD is commonly referred to as the "green plague" of pigeonpea due to the characteristic appearance of infected plants, which exhibit lush green foliage, excessive vegetative growth, and an absence of flowering and pod development. Under favorable conditions, the disease can spread rapidly in an epidemic fashion, resembling a plague. The expression and severity of symptoms are influenced by both the genotype of the plant and the developmental stage at which infection occurs. In highly susceptible genotypes, infection during the early vegetative stage (less than 45 days old) manifest distinct symptoms and complete sterility due to the total suppression of flowering. Conversely, infection occurring at a later stage (after 45 days of growth) leads to delayed symptom development, often seen as a mild mosaic pattern on certain branches or sections of the plant, accompanied by reduced flowering and partial sterility. SMD of pigeonpea was first reported in 1931 by Mitra in the Bihar state of India. The disease manifests as a conspicuous yellow mosaic pattern, with affected plants displaying pale green, bushy growth, reduced leaf size, stunted growth, leaf deformation, and a complete or partial halt in reproductive development. In addition to floral sterility, infected plants exhibit significant reductions in overall plant height, number of branches, number of flowers per plant, and pod length. SMD is transmitted by the eriophyid mite *Aceria cajani* Channabasavanna. This mite has a relatively short life cycle of around two weeks, encompassing an egg stage followed by two nymphal stages. *Aceria cajani* transmits the pigeonpea sterility mosaic virus (PPSMV) in a semi-persistent manner. Although a single mite can achieve up to 40% transmission efficiency, a population of more than five mites is generally required to attain 100% transmission. The minimum acquisition access period (AAP) required for effective virus acquisition is approximately 15 minutes, while the inoculation access period (IAP) necessary for successful virus transfer is about 90 minutes. Following acquisition, the mites remain viruliferous for a period ranging from 6 to 13 hours. Importantly, PPSMV is neither propagative within the mite vector nor transmitted transovarial to progeny.

Another major fungal disease in pigeonpea is blight caused by *Phytophthora cajani*; it holds accountability as the third major disease for yield loss. It is controlled by a single dominant gene, *Pd1* (Saxena, 2008). PB of pigeonpea is sporadic in nature but occasionally assumes epidemic proportions in places of heavy and frequent rain fall. High incidence is usually associated with poor surface drainage and waterlogging (Bisht et al., 1988; Mishra et al., 2025b). The disease typically starts with water-soaked lesions at the collar region or on the lower stem, which rapidly expands and turns to dark brown to black. These lesions may girdle the stem, leading to the collapse of the stem above the site of infection. Infected plants often show sudden wilting of leaves, even when soil moisture is adequate—a classic symptom of vascular disruption. The wilting is generally irreversible and leads to plant death within a few days. In seedlings, symptoms include damping-off, where infected seedlings collapse and die soon after emergence. In mature plants, especially during flowering and podding stages, the disease progresses more aggressively. Roots of affected plants show soft, brown to black rot, with the cortex easily sloughing off when touched. Under humid conditions, whitish mycelial growth of the pathogen can sometimes be observed at the base of the stem or on root surfaces. Infected plants are often scattered in the field but may appear in clusters, especially in low-lying or poorly drained areas where water stagnation occurs, favoring the pathogen's spread (Kannaiyan et al., 1984; Singh and Chauhan, 1985).

Pigeonpea is highly vulnerable to several insect pests and pathogens, including fungal, bacterial, and viral agents, which significantly reduce crop productivity. Among the major insect pests, the pod borer (*Helicoverpa armigera*) is the most economically damaging, capable of causing total crop loss under heavy infestation. Other pests such as the pod fly (*Melanagromyza obtusa*), blue butterfly (*Lampides boeticus*), and leaf hopper (*Empoasca* spp.) also contribute to yield losses (Dhanyakumar et al., 2025). According to Sharma et al. (2003, 2007), host plant resistance to insect pests is quantitatively inherited, involving multiple genes associated with morphological and biochemical traits, including trichome density, leaf toughness, and secondary metabolites like flavonoids and phenols. Another severe fungal disease is dry root rot caused by *Macrophomina phaseolina*, where resistance is under polygenic control and is strongly influenced by environmental factors such as high soil temperatures and drought (Pandey & Basandrai, 2021).

Viral diseases such as pigeonpea yellow mosaic virus (PYMV), transmitted by *Bemisia tabaci*, cause chlorotic mottling, stunting, and significant yield losses. Resistance to PYMV is often conferred by recessive genes, although recent studies suggest a polygenic basis for partial resistance in some cultivars (Patil & Kumar, 2015). Bacterial diseases, though less frequent, include bacterial leaf spot caused by *Xanthomonas campestris* pv. *cajani*, which leads to necrotic lesions and defoliation under high humidity. Resistance mechanisms here are less well understood but are believed to involve R-gene mediated and inducible defense pathways.

In general plant immune networks, cell-surface pattern recognition receptors (PRRs) recognize pathogen-associated molecular patterns (PAMPs) or damage-associated molecular patterns (DAMPs). This leads to PRR-mediated immunity, commonly known as pattern-triggered immunity (PTI). Pathogens secrete virulence molecules, termed effectors, or resistance genes, to inhibit PTI or interfere with plant physiological responses. Some effectors are recognized by intracellular nucleotide-binding domain leucine-rich repeat-containing receptors (NLRs). This results in NLR-mediated immunity, which is commonly known as effector-triggered immunity (ETI). Both PTI and ETI can elevate the biosynthesis of salicylic acid (SA), a defense phytohormone that mediates systemic acquired resistance (SAR). As secondary responses, the hypersensitivity response, ROS accumulation, Ca+2 sensor response, or apoptosis further prevent the disease from spreading (Jones et al., 2024; Ngou et al., 2022).

11.3 APPROACHES TO UNDERSTAND AND USE GENETIC RESISTANCE MECHANISMS IN PIGEONPEA BREEDING

Understanding and utilizing genetic resistance in pigeonpea is crucial for managing biotic stresses (such as diseases and insect pests), including FW, PB, *Helicoverpa armigera*, and SMD. Classical breeding methods, such as pedigree and backcross breeding, have long been employed to introduce resistance traits from diverse landraces and wild relatives into elite cultivars. However, these approaches are often constrained by polygenic inheritance and environmental interactions (Saxena et al., 2010a). With advances in molecular genetics, genome sequencing and genomics-assisted breeding facilitate rapid identification and utilization of disease resistance genes/loci in breeding programs. Molecular markers such as SSRs (simple sequence repeats) and SNPs (single-nucleotide polymorphisms) have been extensively used to tag and track resistance loci in breeding populations (Patil et al., 2017a, 2017b, Saxena et al., 2018). More recently, QTLs associated with resistance to FW and SMD have been targeted for introgression into popular pigeonpea varieties using MAS (Saxena et al., 2020) approach.

Genomics-assisted breeding (GAB) further accelerates the discovery of resistance genes through genetic linkage analysis, genome-wide association studies (GWAS), transcriptomics, and comparative genomics. The availability of the pigeonpea draft genome (Varshney et al., 2013) has significantly facilitated the identification of candidate genes for disease resistance, including NBS-LRR-type genes, which are known for their role in pathogen recognition. Additionally, transcriptome analyses have been used to identify differentially expressed genes under pest or pathogen attack, contributing to the understanding of resistance mechanisms at the molecular level (Priyanka et al., 2010). More recently, genomic selection (GS) and clustered regularly interspaced short palindromic repeats (CRISPR)/Cas-based genome editing are emerging as future tools for enhancing resistance by targeting specific genes or regulatory elements involved in plant defense (Varshney et al., 2021). Collectively, integrating conventional breeding with modern omics and gene-editing approaches offers a comprehensive pathway to develop pigeonpea cultivars resilient to biotic stresses, ultimately improving yield stability and food security in semi-arid tropics.

11.4 GERMPLASM SURVEY TO IDENTIFY SOURCES OF RESISTANCE

For creation of new genetic diversity, it is imperative to identify the resistance germplasm for use in breeding program. Germplasm harbors resistance genes against FW, PB, SMD and some resistance to pod borer, pod fly, and pod wasp are also being conserved ex-situ in the gene banks at ICRISAT, Hyderabad and ICAR-NBPGR, New Delhi. Extensive multidisciplinary evaluation of core/mini-core collection at ICRISAT has led to the identification of promising accessions for SMD (24), FW (6), FW + SMD (5), pod borer (10), salinity (16), water logging (23), high yield (54), high zinc (15), and iron (15) content, whereas NARS identified trait-specific germplasm for early maturity (8), high seed yield (2), wilt (39), and SMD + wilt (24) from the mini-core collection (Upadhyaya et al., 2012). See Table 11.2 for more germplasm that includes elite cultivars, landraces, and wild relatives.

TABLE 11.2
Germplasm Resources Containing Resistant Traits Against Various Pathogens

Germplasm/ Wild-Relative (Accessions)	Resistance to Biotic Stresses	Origin/Notes	Reference(s)
ICP 7035	Complete resistance to all three races of SMD, also resistance to other pests and diseases	Vegetable pigeonpea landrace from Madhya Pradesh; widely used as SMD donor in India, Myanmar, Nepal	Rangaswamy et al. (2005)
ICP-8863 (Maruti)	FW resistance; moderate resistance to PB	Cultivar derived from landrace pool; used extensively in breeding	Pawar et al. (2024)
ICPL 87119 (Asha)	*FW*, SMD, partial tolerance to PB	Elite Indian variety; commonly used recipient parent in interspecific hybridization	Sharma (2017)
ICP 14819	FW	Identified in ICAR mini-core screening as wilt resistant, also moderate resistance to pests	Kanagaraj Murali-Baskaran et al. (2019)
ICP 7426	Pod fly (*Melanagromyza obtusa*), moderate resistance to pod borers	Mini-core accession evaluated under field screening	Kanagaraj Murali-Baskaran et al. (2019) Singh (2009)
Cajanus scarabaeoides (wild species)	Very high resistance to *Helicoverpa armigera* (pod borer), pod fly, PB, cyst nematode, wilt	Closest wild relative of *C. cajan*; easily crossable; used for introgression pest/disease resistance	Sharma et al. (2022)
Cajanus acutifolius (wild species)	Resistance to PB, SMD, pod borer	Secondary gene pool species used for interspecific lines with ICPL 87119	Sharma (2017)
Cajanus platycarpus (wild species)	Resistance to SMD, PB, pod borer, pod fly	Tertiary gene pool; used via embryo rescue in interspecific breeding	Sharma (2017) prebreeding trials
Cajanus sericeus (wild species)	Resistance to pod fly, PB, SMD	Wild species evaluated for pest/disease resistance	Khoury et al. (2015)

(*Continued*)

TABLE 11.2 (Continued)
Germplasm Resources Containing Resistant Traits Against Various Pathogens

Germplasm/ Wild-Relative (Accessions)	Resistance to Biotic Stresses	Origin/Notes	Reference(s)
Cajanus albicans (wild species)	Resistance to pod borer, pod fly, PB, SMD	Secondary gene pool; strong multitrait donor	Khoury et al., 2015
Cajanus cajanifolius (wild species)	Resistance to SMD, PB, moderate to pod pests	Secondary gene pool; used in crosses with ICPL 87119	Sharma (2017)
Cajanus lineatus (wild species)	Resistance to pod fly, PB, SMD	Wild relative evaluated under field screening	Khoury et al., 2015

Researchers at ICRISAT identified ICPL 20109, ICPL 20096, ICPL 20115, ICPL 20116, ICPL 20102, and ICPL 20094 as resistant genotypes (Sharma et al., 2016) after an extensive screening of 976 breeding and germplasm lines in wilt-sick plots. ICP 8863 (Maruti), the first released wilt-resistant variety, and ICPL 87119 (Asha) are widely cultivated for wilt resistance in India. Similarly, several screening methods have been developed for identifying PB resistance, but screening in sick plots has emerged as the best method for large-scale screening of germplasm (Singh & Chauhan, 1992). *Cajanus platycarpus* and *Cajanus sericeus* were found to be resistant to the P2 isolate but susceptible to the P3 isolate (Reddy et al., 1991). BDN 627, Sehore 1971, ICPL 1871, ICPL 84052, ICPL 84023 and ICPL 88009 emerged as resistant lines after an extensive screening of 258 genotypes (Gupta et al., 1997). Pande et al. (2012) identified ICP 11376-5, ICP 12730, ICP 12751, ICP 12755, ICPL 20093, ICPL 20100, ICPL 20101, ICPL 20104, ICPL 20105, and ICPL 20109 as resistant lines. Gupta et al., (1997) reported the monogenic dominant nature of resistance and the involvement of minor genes in the PB resistance.

However, the same kind of resistance against *Helicoverpa armigera* is not available among the cultivated *Cajanus* species. Gene pyramiding with two different insecticidal genes and tissue-specific expression to reduce the risk of developing insect resistance is another attractive option to achieve durable resistance against this pest. Expression of a chimeric cry1AcF (encoding cry1Ac and cry1F domains) gene in transgenic pigeonpea has been demonstrated to impart resistance to *H. armigera* (Ramu et al., 2012). Apart from this, an advanced generation population from the cross involving *Cajanus acutifolius* (a wild relative from the secondary gene pool) as the pollen parent, has shown considerable resistance for pod-borer damage (Jadhav, et al., 2012; Mallikarjuna et al., 1997). Transferring a strong-effect gene(s)/QTL(s) into susceptible genetic backgrounds is a straightforward method to harness the potential of germplasm having resistant traits. However, the method faces challenges in long term because of the constant evolution of the pathogen, resulting in the breakdown of the resistance.

11.5 UNDERSTANDING THE GENETIC MAKEUP OF RESISTANCE

The availability of whole-genome sequence information and a plethora of new generation molecular markers has led breeders to rapidly map quantitative trait loci (QTLs) responsible for several traits of economic significance. The recent development of modern genomic tools and technologies, including next-generation sequencing (NGS) technologies, genomic selections, and genome-assisted breeding, has opened new avenues in pigeonpea breeding. Crop wild relatives, landraces, and diverse accessions available in gene banks have also significantly contributed to identify new variation for breeding and research (Bohra et al., 2022a, 2022b). From both the convention breeding methods and marker-assisted breeding, have helped breeders to gather knowledge about the genetic resistance. For example, in *FW*, resistance is controlled by the single recessive gene (Jain & Reddy, 1995). Later, it was confirmed that two complementary gene interactions predominate the susceptivity (Saxena et al., 2012). In case of SMD, two genes govern resistance, but they are two distinct nonallelic genes with complementary gene interaction. But later, the existence of recessive genes controlling resistance to SMD was verified by Patil and Kumar (2015). Limited research has been done on inheritance of resistance against PB; inheritance of the P2 resistance race established that two homozygous resistance gene exist, whereas in P3, race is regulated by a single dominant gene along with few minor genes (Singh et al., 2003; Gupta et al., 1997). Similarly, for *Alternaria* blight only single recessive gene is determining the fate of resistance (Ojha & Choudhary, 1993).

To understand the genetic basis of plant traits, two methods-association studies and QTL mapping are used (Bohra, 2013). This started with biparental mapping population, and later, due to some disadvantages in QTL mapping, association mapping with multiparent population has been developed. Concerning SMD, research based on SSR markers has facilitated mapping of QTLs in the population (Gnanesh et al., 2011). With the advent of NGS tools, SNPs have been used to map the loci associated with target traits, In recent years MTAs identified through GWAS has helped comprehending the disease resistance mechanism more precisely.

Six hypervariable SSRs were found to be linked with FW resistance among which two were additionally verified in an F2 population (Patil et al., 2017a, 2017b). Using sequencing-based bulk segregant analysis (Seq-BSA) extreme bulks of the RILs (ICPL 20096 × ICPL 332) were analyzed, and four SNPs were found to be linked with the FW and SMD. On the other hand, Saxena et al. (2017a, 2017b) identified three QTLs, qFW11.1, qFW11.2, and qFW11.3, using a genotyping-by-sequencing (GBS) analysis of three mapping populations segregating for FW.

Identification and introgression of genomic segments attributing disease resistance through a GAB approach is a sustainable approach for the development of disease-resistant pigeonpea varieties (Saxena et al., 2017a). Genotyping-by-sequencing (GBS) was used for simultaneous identification and genotyping of SNPs, and the candidate genomic region identified on CcLG11 was a promising QTL for molecular breeding in developing superior lines with enhanced resistance to SMD (Saxena et al., 2017a). In an earlier study, six QTLs explaining phenotypic variation were identified on LG7 and LG9 following analysis of extensive phenotyping and SSR data on population segregating for SMD resistance (Gnanesh et al., 2011).

In another study, Saxena et al., (2021) developed an effective molecular diagnostic kit comprising ten competitive allele specific PCR (KASP) SNP markers each for FW and SMD resistance. Investigation on the diversity of these loci has uncovered different alleles in different resistance genotypes. Two genes (*C. cajan*_03691 and *C. cajan*_18888) for FW resistance and four genes (*C. cajan*_07858, *C. cajan*_20995, *C. cajan*_21801, and *C. cajan*_17341) for SMD resistance were identified. Development of a customized and cost-effective KASP genotyping assay paves the way for accelerated trait improvement in pigeonpea breeding programs. Table 11.3 shows a list of putative genes playing roles in various biotic stresses.

TABLE 11.3
Putative Genes and Their Function in Various Biotic Stresses

Gene	Type	Function	Key Reference(s)
	a) Putative Function in Relation to FW		
NBS-LRR (CcNBS-LRR) family members	Resistance gene family (R-genes)	Several CcNBS-LRR genes are upregulated in resistant cultivars early after *F. udum*	Bajpai et al. (2022), Purohit et al. (2021)
Pathogenesis-related (PR) genes: β-1,3-glucanase, chitinase, PR1, thaumatin-like proteins	Defense (PR) proteins	Reported upregulation in resistant pigeonpea lines after *F. udum* challenge	Biswas et al. (2020)
Antioxidant enzymes: SOD, APX (ascorbate peroxidase), peroxidases	ROS-scavenging/ oxidative burst modulators	ROS signaling and detoxification	Biswas et al. (2020)
Transcription factors and signaling genes: WRKYs, NPR1, MAPKs, JA/SA pathway genes (LOX, PAL)	Regulatory/signaling genes	TFs and hormone-pathway genes coordinate defense responses (JA/SA cross talk)	Kumar et al. (2019)
Cell-wall modification genes: pectin methylesterase, expansins, cellulose synthases, lignin/ peroxidase enzymes	Cell wall strengthening/ vascular obstruction	Modifications to cell wall can limit pathogen spread in vascular wilt diseases	Soren et al. (2021)
	b) Putative Function in Relation to SMD		
C. cajan_01839– Probable serine–threonine protein phosphatase	Regulator for signal transduction, including defense signaling	One damaging substitution was identified in the susceptible cultivar via BSA and nsSNP analysis.	Singh et al. (2016)
C. cajan_07858- Putative ABC transporter (ABCG sub-family)	Movement of defense compounds and promotes virus interactions	This locus was also used for a diagnostic kit for SMD via KASP assays.	Saxena et al. (2021)
C. cajan_20995- target of Myb protein 1-like isoform X6	Intracellular protein transport	This locus was also used for a diagnostic kit for SMD via KASP assays.	Saxena et al. (2021)

(*Continued*)

TABLE 11.3 (Continued)
Putative Genes and Their Function in Various Biotic Stresses

Gene	Type	Function	Key Reference(s)
C. cajan_21801- myb transcription factor	MYB TFs predominantly play a significant role in defense against various biotic stresses, including viruses	This locus was also used for a diagnostic kit for SMD via KASP assays.	Saxena et al. (2021)
C. cajan_17341- Putative receptor-like protein kinase (RLK / RLK-type)	Core components of pathogen recognition and part of the defense signaling	One-bp indel in susceptible vs. resistant genotypes.	Saxena et al. (2021)
C. cajan_05815/ *C. cajan_05816*- Annotated as conserved oligomeric Golgi (COG) complex subunit 5 (putative)	Annotated as COG complex subunit 5 (putative)	Indel-seq found a one-bp deletion/insertion at a CcLG02 position associated with SMD.	Singh et al. (2017b)
C. cajan_15032	Uncharacterized protein (Indel-seq reported frame-shift effect for an SMD-associated Indel at CcLG10)	Indel-seq: one-bp insertion/deletion on CcLG10 associated with SMD.	Singh et al. (2017b)
	c) Putative Function in Relation to PB		
Host resistance gene families (candidate sets)—NBS-LRR (R-genes)	These genes are established effectors/recognition receptors that mediate ETI against bacterial pathogens in other crops.	Indirect but high-priority candidates, but not PB-specific. Good starting point for mapping/functional screens.	Singh et al. (2012), Wu et al. (2017)
Pathogenesis-related (PR) genes (PR1, chitinases, β-1,3-glucanases, thaumatin-like proteins)	PR genes are induced in defense responses (SA/ETI/ biotic stress).	These genes are general defense effectors and are often induced in bacterial infections in other legumes.	Purohit et al. (2021), Mahato et al. (2019)
Transcription factors and signaling (WRKY, MAPK, NPR1, JA/SA pathway genes)	TFs coordinate defense gene expression.	Useful regulators to test functionally.	Purohit et al. (2021)
	d) Putative Function in Relation to *H. armigera*		
Protease inhibitors (Kunitz-type trypsin inhibitors, Bowman-Birk family)—*endogenous* pigeonpea proteins	Biochemical isolation; gene cloning/family catalogs; functional rationale from insect physiology	Seed/leaf trypsin and chymotrypsin inhibitors inhibit digestive proteases in lepidopteran larvae.	Swathi et al. (2016)

(*Continued*)

TABLE 11.3 (Continued)
Putative Genes and Their Function in Various Biotic Stresses

Gene	Type	Function	Key Reference(s)
Lectins/amylase inhibitors/other seed defense proteins—*endogenous*	Biochemical and proteomic evidence	Some lectins and amylase inhibitors reduce insect feeding and digestion.	Proteomic and biochemical studies. (CAFNR Faculty, ScienceDirect)
Secondary-metabolite/ flavonoid pathway genes (from wild relatives)	Comparative metabolomics/ transcriptomics (wild *Cajanus platycarpus*/ *Cajanus scarabaeoides*)	Wild relatives with lower H. armigera survival show different profiles of flavonoids/ phenolics.	Tyagi et al. (2022)
Host transcriptional responses (JA/defense pathway genes, wound-responsive genes)	Comparative transcriptomics (resistant vs susceptible) and wounding/OS (oral secretion) experiments	Herbivore wounding induces JA -JAZ signaling activating defense genes and secondary-metabolism to regulate plant defense responses.	Njaci et al. (2020)
	e) Putative Function in Relation to Pod Fly		
Protease inhibitors – Kunitz and Bowman-Birk (PI/BBI/KTI)	Inhibit insect gut proteases reduce larval digestion, growth and survival (antibiosis).	Pigeonpea seeds and wild relatives contain active PIs; differential inhibition of insect gut proteinases shown.	Bonturi et al. (2022), Mohanraj et al. (2019)
Trichome regulatory TFs (R2R3-MYB e.g., GL1-like, bHLH GL3/EGL3, TTG1 complex)	Increase density/ length of (nonglandular) trichomes on pods, physical barrier, deter oviposition/ larval entry (antixenosis).	Higher nonglandular trichome density/ longer trichomes associated with lower pod fly/pod borer damage.	Khetnon et al. (2022)
Pod wall/cell wall and lignin pathway genes (CESA, MYB46, PAL, CCR, CCoAOMT)	Thicker/stronger pod wall makes penetration and larval establishment harder. Lignification and cellulose deposition reinforce tissue.	Pod-wall thickness regulated by cell wall/MYB genes is negatively correlated with pod fly damage.	Dabravolski and Isayenkov (2023)
Terpene synthases (TPS family)	Make volatile terpenes or contact terpenoids that repel adults or attract parasitoids (antixenosis/ indirect defense).	Terpene roles in insect-plant interactions well documented.	Boncan et al. (2020)

(Continued)

TABLE 11.3 (Continued)
Putative Genes and Their Function in Various Biotic Stresses

Gene	Type	Function	Key Reference(s)
Secondary-metabolism genes (flavonoid/ phenolic biosynthesis: CHS, F3H, UGTs)	Produce phenolics or flavonoids that reduce feeding or oviposition (antibiosis/ deterrence).	Biochemical differences among pigeonpea genotypes correlate with pest damage.	Sharma et al. (2022)
JA-pathway regulators and defense TFs (JAZ, MYC, WRKY)	Control induced anti-herbivore responses (protease inhibitors, volatile emission).	Herbivore/wounding responses in legumes show JA/TFs regulate armory.	Choudhary et al. (2013)

With regard to insect pests, transgenic research has led to the development of resistance against *Helicoverpa*, by introgression of the *Bacillus thuringiensis* insecticidal crystal proteins Cry2Aa and Cry1AcF, which showed reduced damaged to the pod-borer insect (Singh et al., 2018).

11.6 GENE-EDITING FRAMEWORK IN PIGEONPEA

Conventional or traditional breeding methods employ mutagens and hybridization techniques to create new variations in a breeding program. With the discovery of genome editing tools, manipulation of the target gene(s) at a precise nucleotide position could be further used to generate new genetic diversity. This approach has been widely used in many crops including cereals to incorporate genetic resistance in the plants (Kumar et al. 2023). In this context, several genome editing tools such as zinc finger nucleases (ZFNs), transcription activator-like effector nucleases (TALENs), and CRISPR have been used for accelerating the pace of plant improvement. In a review by Rasheed et al., (2022) all the editing tools from meganucleases (MN) to modern CRISPR have been summarized in a comparative way. They discuss various advantages of CRISPR over other tools (MN, TALEN, and ZFN), such as higher-target recognition, low cost, few off targets, easy multiplexing, and higher editing efficiency. Transcription activator-like effectors (TALEs) are naturally occurring DNA-binding proteins derived from plant-pathogenic bacteria belonging to the genus *Xanthomonas*. These proteins are characterized by their central repeat domains, each consisting of 33–35 amino acids, with each repeat specifically recognizing a single nucleotide within the target DNA sequence. The DNA-binding specificity of each repeat is determined by two hypervariable residues known as repeat-variable di-residues (RVDs), which directly interact with the nucleotide bases (Boch et al., 2009; Moscou & Bogdanove, 2009). TALENs have been successfully used to target fatty desaturase gene—namely, the Fatty Acid Desaturase 2 (FAD2) gene that enhances oleic acid production (Haun et al., 2014). In another study, TAL effectors were used to develop dwarf and albino phenotypes by altering the GmPDS11 and GmPDS18 genes in soybean (Du et al., 2016). ZFN technology integrates a zinc finger protein domain along with a nuclease domain to enable gene editing. Similar to the TALEN method, ZFNs function as pairs to create double-strand breaks in DNA, utilizing the FokI enzyme for cutting. The zinc finger (ZF) domain is made up of as many as six protein subunits, each designed to bind to a specific DNA sequence. These zinc finger proteins (ZFPs) are engineered to interact with Zn^{2+} ions, which stabilize their structure, enabling them to bind effectively to three DNA base pairs. ZFN has been used in soybean for traits for the production of lipoxygenase enzyme (*GmLOX1 and GmLOX2*) and dicer-like protein DCL4a/b that enhanced lateral root growth (Curtin et al., 2011; Menz et al., 2020). However, unlike zinc fingers, TALE arrays do not require engineered inter-domain linkers, allowing for straightforward construction of long repeat arrays capable of targeting specific genomic loci with high precision (Boch & Bonas, 2010). TALE-based platforms have become invaluable tools in genome engineering, as their modularity enables fusion to diverse effector domains, such as FokI nucleases (creating TALENs), transcriptional activators, and recombinases, thereby facilitating a range of targeted genetic modifications (Christian et al., 2010). Despite their high target specificity and single-base resolution, the repetitive nature of TALE arrays poses significant challenges in molecular cloning due to sequence redundancy. To address this, various efficient assembly methods have been developed, including Golden Gate cloning, high-throughput solid-phase synthesis, and ligation-independent cloning techniques. These innovations allow rapid and reliable construction of custom TALE arrays, thereby enhancing their utility in research and diagnostic kits applications (Cermak et al., 2011; Doyle et al., 2012).

CRISPR-associated protein (Cas) is an adaptive immune system of prokaryotes and archaea that copes with foreign genetic material, such as viral DNA. Upon initial invasion by phages or plasmids, exogenous nucleic acids (protospacers) are incorporated into the CRISPR array, yielding spacers using Cas effectors. This step is one of the three steps, known as acquisition, which is further followed by the expression where CRISPR locus is transcribed into a long RNA molecule, which is then processed into smaller RNA molecules called CRISPR RNAs (crRNAs). These crRNAs contain spacer sequences that are crucial for target recognition. The third and final step is known as interference, where crRNAs, along with Cas proteins, form a Cas effector complex that searches for and binds to complementary sequences in the invading DNA or RNA. Once bound, the Cas proteins can then degrade the foreign DNA or RNA, effectively neutralizing the threat. This mechanism creates DNA double-strand breaks (DSBs) at specific locations within the genome, directed by a particular RNA known as guide RNA (gRNA) that associates with the Cas9 endonuclease (Figure 11.1). Notably, Cas protein has two domains, HNH and RuvC, RuvC nuclease domain found in many Cas proteins, including Cas9 and Cas12, that cleaves non-target DNA, whereas HNH domain cleaves the target DNA (Jinek et al., 2012; Makarova et al., 2013). This creates the DSBs that were further repaired by endogenous repair mechanism, nonhomologous end joining (NHEJ) and homology-dependent repair (HDR). Nonhomologous end-joining (NHEJ) repair pathway is prone to errors, frequently resulting in the insertion or deletion of several nucleotides at the intended target site (Jinek et al., 2012). The CRISPR/Cas9 technology has been used successfully for

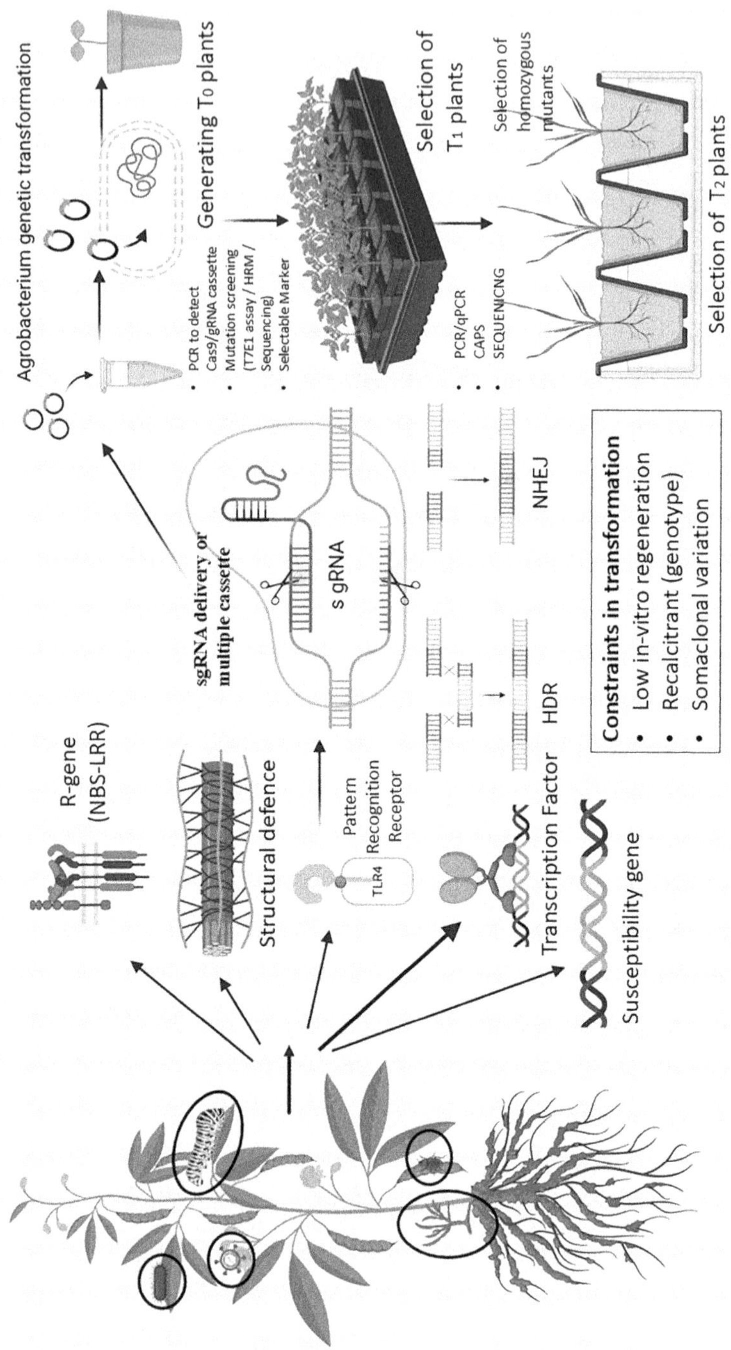

FIGURE 11.1 Gene editing in pigeonpea.

precise gene editing in a wide range of crops, including model plants such as Arabidopsis (Jiang et al., 2014), in legumes like Medicago (Meng et al., 2017), soybean (Lu & Tian, 2022) and cowpea (Che et al., 2021; Ji et al., 2019). The primary challenge is their recalcitrance to *in vitro* regeneration, which obstructs the extensive use of gene editing in legumes (Ochatt et al., 2018; Pratap et al., 2018). As shown in Figure 11.1, researchers have attempted to create an efficient Agrobacterium-mediated genetic transformation system for pigeonpea (Ghosh et al., 2017; Sharma et al., 2006) and groundnut (Mehta et al., 2013; Prasad et al., 2013).

Recently, Prasad et al. (2024) successfully edited the *phytoene desaturase (PDS)* gene, essential for carotenoid biosynthesis, as its disruption results in albino phenotypes and dwarfism in pigeonpea. With an Agrobacterium-mediated transformation efficiency of 15.2%, mutations in the PDS gene in pigeonpea led to an albino phenotype, with editing efficiencies between 4% and 6%. Sequence analysis revealed a nucleotide deletion of base (A) in pigeonpea resulting in premature stop codon, truncated gene that further resulted in an albino phenotype. More recently, Senthil et al. (2025) reported transformation efficiency to be enhanced by 8.80% and 9.16% in *in planta* and *in vitro* transformations, respectively, by targeting the PDS gene. Another research elucidated genetic changes in herbicide resistance gene in pigeonpea by modifying the PEP-binding site within the native 5-*enolpyruvylshikimate-3-phosphate synthase (EPSPS)* enzyme at positions 182G-to-A, 183T-to-I, and 187P-to-S, for reducing glyphosate binding affinity in EPSPS via HDR efficiently confers glyphosate resistance in pigeonpea (Bharti et al., 2025).

11.7 CONSTRAINTS IN TRANSFORMATION AND DELIVERY OF VECTORS

The most critical problem faced during genome editing of legumes is their transformation recalcitrance. Reports on genome editing of plants such as rice and tomato are growing in number (Jaganathan et al., 2018; Ukhatovaa et al., 2023) because they are amenable to transformation. Even soybean, whose genetically modified versions are commercially cultivated in some countries, is known for its recalcitrance to transformation like other legumes (Xu et al., 2022). Only certain cultivars of soybean have generated an appreciable transformation efficiency. Amenability to transformation is desirable since it not only reduces the labor of handling more explants but also generates more edited individuals, thereby increasing the probability of obtaining desirable and clean individuals without off-target mutations. For the commercial release of an edited mutant, it is desirable that the best out of many is chosen after thorough genotype/phenotype screening. To attempt a successful transformation and regeneration various methods have been developed, which were reviewed by Nivya and Shah (2023), and they mentioned that although the transformation efficiency is very high when transformed, this is counted based on the expression of various reporter genes GUS/GFP/YFP, the number of transgenic plants drastically reduced after the antibiotic-containing media selection. It should also be noted that the true transformation efficiency can be calculated after thorough molecular screening. Although PCR confirms the transgenic nature of plants, the occurrence of clones due to the same transgenic events can be identified after junction fragment analysis by southern hybridization. To mention the reason of the transformation recalcitrant, we need to revisit all the events in transformation. The first step is the entry of the transgene, and second is the transgene integration into the nuclear genome. Editing can be done in this step by using suitable plant vectors using different selectable markers and reporter genes. The third major and crucial step is selection and regeneration; legumes face major challenges in this step only; regeneration from the transformed cells does not occur frequently. While the *in vitro* regeneration is more efficient without selection pressure, the subsequent focus shifts intensely to robust molecular screening to guarantee the stable inheritance and non-chimeric nature of the desired genetic alterations in future generations.

Using various sources of explant such as cotyledonary node, embryonic axis, plumule node, axillary meristem, embryonic axis attached with cotyledon, they have successfully been transformed and regenerated. All these methodologies have been reviewed by Nivya and Shah (2023), they have mentioned that the trait inheritance is confirmed by the PCR confirmed in the T_0 generation which is further validated by southern blot. In further stages, the inheritance is not stable, as some of the PCR steps confirmed the gene, but in blotting, the edited loci inheritance is lost.

When it comes to legumes, the transformation efficiency is directly proportional to regeneration efficiency. It is known that regeneration depends on multiple factors, such as culture media, hormone concentration, explant, and other supplements. In this context, soybean sets an excellent example of an explant-dependent enhancement of transformation. Although soybean is considered a crop recalcitrant to transformation (Xu et al., 2022). From the initial step using a suitable explant, one can increase the efficiency of transformation, such as in soybean; use of imbibed split seed with an attached partial embryonic axis resulted in 18.7% transformation, in comparison to the average efficiency of 8.7% (Pareddy et al., 2020). Similarly, the choice of cultivar and agrobacterium strain for infection also need to be done judiciously. To mention chickpea as an example, the *A. tumefaciens* strain GV3101 performed better with transformation efficiency of 17.56%, in comparison with two other strains EHA105 and LBA4404 with 8.54% and 5.43% efficiencies, respectively (Gupta et al., 2023). Other than that, many *in planta* transformation methods are being adopted in other plants, such as the floral dip method, which has been recently used to reduce chimerism in transformed tissues and to enhance *in planta* gene targeting and frequency of mutagenesis, but so far, it has only been successfully employed in *A. thaliana* and a few close relatives. Some other widely used protocols such as haploid inducer in maize and wheat (Budhagatapalli et al., 2020), *de novo* induction of meristems, virus induced genome editing, grafting, and delivery by nanoparticles have been shown to be promising procedures for the generation of GE plants bypassing *de novo* regeneration in tissue culture (Cardi et al., 2023). In addition to these factors, designing of the guide RNA is also a crucial factor for editing a target sequence, for that first step is to obtain a complete nucleotide sequence of the gene/transcript. Then designing of the guide RNA involves locating the PAM motif, preferably within a first few exons; checking off targets; and designing of primer for validating knockout mutants.

11.8 RECENT DEVELOPMENTS IN EDITING TOOLS

To utilize the full potential of CRIPSR technology, we can use the genome editing tools for a variety of purposes. Such as by combining rapid isothermal amplification (RPA) with the highly specific, "collateral effect" of activated CRISPR enzymes (Cas12/DETECTR for DNA viruses and Cas13/SHERLOCK for RNA viruses), scientists have developed rapid, sensitive, and effective diagnostic kits that utilize real-time fluorescence to diagnose viral infections with high accuracy (Gootenberg et al., 2017). Srivastava et al., (2022) made a successfully diagnostic kit for detecting viruses, begomovirus, *viz.*, Mungbean yellow mosaic India virus, which infects a wide range of leguminous crops and other one Ageratum enation virus is reported to infect a wide range of crops from weed to opium poppy. They harnessed the unique property of Cas protein, i.e., collateral cleavage-independent CRISPR/Cas12a, which means the unintended cleavage of the single stranded of the virus that is labeled with the fluorophore.

To advance the future of genome editing, it is essential to recognize that naturally occurring structural variations (SVs) are fundamental to the phenotypic diversity leveraged in breeding programs. When two or more DSBs occur simultaneously, the DNA ends can be repaired through nonhomologous end joining (NHEJ) or, if flanking sequences are homologous, through HDR. This repair process can generate novel junctions in various configurations, leading to a wide range of inter- and intra-chromosomal SVs. Similar to naturally occurring or mutagen-induced large genomic alterations, not all engineered SVs will be stable or viable. However, genome editing enables the deliberate and precise design of these variations to achieve specific outcomes. Such SVs include deletions, insertions, single-gene, or tandem repeat duplications, and larger-scale rearrangements like translocations, recombinations, and inversions. These modifications contribute to genetic diversity, enabling the development of more resilient and superior genotypes within a gene pool.

Recent advances in genome editing have introduced base editing (BE) and prime editing (PE) platforms, which allow for precise single-nucleotide substitutions, small insertions, deletions, and sequence replacements in plant genomes. BE enables targeted modifications at single-base resolution, allowing for precise alterations in both transcriptomes (RNA) and genomes (DNA) within living cells. Base editors are broadly categorized into RNA BEs, which modify individual ribonucleotides, and DNA BEs, which introduce specific point mutations in DNA. BE was first developed by Dr. David Liu and his team in 2016, marking a significant breakthrough in precision genome editing. BEs are further subclassified into cytosine base editors (CBEs), adenine base editors (ABEs), cytosine-to-guanine base editors (CGBEs), dual-base editors, and organellar BEs. The initial system, known as cytosine BE, was designed to convert cytosine (C) to thymine (T) in DNA without introducing DSBs or requiring a donor template (Komor et al., 2016). This was later followed by the development of ABEs, which convert adenine (A) to guanine (G) (Gaudelli et al., 2017). These tools employ a catalytically impaired CRISPR-Cas9 enzyme fused to a deaminase, enabling precise nucleotide substitutions with minimal undesired mutations or large genomic alterations. The advent of BE has revolutionized the ability to correct point mutations in a wide variety of organisms, including plants, and laid the foundation for further innovations such as dual-base and organellar base editors.

PE, developed by (Anzalone et al., 2019), offers a more flexible and potentially more accurate alternative to traditional CRISPR-Cas9 and BE systems. It facilitates "search-and-replace" style DNA modifications without the need for DSBs or donor DNA templates. Unlike traditional CRISPR-Cas9, which relies on DSBs and donor templates, PE uses a catalytically impaired Cas9 nickase fused to a reverse transcriptase, guided by a prime editing guide RNA (pegRNA). This pegRNA both directs the Cas9 to the target site and encodes the desired edit, enabling precise insertions, deletions, and all 12 types of base substitutions. The versatility, precision, and reduced off-target effects of PE make it especially promising for plant genome engineering.

In addition, transposon-based genome editing has emerged as a promising technique. The TnpB protein, a compact enzyme of approximately 400 amino acids, has been successfully applied for gene editing in both monocot (rice) and dicot (Arabidopsis) plant species. Enhancing the expression system has led to increased editing efficiency. TnpB-mediated cleavage requires the presence of a transposon-associated motif (TAM) located approximately 50 base pairs upstream of the target sequence (Karmakar et al., 2024).

Gene editing holds immense potential for enhancing yield, nutritional quality, stress tolerance, and disease resistance. The traits of economic significance in crops including legumes such as soybean, chickpea, pigeonpea, and cowpea, are of complex nature, regulated by multiple gene(s)/QTLs and are often difficult to improve through conventional breeding. Editing could be used to fine-tune key genes involved in nitrogen fixation, flowering time, drought tolerance, or resistance to pathogens such as *Fusarium* or *Phytophthora*. For example, single-nucleotide changes in promoter or coding regions of transcription factors could regulate gene expression more precisely than previous genome editing approaches. Moreover, traits such as seed coat color, oil content, or anti-nutritional factor reduction can be addressed through highly specific edits.

Although successful implementation in legumes is still limited due to transformation inefficiencies and pegRNA design challenges, advances in delivery systems (e.g., nanoparticle-mediated or viral vectors) optimized expression constructs, and improved in silico gRNA prediction tools are expected to enhance editing efficiency in legume species. As genome assemblies and functional annotations for legume crops continue to improve, the integration of editing tools with omics approaches and speed breeding could accelerate the development of elite, climate-resilient, and nutritionally superior legume varieties. Thus, modification represents a transformative tool for next-generation legume breeding and sustainable agriculture.

11.9 TRANSGENE-FREE EDITED CROPS AND REGULATORY FRAMEWORK

To create transgene-free, gene-edited crops, several strategies can be employed. These can be broadly categorized into eliminating the integrated CRISPR/Cas9 construct through genetic segregation, transient expression of the editor, or using DNA-free editing methods. Genetic segregation is very useful for sexually propagated crops for introducing desired features, and undesirable traits can be readily removed in subsequent generations. But in crops that are propagated asexually by transformation methods, that leads to the illegitimate recombination events, which possess a significant risk. Illegitimate recombination is a nonhomologous process that inserts DNA randomly and unpredictably rather than at the intended target site. These unintended recombination events can sometimes insert exogenous DNA sequences that mimic the intended edited region, leading to false-positive results during initial molecular screening or creating complex, undesirable changes in the plant's genetic makeup that are difficult to characterize precisely. The vector used in the genome editing, CRISPR gene-editing construct, which consists of the Cas9 and gRNA cassettes along with selection signals such the kanamycin-resistant gene. The construct is subsequently inserted into plant cells via either biolistic transformation (gene gun) or Agrobacterium-mediated means. The transgenes can generate gRNAs and CRISPR enzymes for target gene editing after they are incorporated into plant chromosomes. Although some of them are altered at the target loci, the CRISPR transgenes are present in most T_0 plants (Bhattacharjee et al., 2023). To obtain transgene-free and edited plants, selfing or backcrossing T_0 plants to wild-type plants takes an extra generation. Therefore, plants free of transgenes can be isolated from the progeny of T_0 plants or backcrossed populations using genetic segregation, however the process remains laborious and time-consuming. Additionally, a few other methods can also be employed to remove transgenes, such as the usage of the recombinase system, site specific recombinase, Cre lox system, and transposons can be successfully used to excise out the Cas gene and successfully create the transgene-free crops in T_1 and further generation. Another method avoids stable integration of the transgene by delivering the CRISPR/Cas9 components (e.g., Cas9 protein and guide RNA) directly into plant cells without integrating into the plant genome are Ribonucleoprotein (RNP) delivery where Preassembled Cas9 protein-guide RNA complexes are introduced into plant cells, editing the genome without leaving any DNA traces. Transient expression of CRISPR/Cas9 RNA from DNA vectors can be used where plasmids carrying the CRISPR/Cas9 components are introduced into plant cells, but they do not integrate into the genome and are eventually lost during plant development.

The regulation of the genome edited crops, CRISPR technology falling under the categories of SDN-1 and SDN-2, will be exempted from biosafety regulations (Refer Guidelines for the Safety Assessment of Genome Edited Plants, 2022) under India's biosafety regulations for general crops. The progress in DNA-free plant genome editing is directed toward a modification based on reduced off-target and increased transformation efficiency, along with good regeneration efficiency of explants. More recently, two genome edited rice lines have been developed, DRR Rice 100 (Kamla) and Pusa DST Rice 1 by the researchers at ICAR, India. These lines have the potential to bring about revolutionary changes in terms of higher production, climate adaptability, and water conservation. SDN1 involves creating small insertions or deletions (indels) in the DNA sequence without the use of a DNA template, while SDN2 utilizes a DNA template to guide the repair and introduce specific changes, like point mutations. Both SDN1 and SDN2 are considered to result in edits that are indistinguishable from naturally occurring mutations or those created through conventional breeding methods. Therefore, the acceptance of the edited lines under SDN1 and SDN2 can be accepted under the regulatory guidelines, but general public perception on biotech crop products is still not very positive, the situation is gradually changing though.

11.10 CONCLUSION AND PROSPECTS

Genome editing has emerged as a revolutionary approach to improve disease and insect resistance in pigeonpea, a crop severely constrained by biotic stresses such as FW, SMD, PB, pod borers and pod flies. By enabling precise modifications in key defense genes, editing tools such as CRISPR/Cas9, base editors, and prime editors open new possibilities for developing durable and broad-spectrum resistance. One of the most promising categories of targets includes NBS-LRR genes, which encode intracellular immune receptors responsible for pathogen recognition and activation of ETI. In pigeonpea, several NBS-LRR homologs were reported to be upregulated in resistant genotypes in response to *Fusarium udum* infection, highlighting their role in defense (Kumar et al., 2014). Similarly, central signaling regulators such as EDS1 and PAD4, which play a key role in SA-mediated defense, show elevated expression in SMD-resistant pigeonpea lines, making them attractive targets for BE (Ranjan et al., 2019). WRKY transcription factors also represent valuable breeding targets; for example, CcWRKY28 differentially expressed in Fusarium-responsive pigeonpea lines, implicating its role in disease defense (Singh et al., 2017b).

In viral resistance, host translation initiation factors such as eIF4E and eIF(iso)4E are increasingly recognized as susceptibility genes. Disruption of these genes has already conferred virus resistance in crops like cucumber and soybean (Chandrasekaran et al., 2016), and editing their homologs in pigeonpea could offer resistance to PPSMV. Another important group of targets includes SWEET sugar transporter genes, which are often hijacked by pathogens for nutritional gain. Knockout of SWEET genes has been shown to enhance resistance in rice (Chen et al., 2010), and similar strategies could be tested in other crops. Structural genes involved in lignin biosynthesis, such as PAL (Phenylalanine ammonia lyase) and CCoAMT (Caffeoyl-CoA O-methyltransferase), were reported to be differentially expressed during *Fusarium* infection, suggesting their potential for enhancing plant resistance (Varshney et al., 2013). In addition, the RNA silencing machinery, including DCL (dicer-like), AGO (Argonaute), and RDR (RNA-dependent RNA polymerase), is vital for antiviral defense, particularly against PPSMV. These genes are strongly activated in resistant pigeonpea lines, and genome editing tools could enhance their expression for stronger RNAi-mediated immunity (Srivastava et al., 2021).

Beyond disease resistance, genome editing also provides opportunities to mitigate yield losses caused by insect pests such as *Helicoverpa armigera* (pod borer), *Etiella zinckenella* (spotted pod borer), and *Melanagromyza obtusa* (pod fly). Proteinase inhibitors (PIs), including trypsin and chymotrypsin inhibitors, are naturally expressed in pigeonpea during insect attack and interfere with pest digestion and editing could help manipulate their expression or regulatory control (Srinivasan et al., 2005). The jasmonic acid (JA) signaling pathway, mediated by COI1, JAR1, and MYC2, is another promising defense system that could be fine-tuned using CRISPR-based tools, while WRKY transcription factors such as WRKY53 and WRKY70 may be edited to optimize cross talk between JA and SA pathways (Singh et al., 2017b). Furthermore, insecticidal proteins such as lectins and ribosome-inactivating proteins (RIPs) have been shown to disrupt insect gut cells.

Their transgenic expression in legumes like chickpea and cowpea provided resistance to *H. armigera* (Sarmah et al., 2004), and editing of native pigeonpea homologs could offer similar protection without transgenes.

Genome editing can also reduce pest susceptibility by disrupting host susceptibility (S) genes. For instance, SWEET transporters, which provide sugars to both pathogens and insects, can be targeted to reduce pest advantage (Chen et al., 2010). Similarly, enhancing cuticular wax biosynthesis genes such as CER1 and WAX2 could strengthen physical barriers to insect feeding. Secondary-metabolite pathways, including those regulated by chalcone synthase (CHS) and isoflavone synthase (IFS), are also potential targets, as their products (flavonoids, alkaloids, tannins) deter insect feeding. Transcriptomic analyses in pigeonpea confirm the differential regulation of these pathways under insect stress, supporting their functional relevance (Priyanka et al., 2010).

Together, these studies highlight a broad array of genome editing targets—from immune receptors, signaling regulators to structural and susceptibility genes—that can be precisely engineered to provide pigeonpea with durable resistance against both diseases and insect pests. Importantly, integrating these strategies through CRISPR-based multiplex editing and gene stacking has the potential to create modern pigeonpea cultivars with broad-spectrum resistance, thereby reducing reliance on chemical pesticides and promoting sustainable pigeonpea production.

REFERENCES

Anzalone AV, Randolph PB, Davis JR, Sousa AA, Koblan LW, et al. (2019) Search-and-replace genome editing without double-strand breaks or donor DNA. *Nature* 576(7785):149–157.

Arnoldi A, Zanoni C, Lammi C, Boschin G (2015) The role of grain legumes in the prevention of hypercholesterolemia and hypertension. *CRC Crit Rev Plant Sci* 34: 144–168.

Bajpai R, Ankita S, Birinchi S (2022) Identification and Characterization of Fusarium Udum Responsive NBS-LRR Genes in Pigeonpea (*Cajanus Cajan (L.) Millsp.*) Available at SSRN 4200658.

Bakala HS, Singh G, Srivastava P (2020) Smart breeding for climate resilient agriculture. In: *Plant breeding-current and future views*. IntechOpen, London, UK.

Bharti J, Verma R, Thangaraj A, Sony SK, Fathy K, Rawat B, et al. (2025) Combating aggressive weeds: Reinforcing herbicide resistance in pigeonpea (*Cajanus cajan L.*) through genome editing. *Plant Physiol Biochem PPB*, 222:109550.

Bhattacharjee S, Bhowmick R, Kant L, Paul K (2023) Strategic transgene-free approaches of CRISPR-based genome editing in plants. *Mol Genet Genomic* 298(3):507–520.

Bisht VS, Kannaiyan J, Nene YL (1988) Methods of metalaxyl application to control of Phytophthora blight in pigeonpea. *International Pigeonpea Newsletter* 8: 9–10.

Biswas K, Tarafdar A, Kumar R, Singhvi N, Ghosh P, Sharma M, et al. (2020) Molecular analysis of disease-responsive genes revealing the resistance potential against Fusarium wilt (*Fusarium udum Butler*) dependent on genotype variability in the leguminous crop pigeonpea. *Front genet* 11: 862

Boch J, Bonas, U (2010) Xanthomonas AvrBs3 family-type III effectors: discovery and function. *Annu Rev Phytopathol* 48: 419–436.

Boch J, Scholze H, Schornack S, Landgraf A, Hahn S, Kay S, et al. (2009) Breaking the code of DNA binding specificity of TAL-type III effectors. *Science (N.Y.)* 326(5959): 1509–1512.

Bohra A (2013) Emerging paradigms in genomics-based crop improvement. *The Scientific World Journal* 2013:585467.

Bohra A, Saxena KB, Varshney RK, Saxena RK (2020) Genomics-assisted breeding for pigeonpea improvement. *Theor Appl Genet* 133(5): 1721–1737.

Bohra A, Tiwari A, Kaur P, Ganie SA, Raza A, Roorkiwal M et al. (2022a) The Key to the Future Lies in the Past: Insights from Grain Legume Domestication and Improvement Should Inform Future Breeding Strategies. *Plant Cell Physiol* 63(11): 1554–1572.

Bohra A, Kilian B, Sivasankar S, Caccamo M, Mba C, McCouch SR, Varshney RK (2022b) Reap the crop wild relatives for breeding future crops. *Trends in Biotechnology* 40:412–431.

Boncan DAT, Tsang SS, Li C, Lee IH, Lam HM, Chan TF, et al. (2020) Terpenes and terpenoids in plants: Interactions with environment and insects. *Int J Mol Sci* 21(19): 7382.

Bonturi CR, Silva Teixeira AB, Rocha VM, Valente PF, Oliveira JR, Filho CMB, et al. (2022) Plant kunitz inhibitors and their interaction with proteases: Current and potential pharmacological targets. *Int J Mol Sci* 23(9): 4742.

Budhagatapalli N, Halbach T, Hiekel S, Büchner H, Müller AE, Kumlehn J (2020) Site-directed mutagenesis in bread and durum wheat via pollination by cas9/guide RNA-transgenic maize used as haploidy inducer. *Plant Biotechnol J* 18:2376–2378. doi: 10.1111/pbi.13415

Cardi T, Murovec J, Bakhsh A, Boniecka J, Bruegmann T, Bull SE, et al. (2023) CRISPR/Cas-mediated plant genome editing: outstanding challenges a decade after implementation. *Trends Plant Sci 28*(10): 1144–1165.

Cermak T, Doyle EL, Christian M, Wang L, Zhang Y, Schmidt C et al. (2011) Efficient design and assembly of custom TALEN and other TAL effector-based constructs for DNA targeting. *Nucl Acids Res* 39(12): e82.

Chandrasekaran J, Brumin M, Wolf D, Leibman D, Klap C, Pearlsman M et al. (2016) Development of broad virus resistance in non-transgenic cucumber using CRISPR/Cas9 technology. *Mol Plant Path* 17(7): 1140–1153.

Che P, Chang S, Simon MK, Zhang Z, Shaharyar A, Ourada J, O'Neill D, Torres-Mendoza M, Guo Y, Marasigan KM (2021) Developing a rapid and highly efficient cowpea regeneration, transformation and genome editing system using embryonic axis explants. *Plant J* 106:817–830

Chen LQ, Hou BH, Lalonde S, Takanaga H, Hartung ML, Qu XQ, et al. (2010) Sugar transporters for intercellular exchange and nutrition of pathogens. *Nature* 468(7323):527–532.

Choudhary AK, Raje RS, Datta S, Sultana R, Ontagodi T (2013) Conventional and molecular approaches towards genetic improvement in pigeonpea for insects resistance. *Am J Plant Sci* 4(2): 372–385.

Christian M, Cermak T, Doyle EL, Schmidt C, Zhang F, Hummel A et al. (2010) Targeting DNA double-strand breaks with TAL effector nucleases. *Genetics* 186(2): 757–761.

Curtin SJ, Zhang F, Sander JD, Haun WJ, Starker C, Baltes NJ, Voytas DF, Stupar RM (2011) Targeted mutagenesis of duplicated genes in soybean with zinc-finger nucleases. *Plant Physiol* 156:466–473.

Dabravolski SA, Isayenkov SV (2023) The regulation of plant cell wall organisation under salt stress. *Front Plant Sci* 14:1118313.

Dhanyakumar O, Ranjitha M, Yogendra K, Revanayya G, Sharma HC, Taggar GK, … Jaba J (2025) Decoding plant resistance mechanism in pigeonpea (Cajanus cajan) against major insect pests: decade of efforts and emerging directions. *Plant Stress* 101089. https://doi.org/10.1016/j.stress.2025.101089

Doyle EL, Booher NJ, Standage DS, Voytas DF, Brendel VP, Vandyk JK, Bogdanove AJ (2012) TAL Effector-Nucleotide Targeter (TALE-NT) 2.0: tools for TAL effector design and target prediction. *Nucleic Acids Res* 40:W117–W122.

Du H, Zeng X, Zhao M, Cui X, Wang Q, Yang H, Cheng H, Yu D (2016) Efficient targeted mutagenesis in soybean by TALENs and CRISPR/Cas9. *J Biotechnol* 217:90–97.

Gaudelli NM, Komor AC, Rees HA, Packer MS, Badran AH, Bryson DI, Liu DR (2017) Programmable base editing of A•T to G•C in genomic DNA without DNA cleavage. *Nature* 551:464–471.

Ghosh G, Ganguly S, Purohit A, Chaudhuri RK, Das S, Chakraborti D (2017) Transgenic pigeonpea events expressing Cry1Ac and Cry2Aa exhibit resistance to Helicoverpa armigera. *Plant Cell Rep* 36:1037–1051.

Gnanesh BN, Bohra A, Sharma M, Byregowda M, Pande S, Wesley V, Saxena RK, Saxena KB, Kavi Kishor PB, Varshney RK (2011) Genetic mapping and quantitative trait locus analysis of resistance to sterility mosaic disease in pigeonpea [Cajanus cajan (L.) Millsp.]. *Field Crops Res* 123:56–61.

Gootenberg JS, Abudayyeh OO, Lee JW, Essletzbichler P, Dy AJ, Joung J, Verdine V, Donghia N, Daringer NM, Freije CA, Myhrvold C, Bhattacharyya RP, Livny J, Regev A, Koonin EV, Hung DT, Sabeti PC, Collins JJ, Zhang F (2017) Nucleic acid detection with CRISPR-Cas13a/C2c2. *Science* 356:438–442.

Gupta AK, Singh IS, Reddy MV, Bajpai GC (1997) Genetic resistance of P3 isolate to Phytophthora blight of pigeonpea. *Euphytica* 95:73–76.

Gupta SK, Vishwakarma NK, Malakar P, Vanspati P, Sharma NK, Chattopadhyay D (2023) Development of an Agrobacterium-delivered codon-optimized CRISPR/Cas9 system for chickpea genome editing. *Protoplasma* 260:1437–1451.

Haun W, Coffman A, Clasen BM, Demorest ZL, Lowy A, Ray E, Retterath A, Stoddard T, Juillerat A, Cedrone F, Mathis L, Voytas DF, Zhang F (2014) Improved soybean oil quality by targeted mutagenesis of the fatty acid desaturase 2 gene family. *Plant Biotechnol J* 12:934–940.

Jadhav DR, Mallikarjuna N, Sharma HC, Saxena KB (2012) Introgression of Helicoverpa armigera resistance from Cajanus acutifolius—a wild relative from secondary gene pool of pigeonpea (Cajanus cajan). *Asian J Agric Sci* 4:242–248.

Jaganathan D, Ramasamy K, Sellamuthu G, Jayabalan S, Venkataraman G (2018) CRISPR for crop improvement: an update review. *Front Plant Sci* 9:985.

Jain KC, Reddy MV (1995) Inheritance of resistance to Fusarium wilt in pigeonpea (*Cajanus cajan* (L) Millsp.). *Indian J Genet Pl Breed* 55:434–437.

Ji J, Zhang C, Sun Z, Wang L, Duanmu D, Fan Q (2019) Genome editing in cowpea *Vigna unguiculata* using CRISPR-Cas9. *Int J Mol Sci* 20:2471.

Jiang W, Yang B, Weeks DP (2014) Efficient CRISPR/Cas9-mediated gene editing in *Arabidopsis thaliana* and inheritance of modified genes in the T2 and T3 generations. *PLoS One* 9(6):e99225.

Jinek M, Chylinski K, Fonfara I, Hauer M, Doudna JA, Charpentier E (2012) A programmable dual-RNA-guided DNA endonuclease in adaptive bacterial immunity. *Science* 337(6096):816–821.

Jones JD, Staskawicz BJ, Dangl JL (2024) The plant immune system: From discovery to deployment. *Cell* 187(9):2095–2116.

Kanagaraj P, Murali Baskaran RK, Saxena RK, Singh VK, Parupalli S, Sameerkumar CV, Varshney RK (2019) Genotypic variability for resistance to Fusarium wilt and sterility mosaic disease in a mini-core collection of pigeonpea [*Cajanus cajan* (L.) Millsp.]. *Plant Genet Resour-C* 17(2):102–110.

Kannaiyan J, Nene YL, Reddy MV, Ryan JG, Raju TN (1984) Prevalence of pigeonpea diseases and associated crop losses in Asia, Africa and the Americas. *Int J Pest Manag* 30(1):62–72.

Karmakar S, Panda D, Panda S, Dash M, Saha R, Das P, Avinash SP, et al. (2024) A miniature alternative to Cas9 and Cas12: Transposon-associated TnpB mediates targeted genome editing in plants. *Plant Biotechnol J* 22(10):2950–2953.

Khetnon P, Busarakam K, Sukhaket W, Niwaspragrit C, Kamolsukyeunyong W, Kamata N, Sanguansub S (2022) Mechanisms of trichomes and terpene compounds in indigenous and commercial Thai rice varieties against brown planthopper. *Insects* 13(5):427.

Khoury CK, Castaneda-Alvarez NP, Achicanoy HA, Sosa CC, Bernau V, Kassa MT, … Struik PC (2015) Crop wild relatives of pigeonpea [Cajanus cajan (L.) Millsp.]: Distributions, ex situ conservation status, and potential genetic resources for abiotic stress tolerance. *Biological Conservation* 184:259–270.

Komor AC, Kim YB, Packer MS, et al. (2016) Programmable editing of a target base in genomic DNA without double-stranded DNA cleavage. *Nature* 533(7603):420–424.

Kumar V, Singh A, Amitha Mithra SV, Krishnamurthy SL, Parida SK, Jain S, et al. (2014) Genome-wide association mapping of salinity tolerance in rice (*Oryza sativa*). *DNA Res* 22(2):133–145.

Kumar G, Bajpai R, Sarkar A, Mishra RK, Kumar Gupta V, Singh HB, et al. (2019) Identification, characterization and expression profiles of *Fusarium udum* stress-responsive WRKY transcription factors in *Cajanus cajan* under the influence of NaCl stress and *Pseudomonas fluorescens* OKC. *Sci Rep* 9(1):14344.

Kumar M, Prusty MR, Pandey MK, Singh PK, Bohra A, Guo B, Varshney RK (2023) Application of CRISPR/Cas9-mediated gene editing for abiotic stress management in crop plants. *Front Plant Sci* 14: 1157678.

Lu QSM, Tian L (2022) An efficient and specific CRISPR-Cas9 genome editing system targeting soybean phytoene desaturase genes. *BMC Biotechnol* 22:7.

Mahato AK, Sharma AK, Singh NK (2019) Genome-wide characterization and expression patterns of chitinase genes in the pigeonpea (Cajanus cajan (L.) Millsp.) genome. *Plant Omics* 12:109–119.

Makarova KS, Wolf YI, Koonin EV (2013) Comparative genomics of defense systems in archaea and bacteria. *Nucleic Acids Res* 41:4360–4377.

Mallikarjuna N, Sharma HC, Upadhyaya HD (1997) Exploitation of wild relatives of pigeonpea and chickpea for resistance to Helicoverpa armigera. *J SAT Agric Res* 3:4.

Mehta R, Radhakrishnan T, Kumar A, Yadav R, Dobaria JR, Thirumalaisamy PP, et al. (2013) Coat protein-mediated transgenic resistance of peanut (Arachis hypogaea L.) to peanut stem necrosis disease through Agrobacterium-mediated genetic transformation. *Indian J Virol* 24:205–213.

Meng Y, Hou Y, Wang H, Ji R, Liu B, Wen J, et al. (2017) Targeted mutagenesis by CRISPR/Cas9 system in the model legume Medicago truncatula. *Plant Cell Rep* 36:371–374.

Menz J, Modrzejewski D, Hartung F, Wilhelm R, Sprink T (2020) Genome edited crops touch the market: a view on the global development and regulatory environment. *Front Plant Sci* 11:586027.

Mishra RK, Pandey S, Mishra M, Bohra A, Mishra O, Tiwari S, Satheesh Naik, Naimuddin, Chaudhary RG, Dhar V, Singh IP, Singh NP, Dixit GP (2025a) Pathogenicity, Characterisation and Racial Distribution of Fusarium udum Isolates in India Causing Wilt Disease of Pigeonpea (Cajanus cajan). *Plant Pathology*. https://doi.org/10.1111/ppa.70006

Mishra RK, Bohra A, Kumar D, Parihar AK, Hazra KK, Pandey S, Mishra M, Satheesh Naik SJ, Kumari N, Datta D, Tiwari S, Naimuddin, Dixit GP (2025b) Morpho-cultural variability, pathogenicity and molecular identification of Phytopthora cajani isolates causing blight disease of pigeonpea in India. *Physiological and Molecular Plant Pathology* 138:102674.

Mohanraj SS, Gujjarlapudi M, Lokya V, Mallikarjuna N, Dutta-Gupta A, Padmasree K (2019) Purification and characterization of Bowman-Birk and Kunitz isoinhibitors from the seeds of Rhynchosia sublobata (Schumach.) Meikle, a wild relative of pigeonpea. *Phytochemistry* 159:159–171.

Moscou MJ, Bogdanove AJ (2009) A simple cipher governs DNA recognition by TAL effectors. *Science* 326:1501.

Ngou BPM, Heal R, Wyler M, Schmid MW, Jones JD (2022) Concerted expansion and contraction of immune receptor gene repertoires in plant genomes. *Nat Plants* 8:1146–1152.

Nivya VM, Shah JM (2023) Recalcitrance to transformation, a hindrance for genome editing of legumes. *Front Genome Ed* 5:1247815.
Njaci I, Ngugi-Dawit A, Oduor RO, Kago L, Williams B, Hoang LTM, et al. (2020) Comparative analysis delineates the transcriptional resistance mechanisms for pod borer resistance in the pigeonpea wild. *Int J Mol Sci* 22(1):309.
Ochatt S, Conreux C, Moussa Mcolo R, Despierre G, Magnin-Robert J-B, Raffiot B (2018) Phytosulfokine-alpha, an enhancer of in vitro regeneration competence in recalcitrant legumes. *Plant Cell Tissue Organ Cult* 135:189–201.
Ojha GC, Choudhary SK (1993) Inheritance of resistance to alternaria leaf blight in pigeonpea (*Cajanus cajan* (L.) Millsp.). *Int. Pigeonpea Newsl* 7:4–5.
Pande S, Sharma M, Mangla UN, Ghosh R, Sundaresan G (2012) Phytophthora blight of pigeonpea (*Cajanus cajan* (L.) Millsp.): An updating review of biology, pathogenicity and disease management. *Crop Prot* 30:951–957.
Pandey AK, Basandrai AK (2021) Will *Macrophomina phaseolina* spread in legumes due to climate change? A critical review of current knowledge. *J Plant Dis Prot* 128(1):9–18.
Pareddy D, Chennareddy S, Anthony G, Sardesai N, Mall T, Minnicks T, Karpova O, Clark L, Griffin D, Bishop B, Shumway N, Samuel P, Smith K, Sarria R (2020) Improved soybean transformation for efficient and high throughput transgenic production. *Transgenic Res.* 29(3):267–281.
Patil BL, Kumar PL (2015) Pigeonpea sterility mosaic virus: a legume-infecting *Emaravirus* from South Asia. *Mol Plant Pathol* 16(8):775–786.
Patil PG, Bohra A, Dubey J, Saabale PR, Mishra RK, Pandey G, Das A, Rathore M, Singh F, Singh NP (2017a) Genetic analysis and molecular resistance to race 2 of Fusarium wilt in pigeonpea [*Cajanus cajan* (L.) Millsp.]. *Crop Prot* 100:117–123.
Patil PG, Dubey J, Bohra A, Mishra RK, Saabale PR, Das A, Rathore M, Singh NP (2017b) Association mapping to discover significant marker-trait associations for resistance against Fusarium wilt variant 2 in pigeonpea [*Cajanus cajan* (L.) Millspaugh] using SSR markers. *J Appl Genet* 58:307–319.
Pawar NB, Mayee CD (1986) Reaction of pigeonpea genotypes and their crosses to Fusarium wilt. *Indian Phytopathol* 39:70–74.
Pawar P, Kale SM, Rathore A, Upadhyaya HD, Varshney RK (2024) Identification and deployment of Fusarium wilt and Phytophthora blight resistance in pigeonpea through combined field and genomic approaches. *Theor Appl Genet* 137(1):55–70.
Prasad K, Bhatnagar-Mathur P, Waliyar F, Sharma KK (2013) Overexpression of a chitinase gene in transgenic peanut confers enhanced resistance to major soil borne and foliar fungal pathogens. *J Plant Biochem Biotechnol* 22:222–233.
Prasad K, Gadeela H, Bommineni PR, et al. (2024) CRISPR/Cas9-mediated mutagenesis of phytoene desaturase in pigeonpea and groundnut. *Funct Integr Genomics* 24:57.
Pratap A, Prajapati U, Singh CM, Gupta S, Rathore M, Malviya N, Tomar R, Gupta AK, Tripathi S, Singh NP (2018) Potential, constraints and applications of in vitro methods in improving grain legumes. *Plant Breed* 137:235–249.
Priyanka B, Sekhar K, Reddy VD, Rao KV (2010) Expression of pigeonpea hybrid-proline-rich protein encoding gene (CcHyPRP) in yeast and *Arabidopsis* affords multiple abiotic stress tolerance. *Plant Biotechnol J* 8(1):76–87.
Purohit A, Ghosh S, Ganguly S, Negi MS, Tripathi SB, Chaudhuri RK, et al. (2021) Comparative transcriptomic profiling of susceptible and resistant cultivars of pigeonpea demonstrates early molecular responses during *Fusarium udum* infection. *Sci Rep* 11:22319.
Ramu SV, Rohini S, Keshavareddy G, Neelima MG, Shanmugam NB, Kumar ARV, et al. (2012) Expression of a synthetic cry1AcF gene in transgenic pigeonpea confers resistance to *Helicoverpa armigera*. *J Appl Entomol* 136:675–687.

Rangaswamy KT, Muniyappa V, Lava Kumar P, Saxena KB, Byregowda M, Raghavendra N, Pandurangaiah K, Vijay Kumar R, Waliyar F, Jones AT (2005) ICP7035–A sterility mosaic resistant vegetable and grain purpose pigeonpea variety. *SAT eJournal* 1:13.

Ranjan A, Nigam D, Asif MH, Singh UM, Ranjan S, Madhugiri R, et al. (2019) Genome-wide expression profiling of pigeonpea accessions to understand resistance against sterility mosaic disease. *Sci Rep* 9:18115.

Rasheed A, Barqawi AA, Mahmood A, Nawaz M, Shah AN, Bay DH, et al. (2022) CRISPR/Cas9 is a powerful tool for precise genome editing of legume crops: a review. *Mol Biol Rep* 49:5595–5609.

Reddy MV, Sarkar N, Nene YL, Raju TN (1991) Predisposing factors for Phytophthora blight of pigeonpea. *Indian Phytopathol* 45:268–270.

Saini P, Saini P, Kaur JJ, Francies RM, Gani M, Rajendra AA, et al. (2020) Molecular approaches for harvesting natural diversity for crop improvement. In: *Rediscovery of genetic and genomic resources for future food security*. Springer, Singapore. doi: 10.1007/978-981-15-0156-2_3

Sarmah BK, Moore A, Tate W, Molvig L, Morton RL, Rees DP, et al. (2004) Transgenic chickpea seeds expressing high levels of a bean α-amylase inhibitor. *Mol Breed* 14:73–82.

Saxena KB (2008) Genetic improvement of pigeonpea—a review. *Trop Plant Biol* 1:159–178.

Saxena KB, Kumar RV, Srivastava RK (2010a) Development of cytoplasmic male-sterility, its inheritance, and stability in pigeonpea. *Euphytica* 173:307–314.

Saxena RK, Saxena KB, Kumar RV, Hoisington DA, Varshney RK (2010b) Simple sequence repeat-based diversity in elite pigeonpea genotypes for developing mapping populations to map resistance to Fusarium wilt and sterility mosaic disease. *Plant Breed* 129:235–241.

Saxena KB, Kumar RV, Saxena RK, Sharma M, Srivastava RK, Sultana R, et al. (2012) Identification of dominant and recessive genes for resistance to Fusarium wilt in pigeonpea and their implication in breeding hybrids. *Euphytica* 188:221–227.

Saxena RK, Kale SM, Kumar V, Parupalli S, Joshi S, Singh VK, et al. (2017a) Genotyping-by-sequencing of three mapping populations for identification of candidate genomic regions for resistance to sterility mosaic disease in pigeonpea. *Sci Rep* 7:1813.

Saxena RK, Singh VK, Kale SM, Tathineni R, Parupalli S, Kumar V, et al. (2017b) Construction of genotyping-by-sequencing based high-density genetic maps and QTL mapping for fusarium wilt resistance in pigeonpea. *Sci Rep* 7:1911.

Saxena RK, Rathore A, Bohra A, Yadav P, Das RR, Khan AW, Singh VK, Chitikineni A, Singh IP, Sameer Kumar CV, Saxena KB, Varshney RK (2018) Development and application of high density Axiom®CajanusSNP Array with 56K SNPs for understanding genome architecture of released varieties and founder genotypes for redefining future pigeonpea breeding programmes. *The Plant Genome* 11: 180005

Saxena RK, Hake A, Hingane A, Sameer Kumar CV, Bohra A, Sonnappa M, Rathore A, Kumar A, Mishra A, Tikle AN, Sudhakar C, Rajamani S, Patil DK, Singh IP, Singh NP, Varshney RK (2020) Translational Pigeonpea Genomics Consortium for accelerating genetic gains in pigeonpea (Cajanus cajan L.). *Agronomy* 10:1289.

Saxena RK, Hake A, Bohra A, Khan AW, Hingane A, Sultana R, et al. (2021) A diagnostic marker kit for Fusarium wilt and sterility mosaic diseases resistance in pigeonpea. *Theor Appl Genet* 134:367–379.

Senthil K, Rathinam M, Parashar M, Dokka N, Tyagi S, Mathur V, et al. (2025) Establishing a CRISPR/Cas9 genome editing framework in pigeonpea (*Cajanus cajan* L.) by targeting phytoene desaturase (PDS) gene disruption. *J Genet Eng Biotechnol* 23:100465.

Sharma M (2017) Screening techniques and resistance sources for wilt, sterility mosaic disease, and Phytophthora blight in pigeonpea. *Phytopathology* 107:187–197.

Sharma D, Rani TS, Upadhyaya HD (2003) Resistance to pod borer, pod fly, and diseases in wild relatives of pigeonpea. *Int Chickpea Pigeonpea Newsl* 10:29–32.

Sharma K, Sreelatha G, Dayal S (2006) Pigeonpea (*Cajanus cajan* L. Millsp). *Methods Mol Biol* 343:359–367.

Sharma HC, Gowda CLL, Stevenson PC, Ridsdill-Smith TJ, Clement SL, Rao GR, et al. (2007) Host plant resistance and insect pest management in chickpea. In: *Chickpea breeding and management.* CABI, Wallingford, UK, pp 520–537.

Sharma M, Ghosh R, Telangare R, Rathore A, Saifulla M, Mahalinga DM, et al. (2016) Environmental influences on pigeonpea-*Fusarium udum* interactions and stability of genotypes. *Front Plant Sci* 7:253.

Sharma, S., Jaba, J., Rao, P. J., Prasad, S., Gopal, N. T. V. V., Sharma, H. C., & Kilian, B. (2022) Reaping the potential of wild *cajanus* species through pre-breeding for improving resistance to pod borer, Helicoverpa armigera, in cultivated pigeonpea *(Cajanus cajan (l.) millsp.). Biology* 11(4): 485.

Singh UP, Chauhan VB. (1985) Relationship between field levels and light and darkness on the development of Phytophthora blight of pigeonpea (Cajanus cajan (L.) Millsp.). *Phytopathol* 114:1.

Singh KB, Chauhan VB (1992) Screening methods for resistance to Phytophthora blight in pigeonpea. *Journal of Mycology and Plant Pathology* 22:62–65.

Singh S, Kumar SSBV, Reddy MPG. (2003) *Sterility mosaic resistant genotypes of pigeonpea.* In: *Proc Nat Symp Pulses Crop Diversif Nat Resour Manage*, Dec 20–22, Kanpur, India. p 235.

Singh, I. P., Katiyar, P. K., & Singh, S. K. (2009) Inheritance of genes imparting resistance to pod fly (Melanagromyza obtusa) in pigeonpea (Cajanus cajan). *Indian J Agric Sci* 79(8): 628.

Singh NK, Gupta DK, Jayaswal PK, Mahato AK, Dutta S, Singh S, et al. (2012) The first draft of the pigeonpea genome sequence. *J Plant Biochem Biotechnol* 21(1):98–112.

Singh VK, Khan AW, Saxena RK, Kumar V, Kale SM, Sinha P, (2016) Next generation sequencing for identification of candidate genes for fusarium wilt and sterility mosaic disease in pigeonpea (*Cajanus cajan*) *Plant Biotechnol J* 14:1183–1194.

Singh VK, Garg R, Jain M, Kumar A. (2017a) Transcriptome analysis identifies differentially expressed genes associated with resistance against Fusarium wilt in pigeonpea (Cajanus cajan). *Front Plant Sci* 8:1951.

Singh VK, Khan AW, Saxena RK, Sinha P, Kale SM, Parupalli S, et al. (2017b) Indel-seq: a fast-forward genetics approach for identification of trait-associated putative candidate genomic regions and its application in pigeonpea (Cajanus cajan). *Plant Biotechnol J* 15(7):906–914.

Singh S, Kumar NR, Maniraj R, Lakshmikanth R, Rao KYS, Muralimohan N, Arulprakash T, Karthik K, Shashibhushan NB, Vinutha T, Pattanayak D, Dash PK, Kumar PA, Sreevathsa R (2018) Expression of Cry2Aa, a Bacillus thuringiensis insecticidal protein in transgenic pigeon pea confers resistance to gram pod borer, *Helicoverpa armigera. Scientific Reports* 8:8820.

Smýkal P, Konečná E. (2014) Advances in pea genomics. In: *Legumes in the omic era.* Springer, New York, NY.

Soren KR, Tripathi S, Pareek S, Hembram M, Gangwar P, Abrol S, et al. (2021) Comparative time series RNA-seq analysis of pigeonpea root tissues in response to Fusarium udum infection. *Front Fungal Biol* 2:664953.

Srinivasan T, Mohan S, Subramanian S. (2005) Proteinase inhibitors from pigeonpea and their effect on the digestive proteases and development of Helicoverpa armigera. *Phytochemistry* 66(14):1654–1660.

Srivastava RK, Verma V, Chauhan A. (2021) Role of RNA silencing in plant-virus interactions: Insights from pigeonpea sterility mosaic virus infection. *Physiol Mol Plant Pathol* 116:101708.

Srivastava A, Gupta T, Srivastava S, Dhir S, Kumar P, Singhal T, et al. (2022) Development of a new Collateral Cleavage-independent CRISPR/Cas12a based easy detection system for plant viruses. *J Virol Methods* 300:114432.

Swathi M, Mishra PK, Lokya V, Swaroop V, Mallikarjuna N, Dutta-Gupta A, Padmasree K. (2016) Purification and partial characterization of trypsin-specific proteinase inhibitors from pigeonpea wild relative *Cajanus platycarpus L.* (Fabaceae) active against gut proteases of lepidopteran pest Helicoverpa armigera. *Front Physiol* 7:388.

Tyagi S, Rathinam M, Shashank PR, Chaudhary N, Shasany AK, Sreevathsa R. (2022) Deciphering of pod borer [*Helicoverpa armigera (Hübner)*] resistance in *Cajanus platycarpus* (Benth.) offers novel insights on the reprogramming and role of flavonoid biosynthesis pathway. *Toxins* 14(7):455.

Ukhatova YV, Erastenkova MV, Korshikova ES, Krylova EA, Mikhailova AS, Semilet TV, et al. (2023) Improvement of crops using the CRISPR/Cas system: new target genes. *Mol Biol* 57:375–397.

Upadhyaya HD, Dronavalli N, Gowda CLL, Singh S (2012) Mini Core Collections for Enhanced Utilization of Genetic Resources in Crop Improvement. *Indian Journal of Plant Genetic Resources* 25(1):111–124.

Varshney RK, Chen W, Li Y, Bharti AK, Saxena RK, Schlueter JA, et al. (2013) Draft genome sequence of pigeonpea (*Cajanus cajan*), an orphan legume crop of resource-poor farmers. *Nat Biotechnol* 30:83–89.

Varshney RK, Saxena RK, Upadhyaya HD, Khan AW, Yu Y, Kim C, et al. (2017) Whole-genome resequencing of 292 pigeonpea accessions identifies genomic regions associated with domestication and agronomic traits. *Nat Genet* 49:1082–1088.

Varshney RK, Bohra A, Yu J, Graner A, Zhang Q, Sorrells ME (2021) Designing future crops: genomics-assisted breeding comes of age. *Trends Plant Sci* 26:631–649.

Vaz Patto MC, Amarowicz R, Aryee ANA, Boye JI, Chung HJ, Martin-Cabrejas MA, et al. (2015) Achievements and challenges in improving the nutritional quality of food legumes. *Crit Rev Plant Sci* 34:105–143.

Wu J, Zhu J, Wang L, Wang S (2017) Genome-wide association study identifies NBS-LRR-encoding genes related with anthracnose and common bacterial blight in the common bean. *Front Plant Sci* 8:1398.

Xu H, Guo Y, Qiu L, Ran Y (2022) Progress in soybean genetic transformation over the last decade. *Front Plant Sci* 13:900318.

Young ND, Bharti AK (2012) Genome-enabled insights into legume biology. *Annu Rev Plant Biol* 63:283–305.

12 Gene Editing in Pigeonpea for Abiotic Stress Tolerance

Rachana Verma, Jyotsna Bharti, Isha Gupta, Puja Chakraborty, R. Shubhra Maithreyi, Arulprakash Thangaraj, Rashmi Kaul, Sugyan Preet, Kunal Tanwar, and Tanushri Kaul

12.1 INTRODUCTION

Securing sustained food and nutrition availability has become a significant challenge for developing and, less economically developed countries particularly in areas such as South Asia and sub-Saharan Africa. The ever-increasing world's population that is projected to hit 10 billion by 2050, the demand for food is expected to surge substantially, reaching to approximately13.6 billion tons (Ahmad et al., 2021). At present, approximately 2.37 billion people globally face food shortages, with 57 million individuals in Asia experiencing hunger in 2020 (FAO, 2021). The United Nations' 2030 Sustainable Development Agenda aims to eradicate poverty and combat environmental degradation. The COVID-19 pandemic further worsened the crisis, disrupted food systems, and limited access to essential nutrition on a global scale. In 2020, nearly 7.8 billion individuals were affected by various health-related issues (Population Reference Bureau, 2021). The World Health Organization (WHO) has warned of severe impacts in the future (Ghebreyesus, 2020; Khorsandi, 2020). Pigeonpea (Cajanus cajan L.) is a crucial legume crop extensively cultivated in tropical and subtropical regions (Saxena et al. 2019, 2021). It serves as an essential source of protein, dietary fiber, and vital nutrients for millions, particularly in South Asia and Africa. Moreover, pigeonpea enhances soil fertility by facilitating nitrogen fixation. However, conventional breeding methods have been unable to keep pace with the growing yield demands of modern agriculture. Moreover, the genetic variation within pigeonpea germplasm has diminished over time as a result of domestication and selective breeding methods (Hyten et al., 2006; Song et al., 2015, Bohra et al. 2022). The primary factors that impede pigeonpea growth and decrease agricultural productivity include environmental stresses like cold, drought, and saline-alkali conditions. Traditional breeding has been used to create pigeonpea varieties that can withstand abiotic stress (Bohra et al. 2020). However, this process is time-consuming, taking around a decade to generate stress-resilient crops. In this regard, emerging

DOI: 10.1201/9781003603092-12

molecular breeding technologies, especially genome editing, have become viable alternatives for accurately enhancing agronomically valuable traits related to abiotic stress tolerance (Gao, 2018; Li et al., 2019). Abiotic stresses, comprising drought, heat, cold, salt, pesticides, metals, and metalloids, significantly impact agricultural production by impairing plant growth and development, ultimately affecting yield and quality (Pandey et al., 2017). Plants change according to various stresses through alterations in their morphology, physiology, biochemistry, and molecular mechanisms (Wang et al., 2003; Usman et al., 2022). As the global population is expected to reach 9.7 billion by 2050, agricultural output must increase by a minimum of 85% to meet food demands (Alexandratos and Bruinsma, 2012; Ray et al., 2013). While traditional breeding methods have played a role in improving crop resistance to abiotic stress, advanced technologies are crucial for accelerating the development of resilient crop varieties. Conventional breeding is time-intensive, often requiring several years, which limits the ability to swiftly address the changing needs of farmers, producers, and the agricultural sector (Manavalan et al., 2009). Hence, the adoption of innovative and highly efficient technologies with immediate effects is essential. Genome editing, first introduced by Capecchi in the 1980s, enables precise modifications in the genetic code, including the deletion, substitution, or insertion of specific DNA sequences. Various genome-editing techniques, such as mega nucleases (Puchta et al., 1993), transcription activator-like effector nucleases (TALENs; Zhang et al., 2014), zinc-finger nucleases (ZFNs; Zhang et al., 2010), and clustered regularly interspaced short palindromic repeats—CRISPR-associated protein 9 (CRISPR-Cas9; Jiang et al., 2013), enable precise genetic modifications. Of these, CRISPR-Cas9 has become the most powerful genome-editing technology, celebrated for its accuracy, affordability, rapidity, and consistency in generating specific mutations (Chen et al., 2019).

The CRISPR/Cas9 system is capable of identifying and cutting specific DNA sequences within the genome, enabling accurate genome editing (Kumar et al. 2023). This technology was derived from a natural defense mechanism in bacteria, which safeguards against viral infections by using Cas enzymes linked to CRISPR/Cas9 (Barrangou et al., 2007). The use of CRISPR/Cas9 may alleviate concerns about 'foreign DNAs' in edited crops and food products, facilitating their commercialization. While some countries, such as Argentina, have already adopted genome-edited crops, others are still engaged in discussions about their approval. A recent example of the commercialization of CRISPR/Cas9-edited crops is the release of edited tomatoes with elevated levels of aminobutyric acid (GABA) in Japan (Waltz, 2022). This advancement is expected to drive wider global adoption of genome-edited plants. The CRISPR/Cas9 method has proven effective across multiple plant species, including model organisms like *Nicotiana benthamiana*, *Nicotiana tabacum*, and *Arabidopsis*, as well as crops like wheat, maize, rice, tomato, potato, soybean, and sugarcane (Kumar et al., 2023). Additionally, CRISPR/Cas9-based multiplexing has been used to target multiple genes in a single organism across various crop species, showcasing its potential for producing genome-edited crop plants tolerant to multiple stresses. This technology enables the targeting of numerous stress-sensitive genes concurrently in elite cultivars, allowing for the development of stress-tolerant crop plants. CRISPR-based gene activation can also be utilized to enhance the expression of

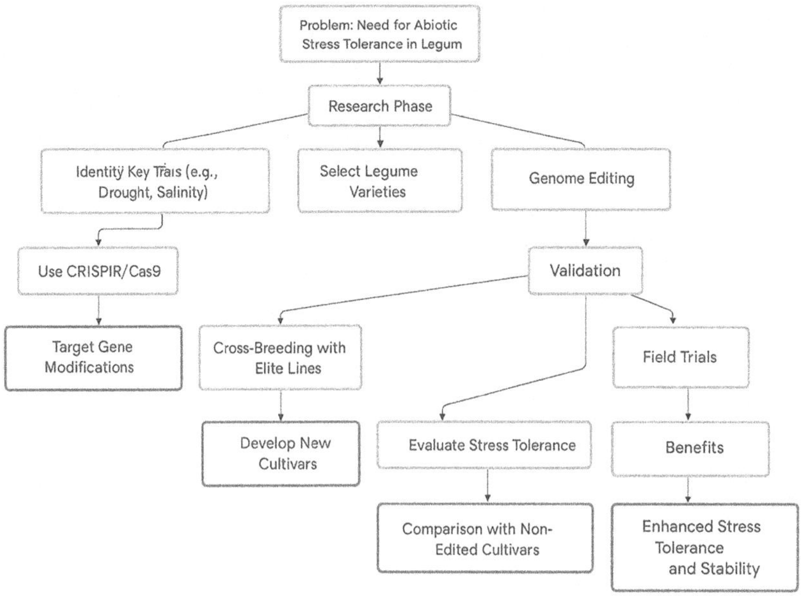

FIGURE 12.1 Flow chart describing the flow of work for developing abiotic stress–tolerant legumes.

tolerance-related genes. Genes involved in stress-responsive regulatory pathways, signaling mechanisms, and biochemical compound production can be modified using CRISPR/Cas9 technology to create crops with improved resilience to environmental challenges. CRISPR/Cas technology has been successfully employed to engineer disease-resistant plants and explore tolerance to abiotic factors such as high temperatures, drought, salinity, and nutrient deficiencies. Research has highlighted the effectiveness of CRISPR/Cas-driven gene editing in targeting complex quantitative genes linked to abiotic stress resistance. Gene modifications using CRISPR/Cas9, including edits to *OsBADH2*, *OsMPK2*, *SAPK2*, and *OsPDS*, have demonstrated promising results in enhancing abiotic stress resilience in rice. Furthermore, alterations in ethylene-responsive factor (ERF), an important transcription factor, have enhanced plants' tolerance to abiotic stress. This review examines the potential uses of CRISPR/Cas9-mediated genome editing in pigeonpea to address various abiotic stresses and predicts future advancements in breeding crops that are resilient to stress (Figure 12.1).

12.2 OVERVIEW OF CRISPR/CAS9-BASED GENOME EDITING

The increasing demand for food security due to population growth and looming threat of climate change has made reliable and efficient methods to guarantee a steady and sufficient diet for the world's population more important than ever (Liu et al., 2021; Gan et al., 2022; Bibi and Rahman, 2023). Therefore, the function of plant biotechnology and its use in engineering plants to meet the demands of global agriculture is now highly desirable. Targeted nucleases simplify the alteration process and offer a more accurate and efficient method of genome editing. "Genome-editing" refers to a collection of technologies that enable the alteration of cellular DNA sequences at specific genomic sites through the generation of altered DNA sequences via nuclease-mediated site-specific DNA breaks. Discovery of homing endonucleases (HEs) or meganucleases (MNs) paved the way for plant genome manipulation through exploitation of the cellular DNA repair pathway by creating a double-strand break (DSB) at the target loci (Antunes et al., 2012). Disruption of *MS26*, a cytochrome P450-like gene, by I-*CreI* homing endonuclease generated male sterile maize lines (Djukanovic et al., 2013). Meganuclease recognizes long DNA sequences, specifically 12–45 bp, to create site-specific DNA cleavage to induce gene targeting (Epinat et al., 2003). Since these cleavage sites are rare and do not occur naturally in the plant genome, a major limitation lies in the need to either introduce the cut site into the region of interest or production of custom mega nuclease for each MN induced gene targeting (Epinat et al., 2003; Silva et al., 2011).

To date, three major next generations genome-editing tools have evolved: first a class of sequence-specific nucleases (SSNs), comprising ZFNs, and TALENs, which offered more flexibility. Alternatively, ZFNs and TALENs generate targeted mutagens through engineered nucleases comprising a common nuclease domain derived from *FokI* endonuclease and the DNA-binding domain of zinc-finger proteins in the case of ZFN and transcription activator-like effector proteins in TALEN (Marintcheva, 2018). ZFNs and TALENs, in comparison to MNs, enable a wider range of genetic modification by permitting customization of the DNA-binding domains to target any sequence in the plant genome (Gaj et al., 2013). ZFNs recognize and bind adjacent to the target site, typically 9–18 bp in length, followed by cleavage and generation of DSB with 5' overhangs (Gupta & Musunuru, 2014). After DNA cleavage, cells repair DSBs predominantly via two pathways—error-prone nonhomologous end joining (NHEJ) or homology-directed repair (HDR; Liu et al., 2019). The third one is CRISPR-Cas9. It originated from the bacterial innate immune system. It introduces cleavage using RNA-guided DNA at specific sites in prokaryotes. The CRISPR/Cas9 is considered a more versatile and improved version in terms of efficiency and specificity for a range of eukaryotic genomes, including plants. In contrast to ZFNs and TALENs, CRISPR/Cas system is capable of targeting numerous genes simultaneously through multiplexing of multiple single guide RNAs (sgRNAs) in the plant vector system (Gupta & Musunuru, 2014). The activity of both ZFNs and TALENs demands the dimerization of a FokI nuclease domain and a sequence-specific, DNA-binding component. Although it ensures specificity, still these interactions are complex, whereas the CRISPR/Cas9 system consists of two components: the Cas9 protein and guide RNA (gRNA). The guide RNA is a short RNA with 18–20 nucleotides that is complementary to the target DNA sequence. This gRNA forms a complex with the RNA-dependent DNA endonuclease Cas9. The Cas9-gRNA complex binds to a protospacer adjacent motif (PAM) at the 3′ end of the target sequence. Streptococcus pyogenes-derived SpCas9 is the most widely utilized Cas9 variant to identify the PAM sequence 5′-NGG-3′. Whenever DSBs are induced, DNA is repaired using high-fidelity, HDR or error-prone nonhomologous end joining (NHEJ). HDR permits accurate editing when a

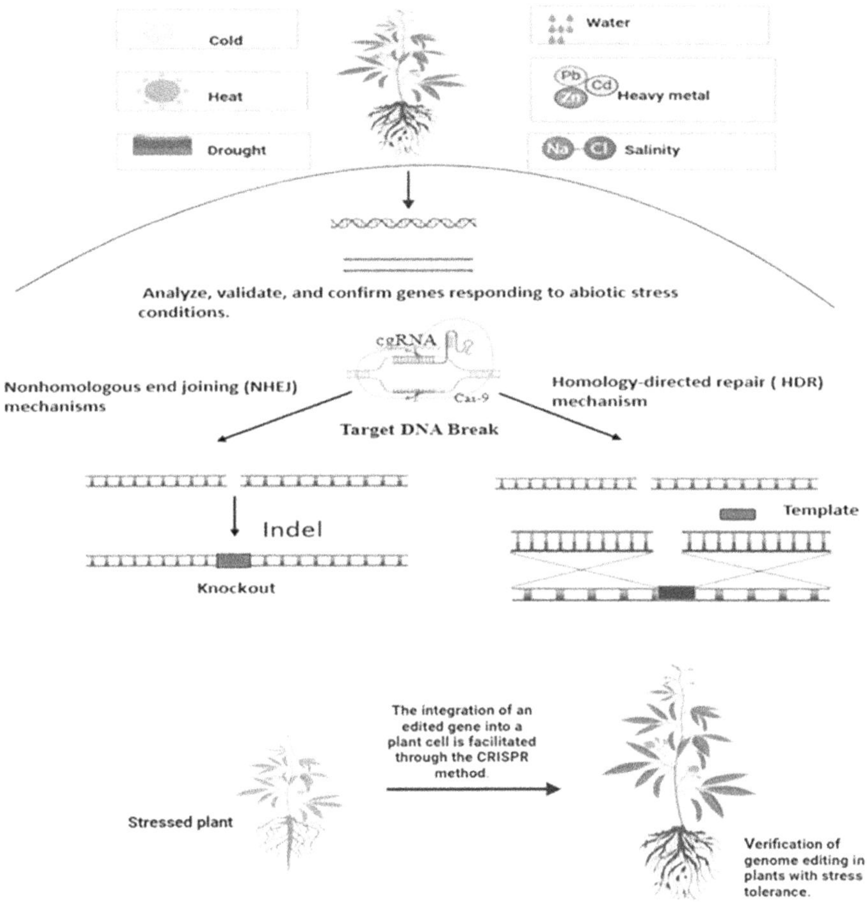

FIGURE 12.2 A comprehensive method is revealed by this schematic illustration of CRISPR/Cas9-based genome editing in plants. The Cas9 enzyme causes a targeted DSB in the DNA under the guidance of a customized single guide RNA (sgRNA). The plant cell then uses either the HDR or NHEJ pathways to fix the break.

repair template is supplied, whereas NHEJ frequently generates indel mutations that can result in frameshifts or premature stop codons for gene knockouts. Larger genomic deletions can also be produced by combining multiple DSBs (Figure 12.2). Despite its effectiveness, it may have off-target effects leading to undesired modifications to genes. To combat this, CRISPR/Cas9 developments have reduced dependency on DSBs and increased specificity (Jinek et al., 2012; Qi et al., 2013; Ran et al., 2013; Baltes et al., 2014; Hsu et al., 2014; Fauser et al., 2014; Sander & Joung, 2014; Wu et al., 2015; Sun et al., 2016a & b; Endo et al., 2016; Li et al., 2016; Haque et al., 2018; Wang et al., 2018; Xu & Li, 2020; Adhikari & Poudel, 2020; Liu et al., 2021; Li et al., 2023).

This advancement has included the development and ongoing improvement of extremely accurate genome-editing tools through the integration of deaminases and activity impaired Cas nucleases. The advent of two primary types of DNA base editors—cytosine base editors (CBEs) and adenine base editors (ABEs) that are intended to precisely introduce particular point mutations—has eliminated the need for donor DNA templates and DSBs. In CBEs, nickase Cas9 and cytidine deaminase APOBEC/AID work together to enable sgRNA-guided targeting, while UGI is incorporated to increase editing precision and efficiency by changing C-G base pairs into T-A base pairs. In ABEs, A-T base pairs are changed to G-C base pairs by fusing a laboratory-evolved TadA deaminase with Cas9 nickase. Similar to this, C to G base editors (CGBEs) increase the editing capabilities of base editors by converting C-G base pairs to G-C base pairs using a combination of uracil-DNA glycosylase (Ung), cytidine deaminase, and Cas9 nickase. Prime editing allows for precise base conversions, as well as small insertions and deletions, according to specific requirements. Prime editors (PEs) consist of a prime editor protein and a prime editing guide RNA, shortly called pegRNA. This PE protein is engineered by fusing a modified Cas9 nickase (H840A) in addition to reverse transcriptase, and it enables the targeted genome modifications directed by pegRNA (Gaudelli et al., 2017; Kantor et al., 2020; Anzalone et al., 2020; Antonioou et al., 2021; Hua et al., 2022; Crop enhancement is one of the major applications of CRISPR-Cas9 technology in agricultural contexts. CRISPR-Cas9 is being explored for improving crop yields, disease resistance, and tolerance to environmental stresses. It's a game-changer for agriculture, allowing for precise edits to enhance desirable traits. Traditional breeding to achieve desired mutations is complicated, time-intensive, and nonspecific. To develop crops with desired traits such as enhanced yield and nutritional value or modify the genome to generate a tolerant/resistant variety, CRISPR/Cas provides promising and unparalleled precision.

12.3 GENOME EDITING–DRIVEN BREEDING FOR ABIOTIC STRESS TOLERANCE

Several important traits in crops, especially in legumes for sustainable agriculture, are enhanced using tools of molecular biology via molecular breeding techniques that are precise and effective for breeding in crops. Molecular breeding includes numerous novel and modern approaches, i.e., marker-assisted backcrossing (MABC), genomics, marker-assisted selection (MAS), genomic selection (GS), and on-the-top genome editing (GE) by accelerating the development of high-performing legume varieties (Bohra et al. 2014a, 2014b, 2020). Despite the fact that legumes face eccentric hurdles and severe challenges as compared to cereals and cash crops like rice or wheat due to the genome complexity and insubstantial availability of genomic tools, in many cases, the presence of polyploidy in genome hampers crop improvement through GE such as in the most important oil-bearing leguminous crop of the world—soybean. Considering all these constraints, genome editing–driven breeding (GEDB) is a modern and alternative approach that improves crops with high-precision genetic modifications with respect to enhancing crucial traits related to increment in yield of crops, enhancing nutritional quality with environmental adaptability and disease resistance. Since the ages several genome-editing approaches like ZFNs, TALENs, and CRIPR-Cas9 have been in practice to revolutionize plant breeding through efficient and targeted alterations (Jaganathan et al., 2018). CRISPR-Cas9 has procured extensive acceptance among all these approaches on account of ease of use, high precision, elevated specificity, and efficiency in making site-specific modifications (Chen et al., 2019). Conventional breeding has also been regularized in crop improvement for a long time but has certain drawbacks, such as the requirement of multiple generations for desired traits. GE has accelerated improvement in crops manyfold by directly modifying desired and specific genes. GEDB overcomes legumes genomic complexity, i.e., polyploidy in soybeans and the effective utilization of wild germplasm diversity. Additionally, GE addresses all the key breeding bottlenecks like low transformation rates, prolonged generation cycles, and linkage drags (Baloglu et al., 2022). It's noteworthy to consider the fact that some GE crops are categorized as non-"genetically modified organisms (GMO)" that do not include foreign DNA under specific regulatory frameworks parallelly commercializing and increasing worldwide public acceptability (Wolt et al., 2016).

Over the last few decades, the pressure to address perpetually escalating food consumption demand while restricting the expansion of land under harvest has powered agricultural intensification. However, unrestricted use of chemicals in the name of agricultural inputs has led to soil deterioration, upsurging greenhouse gas emissions, and environmental degradation stimulating climate changes (Tilman et al., 2011). While catering to food security, nutrition security has taken a major toll. For that reason, diversified cropping systems incorporating legumes alongside cereals will provide several benefits and deal with the issue of nutritional insecurity and can be a key strategy for the sustainable intensification of agriculture, aiding in counteracting these negative effects. Notably, their ability to biologically fix atmospheric nitrogen reduces dependency on synthetic N-fertilizers which promises to enhance soil fertility (Peoples et al., 2009). Additionally, breeding legume varieties that require fewer agricultural inputs such as pesticides and fertilizers reduces the monetary investment, along with environmental impact (Ladha et al., 2005), whereas the development of genome-edited legumes with improved abiotic stress tolerance, for instance heat, drought, and salinity, will further reinforce sustainable climate-smart agriculture (Varshney et al., 2018). Enhancing photosynthetic efficiency and root architecture in legumes will also lead to increased carbon sequestration, and efficient nutrition use efficiency complimented with improved soil health (Lynch, 2019). Moreover, apart from abiotic stressors, biotic stressors cause major qualitative, quantitative, and monetary losses in legumes, thus engineering legumes for enhanced herbicide resilience (HR) (Bharti et al., 2025) along with pest and disease resistance can contribute to integrated pest management (IPM) and crop rotation strategies (Dong & Ronald, 2019).

Recent advancements and database development in plant OMICS technologies—such as genomics, transcriptomics, epigenomics, proteomics, interactomics, metabolomics, and fluxomics—have significantly refined and navigated molecular breeding strategies for crop improvement, especially for legumes. These modern-age tools provide deeper insights at the cellular level into gene expression and protein translation function, post-translational silencing under specific conditions, enabling the identification of key target genes or regulators for breeding. More importantly, high-throughput next-generation sequencing has facilitated the whole genome sequencing (WGS) of many orphan legumes (Varshney et al., 2012; Kaul et al., 2022), and comparative or de novo RNA sequencing (RNA-Seq) has enabled more efficient identification of stress-responsive genes. For example, a recent transcriptomic study in pigeonpea (*Cajanus cajan*) has deciphered numerous key genes linked with salinity stress tolerance, paving the way for developing stress-resilient varieties and extrapolating the information to other crops (Pazhamala et al., 2021). Furthermore, advancements in protein profiling techniques have resulted in the development of a knowledge compendium for an improved understanding of protein function, post-translational modifications, and stress adaptation mechanisms in legumes . In addition, the insights gained into biochemical pathways associated with abiotic stress responses and secondary metabolite biosynthesis have also opened new avenues for system biology endeavors and metabolic engineering in legumes (Katam et al., 2022).

For holistic breeding approaches, the amalgamation of GE and Omics-driven molecular breeding has transformed legume modification strategies. Moving into the revolutionary era of GE for crop breeding, the research orientation should focus on creating multiplex gene-editing tools based on CRISPR to simultaneously introduce many advantageous characteristics. Concurrent modification of multiple genes can allow targeting complex traits like heat and drought resistance. Additionally, legumes are generally recalcitrant, and therefore before genetic modification, it is a prerequisite to have an effective transformation platform; thus, increasing the efficiency of transformation in resistant legume species will allow broader genome-editing deployment (Verma et al., 2022). Leveraging pangenome research and genomic resources can fully capture legume genetic diversity for breeding purposes. Interestingly combining machine learning and artificial intelligence for genotype-phenotype correlations and predictive breeding models for analyzing massive datasets can revolutionize the future of legume breeding (Varshney, 2021). Taking advantage of these improvements, legume breeding can contribute to environmental sustainability, global food and nutrition security, and climate resilience in agriculture.

Having established the significance of GE in molecular breeding for legumes, it is imperative to address the regulatory challenges associated with its implementation. GE frequently falls into a regulatory gray area, with policies varying significantly across different countries. Although technologies such as CRISPR offer high precision, the possibility of unintended or off-target mutations remains a concern. Although genome-edited crops may be classified as non-GMO in some jurisdictions, public disbelief and regulatory barriers persist. Additionally, equitable access to these advancements is essential to ensuring that small-scale farmers and developing countries can benefit from improved legume varieties, promoting sustainable agricultural development (Figure 12.3).

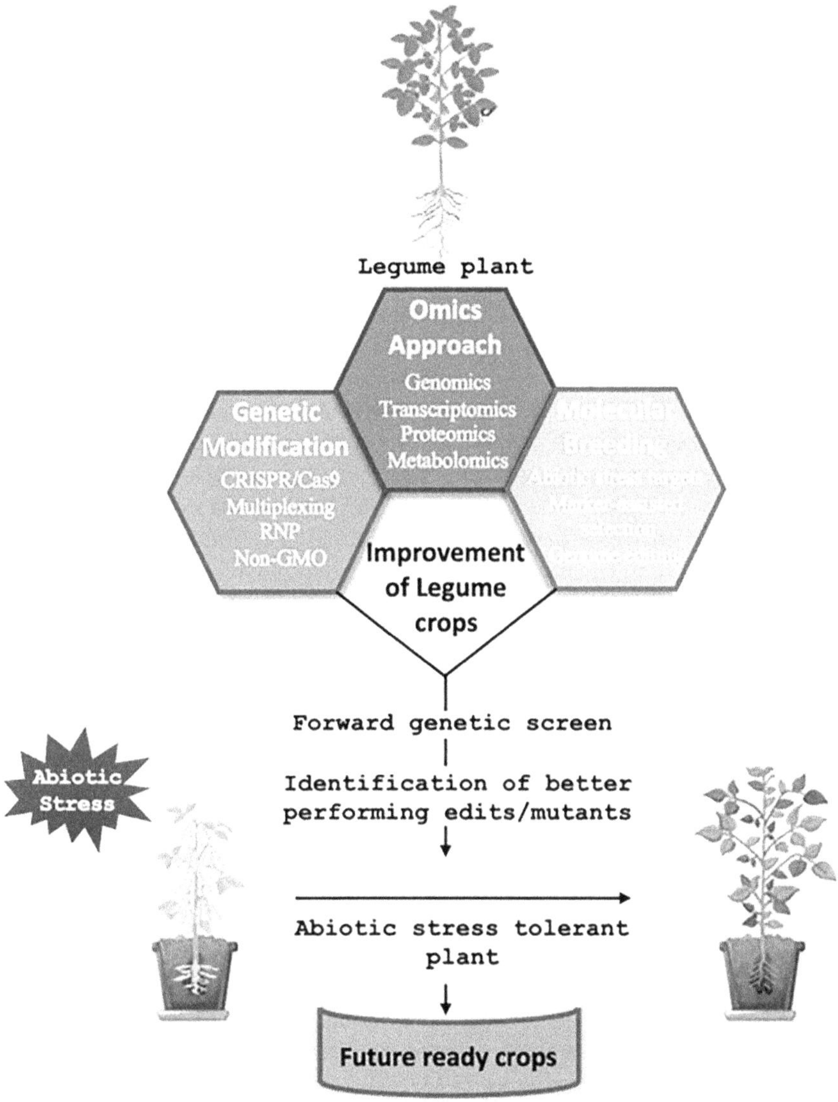

FIGURE 12.3 Role of GE in amalgamation of omics approach for holistic breeding strategy for developing abiotic stress–tolerant legumes.

12.4 ADVANTAGES OF GE APPROACH OVER BREEDING AND TRANSGENIC TECHNOLOGIES

Biofortification of legumes through conventional plant breeding poses several limitations, including laborious introduction of new traits and removal of undesirable ones. In comparison, modern transgenic breeding methods permit speedy introduction of the desired traits from any source to the target plant genome (Anand et al., 2023). For instance, homologous promoter-driven expression of insecticidal crystal protein (*cry1Ac*) conferred resistance against pod borer infestation in chickpea (Chakraborty et al., 2016). Similarly, transgenic lines overexpressing *hrpZm*, a codon-optimized harpin protein-encoding gene, conferred enhanced tolerance to Phytophthora root and stem rot (PRR) causing *P. sojae* (Du et al., 2018a). Furthermore, transgenic lines harboring hygromycin resistance gene was developed using FLP-*FRT* facilitated recombinase-mediated cassette exchange (RMCE) in soybean (Li et al., 2009). Gene targeting through homologous recombination advanced gene modifications through exchange of DNA segments by using site-specific recombinase systems like FLP-*FRT* and Cre-*lox* systems (Gilbertson, 2003; Rao et al., 2010). However, these systems had their limitations of unintended chromosomal rearrangements, low efficiency due to low recombinase expression, and off-target mutagenic recombination (Gaj et al., 2013). While the advent of gene knockdown by RNAi technology paved the way for rapid, high-throughput alternatives to study gene functions in plants, its transient expression, possible off-target silencing, and limited efficiency due to dsRNA degradation reduced its application in agricultural biotechnology (Liu et al., 2020). In like manner, random mutagenesis using chemical agents like ethylmethanesulfonate (EMS) or gamma radiation typically leads to irregular mutations, which may possibly disrupt many in-frame regulatory elements. As a result, achieving the desired trait requires screening large plant populations to identify and eliminate unwanted side effects. In contrast, targeted mutagenesis allows for precise mutations in the desired traits. This method shortens the time needed to create mutated populations and minimizes unwanted off-target effects (Jung & Till, 2021).

ZFN based mutagenesis of *DICER-LIKE* (*DCL*) genes was performed in soybean to create mutations in duplicate genes (Curtin et al., 2011). Similarly, transgenic soybean lines were generated with targeted integration of multiple transgenes to the endogenous *Fatty Acid Desaturase* (FAD2-1) locus using ZFN (Bonawitz et al., 2019). TALENs, on the similar principle, generate DSB by recognizing a single base pair of the desired target site with the DNA-binding domain of TALE proteins (Becker & Boch, 2021). TALEN mediated mutagenesis of FAD2 gene dramatically improved soybean oil quality (Haun et al., 2014). Likewise, direct mutation stacking of FAD2-1A & 1B soybean lines with FAD3A TALEN induced mutation greatly reduced levels of polyunsaturated linolenic acid (Demorest et al., 2016). Luo et al. demonstrated improved stress tolerance in soybean through CRISPR/Cas9 mediated editing of *GmARM* (Luo et al., 2024). The type II CRISPR/Cas9 system specificity is dependent on the 18–20 bp long gRNA sequence and the downstream PAM sequence (5'-NGG-3'). The recognition and nuclease lobe of the Cas9 protein identifies and cleaves 3–4 bp upstream to the PAM site to create blunt ends in the target region (O'Geen et al., 2015; Weeks et al., 2016; Jiang & Doudna, 2017; Liu et al., 2020). Thus, signifying that the CRISPR/Cas9 system is highly specific and efficient. Cellular repair mechanisms typically repair double stranded breaks through either error-prone NHEJ pathway which introduces short insertion/deletion (indels) at the target site or the HDR where the donor template with the desired sequence is used to guide precise insertion at the cutting site (Wu et al., 2014). Recently, a study by Zhang et al. reported a multiplex gene-editing system for the improvement of soybean isoflavone content and enhanced resistance to soybean mosaic virus (SBMY) (Zhang et al., 2020). A CRISPR/Cas9-mediated knockout system was established to simultaneously target *GmF3H1*, *GmF3H2*, and *GmFNSII-1*, the three genes in the isoflavone competition pathway. "Knockout" refers to disruption of a gene's function by introduction of indels within the coding region resulting in frameshift mutations, while "knock-in" specifies precise insertion of the DNA sequence coming from exogenous DNA template using the HDR technique to generate gene modifications (Baloglu et al., 2022; Gan & Ling, 2022).

Increased development of a variety of CRISPR/Cas-systems has warranted time-based, precise, and efficient modifications to the plant genome. The CRISPR/Cas12a system (alternatively known as Cpf1) eases multiplex editing with the help of 3–4 nt T-rich PAM sequence and a standalone crRNA processed through intrinsic RNase activity (Paul & Montoya, 2020). Cpf1 proteins with FAD2 specific crRNAs were targeted through PEG mediated protoplast transformation to create DNA-free improved oleic acid varieties in soybean (Kim et al., 2017). CRISPR-Cas13 system offers the unique advantage of manipulating gene expression without altering the plant's genome. Likewise, Cas14 recognizes and cleaves single-strand ssDNA in place of dsDNA without the need for a PAM sequence, thereby increasing its efficiency. In recent times, it has been used for single nucleotide polymorphism (SNP) genotyping and CRISPR diagnostics (Hillary and Ceasar, 2023). In addition, transcriptional and translational control through activation and repression, along with epigenetic modifications has been made possible with the advent of catalytically inactive dCas proteins (Pan et al., 2021). CRISPR mediated base editing creates single base changes (C to T or A to G) with the help of catalytically inactive Cas9 domain (Cas9 nickase) and Adenine/Cytosine deaminase with minimum indel formation (Mishra et al., 2020). Flowering induction genes, *GmFT2a* and *GmFT4*, were targeted using Cas9n nickase, along with cytosine deaminase to create C to G conversion in soybean plants (Cai et al., 2020). Similar to base editing, prime editing ensures greater control over gene modification through precise base-to-base substitutions or indel production without creating DSBs. Prime editing system ensures greater flexibility by virtue of all 12 base conversions in contrast to base editing which offers only single base reversals (Tingting et al., 2023). Restoration of GFP activity was demonstrated in Chickpea and Cowpea protoplasts through prime editing (Biswas et al., 2022). Emerging DNA-free CRISPR/Cas-systems circumvented the limitation of vector-based systems by skipping the integration of the foreign DNA into the cellular genome and creating modified non-GMOs. These novel techniques included the utilization of virus-like particles, bacterial secretory systems, and in-vitro assembled ribonucleoprotein complex for making precise modification without any permanent change in the genome (Tsanova et al., 2021).

Gene-editing approach using site-directed nucleases (SDNs) has been classified under three categories. SDN-1 includes alterations created in the cellular genome through nuclease and intrinsic repair mechanisms, excluding introduction of exogenous DNA material. These include techniques like CRISPR/Cas9, ZFNs, TALENs, and MNs, often leading to gene knockouts, disruptions or silencing. These alterations are akin to those created with random mutagenesis or through natural mutations achieved through conventional breeding approaches. The SDN-2 approach involves an intrinsic DNA repair mechanism using a 6–20 nt homologous template to generate point mutations through the HDR approach (Ahmad et al., 2021). In March 2022, SDN-1 and SDN-2 generated plants sans foreign gene(s) were exempted from GMO regulations by the Ministry of Environment, Forests and Climate Change of India (Tachikawa & Matsuo, 2024). Similarly, the US Department of Agriculture (USDA) has deregulated SDN-1 and SDN-2 modified crops, treating them the same as natural breeding methods since they do not contain any transgenes (Ahmad et al., 2023). SDN-3, in contrast, involves using a full gene sequence (500 bp or more) as a template to incorporate foreign DNA sequences, leading to establishment of new genetic material. CRISPR-modified SDN-1 and SDN-2 plants are generally considered more acceptable than transgenic plants, which are classified as GMOs because they involve the insertion of foreign genes not naturally found in the plant's genome (Jones, 2015).

12.5 APPLICATION OF GE FOR ABIOTIC STRESS TOLERANCE IN LEGUMES

Genome-editing technology CRISPR/Cas9 particularly has transformed the plant breeding field by targeted modification of genes that control important stress response pathways, including ion homeostasis, water retention, osmotic regulation, and antioxidant defense, in the context of abiotic stress tolerance. Leguminous crops may be made more resilient to environmental stress by modifying genes and introducing the desired traits that affect these characteristics. Abiotic stress, such as drought, salinity, heat, cold, and flooding, significantly reduces the yield and productivity of the important legumes, including peas, soybean, pigeonpea, lentils, chickpea, etc., for global food security. Abiotic stress affects the nutritional quality. Hence there is a need for GE to manipulate the targeted key genes responsible for various abiotic stress in these crops. Since ages, in crucial legumes, several genome-editing practices have been utilized for abiotic stress—drought, salinity, temperature, and heavy metals for enhancing and increasing resistance to various stress (Table 12.1).

Across the world, one of the major challenging stresses in legumes is drought stress, which significantly impacts yield and productivity of legumes by hindering the process of photosynthesis, reducing nitrogen fixation, and disrupting various biochemical, physiological, and morphological factors (Yadav et al., 2023). Scarcity of water has negative influence on yields of legumes by drop down of relative water

TABLE 12.1
A Nonexhaustive List of Genome-Editing Studies Done Hitherto for Developing Abiotic Stress Resilience in Legumes

Stress	Crop (Legumes)	Genome-Editing Tool	Target Gene	Improved Trait	Reference
Heat Stress	Soybean	CRISPR/Cas9	*GmHSP17.9*	Heat shock protein countered heat stress by enhancing nitrogenase activity in root nodules	Yang et al. (2022), Jianing et al. (2022)
			GmHsp90A2	Molecular chaperone and heat shock protein might respond to heat stress; increased malondialdehyde and enhanced thermo-sensitivity	Huang et al. (2019), Erdoğan et al. (2023)
Salinity	Soybean	CRISPR/Cas9	*GmMYB118*	Decreased tolerance to salinity and drought	Du et al. (2018b)
			GmDrb2a, GmDrb2b	Decreased tolerance to salinity and drought	Curtin et al. (2018)
			GmAITRs	Transcription factor knockouts Enhanced tolerance to salinity with longer roots and leaves	Wang et al. (2021), Kumar et al. (2023)
			SOS1	Increased salt sensitivity	Zhang et al. (2022)
Drought	Soybean	CRISPR/Cas9	*DREB*	Decreased tolerance to salinity and drought	Du et al. (2018b)
			GmHdz4	Higher drought tolerance	Zhong et al. (2022)
			GmMYB118	Decreased tolerance to salinity and drought	Du et al. (2018b)
	Medicago truncatula	CRISPR/Cas9	*MtP5CS, MtProDH*	Osmo protectant genes , enhanced drought tolerance	Verdoy et al. (2006), Zhang et al. (2014), and Zhang et al. (2022)
	Chickpea	CRISPR/Cas9	4-coumarate ligase (*4CL*), Reveille 7 (*RVE7*)	Enhanced tolerance to drought	Badhan et al. (2021), Razzaq et al. (2022), and Angon et al. (2023)

potential, leading to closure of stomata as drought alone reduces the yield by 50%–70% (Luo et al., 2016; Kumar, 2020). GE is a platform to address all these issues and focus is to enhance drought tolerance in legumes via targeting key genes that regulate osmo-tolerance, root growth, and development and improve water use efficiency (WUE). According to reports, in crucial legumes like soybean and chickpea genes like DREB (dehydration-responsive element binding) have been edited to make them drought tolerant. Osmo protectant genes like MtProDH (encoding proline dehydrogenase) and MtP5CS (Delta-1-pyrroline-5-carboxylate synthetase) has been altered for enhancing drought tolerance in Medicago truncatula (Verdoy et al., 2006; Zhang et al., 2014; Zhang et al., 2022).GmHdz4 (homeodomain-leucine zipper gene) was overexpressed and gene editing via CRISPR/Cas9 in hairy roots improving the root system architecture, along with total root length and the number of root tips was increased and hence promoted tolerance to drought stress (Zhong et al., 2022). In soybean, GmMYB118 was edited and resulted in improved drought tolerance in addition to salt stress by reducing the reactive oxygen species (ROS) and malondialdehyde (MDA) with two approaches, i.e., CRIPSR and OE-overexpression (Du et al., 2018a). Moreover, several drought resistant varieties of chickpea, faba bean, pigeonpea and common bean are developed through various institutions. In chickpea, knockout of the two protoplasmic genes, i.e., 4-coumarate ligase (4CL) and Reveille 7 (RVE7) edited via CRISPR/Cas9 enhanced drought tolerance as 4CL is one important enzyme in the lignin biosynthesis pathway's phenylpropanoid metabolism, and in a number of plants, it controls the accumulation of lignin under stress and the plant circadian rhythm is regulated by the MYB transcription factor RVE7(Badhan et al., 2021; Razzaq et al., 2022; Angon et al., 2023). In groundnut (*Arachis hypogaea*), for example, CRISPR/Cas9 has been used to edit multiple genes involved in fatty acid biosynthesis, resulting in varieties with enhanced drought tolerance and improved oil quality (Yao et al. 2023). These developments not only contribute to stress resilience but also add value to the crop's nutritional profile (Mohamed et al., 2024). When it comes to worldwide restrictions that restrict crop yield and agricultural land use, salinity stress comes in second only to drought (Gong et al., 2020). Using the CRISPR/Cas9 enhanced sensitivity to salt technique, the loss of function in the gene SOS1 (Salt Overly Sensitive 1) induced increased salt sensitivity in legumes like soybeans, which correlates with excessive Na+ intake and accumulation in roots and leaves (Zhang et al., 2022). The soybean (*Glycine max*) plant was able to withstand salt stress by utilizing CRISPR/Cas9-based knockout mutants of the genes encoding abscisic acid (ABA)-induced transcription repressors (AITRs; Wang et al., 2021). Compared to WT plants, these mutant plants exhibited greater ABA sensitivity and longer roots and shoots (Kumar et al., 2023). Other gene knockouts, such as GmDrb2a and GmDrb2b, also made soybean susceptible to salinity stress (Curtin et al., 2018).

Legumes growth and development are significantly affected by extreme weather conditions, including both heat and cold stress, as plants require optimum temperature for their proper growth and development. Cold stress in plants shows detrimental effects such as membrane destabilization, along with hampering several metabolic activities, whereas thermal stress results in enzyme deactivation, protein denaturation, and oxidative damage to plants. To improvise and overcome temperature stress (cold and thermal) techniques like GE has been used in legumes. GE smartly and precisely has transformed plant breeding and modifying legumes by the introduction of desirable traits that show resistance to heat stress, breaking the barriers over transgenic or conventional breeding methods (Sita et al., 2017; Liu et al., 2019). Targeted and precise mutagenesis via CRISPR/Cas9 has immense potential to produce elite legume cultivars harboring long-term climatic resilience with respect to thermal stresses. GmHSP17.9, a heat shock protein, was edited in soybean using CRISPR/Cas9 to counter heat stress by enhancing nitrogenase activity with variation in nodule fresh weight and number in mutants (Yang et al., 2022; Jianing et al., 2022).

Despite the promising advancements, several challenges persist in the application of GE for abiotic stress tolerance in legumes. One of the major hurdles for GE in legumes is low transformation rate and recalcitrance. Furthermore, off-target effects and concerns related to regulatory aspects and commercialization necessitate the discovery and development of high-precision GE crops. Moreover, swift public acceptance of these GE crops is all the more crucial. Hence, future research must focus on improving delivery methods, such as nanoparticle-mediated delivery systems, and enhancing the specificity of genome-editing tools to minimize unintended modifications. Additionally, the amalgamation of GE with methods of traditional breeding and precocious phenotyping techniques promises the proliferating development of legume varieties with robust abiotic stress tolerance (Şimşek et al., 2023).

To further enhance the impact of GE in legumes, developing region-specific varieties that thrive in local climate, soil, and pest conditions could be a promising approach. This could involve collaboration with local farmers, researchers, and policymakers to create tailored solutions. Additionally, establishing robust biosafety assessments and regulatory frameworks will be crucial for safe deployment. Addressing intellectual property rights and ensuring access to GE technologies for small-scale farmers will also be important. Public engagement and education on GE benefits and risks can help increase adoption rates. Emerging gene-editing tools like base editing or prime editing, microbiome research, and integration with digital agriculture and precision farming practices are also exciting areas for future exploration. Furthermore, exploring the potential of GE legumes to address specific nutritional deficiencies prevalent in regions where legumes are a staple crop could add value. For instance, biofortifying legumes with micronutrients like iron, zinc, or vitamins could improve human health outcomes. Additionally, leveraging gene editing to enhance the nutritional profile of legumes, such as increasing protein content or improving amino acid balance, could have significant impact on global nutrition and health. Moreover, the environmental benefits of GE legumes, such as reduced pesticide use, improved water retention, and increased carbon sequestration, could be further emphasized. Life cycle assessments and environmental impact studies can help quantify these benefits, informing policy decisions and promoting the adoption of GE legumes as a sustainable agricultural practice.

In addition, GE legumes could be engineered to produce nutrient-rich biofertilizers, promoting soil health and reducing synthetic fertilizer use. Or, they could be designed to enhance ecosystem services like pollinator support, biodiversity conservation, or even phytoremediation of contaminated soils. These multifunctional benefits could amplify the value proposition of GE legumes, making them a key component of regenerative agriculture strategies. Notwithstanding, GE legumes could also be engineered to produce novel compounds, like pharmaceuticals or nutraceuticals, providing new revenue streams for farmers and improving healthcare outcomes. For example, legumes producing edible vaccines, antimicrobial peptides, or anti-inflammatory compounds could revolutionize disease prevention and treatment, especially in resource-poor settings. Besides this, GE legumes could be designed to serve as sentinel plants, detecting environmental pollutants or pathogens, and alerting farmers or authorities to potential threats. This could enable early warning systems for agricultural diseases, improving crop yields and reducing losses. What's more, GE legumes could also be engineered for space agriculture, providing a sustainable food source for future space missions or settlements. Their nitrogen-fixing abilities and nutritional profile make them an attractive candidate for Martian or lunar farming. This could be a game-changer for long-duration space exploration.

In conclusion, its noteworthy that novel and innovative genome-editing techniques, especially CRISPR/Cas9 technology, has opened umpteen new avenues and opportunities for breeding legumes with enhanced resilience to abiotic stress along with better nutrition to counteract global climatic challenges as well as micronutrient- & protein-energy malnutrition. Recent developments demonstrate the potential of this technology to address the challenges posed by environmental stresses, thereby contributing to sustainable agriculture, nutritional, food, fiber, energy, and space security. While challenges remain, the progress made thus far demonstrates the transformative potential of GE in achieving nutritional and food security and sustainable agriculture in the face of climate change.

12.6 CONCLUSION

Legumes are renowned for their health and nutritional value, along with their role in promoting the sustainability of agricultural systems. Additionally, they play a pivotal role as the primary source of plant-based protein for humans as well as livestock. They enhance farm productivity and income through intercropping and crop rotation while reducing pests, diseases, and weeds. In terms of enhancing agricultural output, legumes have drawn less attention than cereals, irrespective of their commercial and economic significance. However, legumes are prone to several abiotic stresses that significantly impact their productivity. Pigeonpea, a nutrient-dense legume, facilitates sustainable farming in tropical and subtropical regions. However, it is prone to abiotic stresses such as mineral deficiencies, moisture, temperature, and photoperiod fluctuations. Moreover, the stress factors hinder plant development while affecting their physiology and morphology. Abiotic stresses are governed by several factors such as different stress signal transduction pathways and transcription factors, along with phytohormones' metabolic pathway. CRISPR/Cas-systems are employed to generate climate-smart crops via enhancing the yield and nutritional value. Further, development of plant tolerance to abiotic stress via CRISPR-Cas is significantly increasing, even though its utilization in generating abiotic stress–tolerant crops is still limited. Precise GE by CRISPR/Cas9 enables multiplexed sgRNAs to target multiple genes simultaneously and promotes crop resilience through modifying structural and regulatory genes linked to abiotic stress. As discussed, in this chapter, the representative cases in genome-edited, abiotic stress–tolerant/resistant legumes are still comparatively less, especially no instances in pigeonpea. One potential aspect to consider is the intersection of GE legumes with digital agriculture and data-driven farming practices. For instance, integrating GE legumes with precision agriculture technologies, such as IoT sensors, drones, or satellite imaging, could optimize crop management, predict yields, and detect stress early on. This convergence could amplify the benefits of GE legumes and create more resilient agricultural systems. It would also be interesting to explore is the potential of GE legumes in circular economy models, for instance; designing GE legumes for biodegradable plastics, bioproducts, or bioenergy could create new market opportunities and reduce environmental waste. This could also incentivize sustainable agriculture practices and promote a more regenerative food system. Hence, developing climate resilient, nutrient-rich lines via CRISPR/Cas can revolutionize pigeonpea breeding programs and ensure food and energy security and making pigeonpea farming a more practical and sustainable option for farmers worldwide.

REFERENCES

Adhikari P, Poudel M (2020) CRISPR-Cas9 in agriculture: approaches, applications, future perspectives, and associated challenges. *Malays J Halal Res* 3(1): 6–16.

Ahmad A, Jamil A, Munawar N (2023) GMOs or non-GMOs? the CRISPR conundrum. *Front Plant Sci* 14: 1232938.

Ahmad A, Munawar N, Khan Z, Qusmani AT, Khan SH, Jamil A, Ashraf S, Ghouri MZ, Aslam S, Mubarik MS, Munir A (2021) An outlook on global regulatory landscape for genome-edited crops. *Int J Mol Sci* 22(21): 11753.

Alexandratos N, Bruinsma J (2012) *World agriculture towards 2030/2050: the 2012 revision*. Rome: ESA Working Paper No. 12-03, FAO.

Anand A, Subramanian M, Kar D (2023) Breeding techniques to dispense higher genetic gains. *Front Plant Sci* 13: 1076094.

Angon PB, Mondal S, Akter S, Sakil MA, Jalil MA (2023) Roles of CRISPR to mitigate drought and salinity stresses on plants. *Plant Stress* 8: 100169.

Antonioou P, Miccio A, Brusson M (2021) Base and prime editing technologies for blood disorders. *Front Genome Ed* 3: 618406.

Antunes MS, Smith JJ, Jantz D, Medford JI (2012) Targeted DNA excision in *Arabidopsis* by a re-engineered homing endonuclease. *BMC Biotechnol* 13: 86.

Anzalone AV, Koblan LW, Liu DR (2020) Genome editing with CRISPR–Cas nucleases, base editors, transposases and prime editors. *Nat Biotechnol* 38: 824–844.

Badhan S, Ball AS, Mantri N (2021) First report of CRISPR/Cas9 mediated DNA-free editing of 4CL and RVE7 genes in chickpea protoplasts. *Int J Mol Sci* 22(1): 396.

Baloglu MC, Celik Altunoglu Y, Baloglu P, Yildiz AB, Türkölmez N, Özden Çiftçi Y (2022) Gene-editing technologies and applications in legumes: progress, evolution, and future prospects. *Front Genet* 13: 859437.

Baltes NJ, Gil-Humanes J, Cermak T, Atkins PA, Voytas DF (2014) DNA replicons for plant genome engineering. *Plant Cell* 26(1): 151–163.

Barrangou R, Fremaux C, Deveau H, Richards M, Boyaval P, Moineau S, Romero DA, Horvath P (2007) CRISPR provides acquired resistance against viruses in prokaryotes. *Science* 315(5819):1709–1712. https://doi.org/10.1126/science.1138140

Becker S, Boch J (2021) TALE and TALEN genome editing technologies. *Gene Genome Ed* 2: 100007.

Bharti J, Verma R, Thangraj A, Sony SK, Fathy K, Rawat B, Kaul R, Chakraborty P, Gupta I, Kaul T (2025) Combatting aggressive weeds: reinforcing herbicide resistance in pigeonpea (*Cajanus cajan* L.) through genome editing. *Plant Physiol Biochem* 222: 109550.

Bibi F, Rahman A (2023) An overview of climate change impacts on agriculture and their mitigation strategies. *Agriculture* 13: 1508.

Biswas S, Bridgeland A, Irum S, Thomson MJ, Septiningsih EM (2022) Optimization of prime editing in rice, peanut, chickpea, and cowpea protoplasts by restoration of GFP activity. *Int J Mol Sci* 23(17): 9809.

Bohra A, Jha UC, Kavi Kishor PB, Pandey S, Singh NP (2014a) Genomics and molecular breeding in lesser explored pulse crops: current trends and future opportunities. *Biotechnol Adv* 32: 1410–1428.

Bohra A, Pandey MK, Jha UC, Singh B, Singh IP, Datta D, Chaturvedi SK, Nadarajan N, Varshney RK (2014b) Genomics-assisted breeding in the four major pulse crops of developing countries: present status and prospects. *Theor Appl Genet* 127: 1263–1291.

Bohra A, Saxena KB, Varshney RK, Saxena RK (2020) Genomics assisted breeding for pigeonpea improvement. *Theor Appl Genet* 133: 1721–1737.

Bohra A, Tiwari A, Kaur P, Ganie S, Raza A, Roorkiwal M, Mir R, Fernie A, Smykal P, Varshney R (2022) The key to the future lies in the past: insights from grain legume domestication and improvement should inform future breeding strategies. *Plant Cell Physiol* 63(11): 1554–1572.

Bonawitz ND, Ainley WM, Itaya A, Chennareddy SR, Cicak T, Effinger K, Jiang K, Mall TK, Marri PR, Samuel JP, Sardesai N (2019) Zinc finger nuclease–mediated targeting of multiple transgenes to an endogenous soybean genomic locus via non-homologous end joining. *Plant Biotechnol J* 17(4): 750–761.

Cai Y, Chen L, Zhang Y, Yuan S, Su Q, Sun S, ... Hou W (2020) Target base editing in soybean using a modified CRISPR/Cas9 system. *Plant Biotechnol J* 18(10):1996.

Chakraborty J, Sen S, Ghosh P, Sengupta A, Basu D, Das S (2016) Homologous promoter derived constitutive and chloroplast targeted expression of synthetic cry1Ac in transgenic chickpea confers resistance against *Helicoverpa armigera*. *Plant Cell Tiss Org* 125: 521–535.

Chen K, Wang Y, Zhang R, Zhang H, Gao C (2019) CRISPR/Cas genome editing and precision plant breeding in agriculture. *Annu Rev Plant Biol* 70(1): 667–697.

Curtin SJ, Xiong Y, Michno JM, Campbell BW, Stec AO, Čermák T, et al. (2018) CRISPR/Cas9 and TALENs generate heritable mutations for genes involved in small RNA processing of *Glycine max* and *Medicago truncatula*. *Plant Biotechnol J* 16: 1125–1137.

Curtin SJ, Zhang F, Sander JD, Haun WJ, Starker C, Baltes NJ, Reyon D, Dahlborg EJ, Goodwin MJ, Coffman AP, Dobbs D (2011) Targeted mutagenesis of duplicated genes in soybean with zinc-finger nucleases. *Plant Physiol* 156(2): 466–473.

Demorest ZL, Coffman A, Baltes NJ, Stoddard TJ, Clasen BM, Luo S, Retterath A, Yabandith A, Gamo ME, Bissen J, Mathis L (2016) Direct stacking of sequence-specific nuclease-induced mutations to produce high oleic and low linolenic soybean oil. *BMC Plant Biol* 16: 1–8.

Djukanovic V, Smith J, Lowe K, Yang M, Gao H, Jones S, Nicholson MG, West A, Lape J, Bidney D, Carl Falco S (2013) Male-sterile maize plants produced by targeted mutagenesis of the cytochrome P450-like gene (MS 26) using a re-designed I-CreI homing endonuclease. *Plant J* 76(5): 888–899.

Dong OX, Ronald PC (2019) Genetic engineering for disease resistance in plants: recent progress and future perspectives. *Plant Physiol* 180(1): 26–38.

Du Q, Yang X, Zhang J, Zhong X, Kim KS, Yang J, Xing G, Li X, Jiang Z, Li Q, Dong Y (2018a) Over-expression of the *Pseudomonas syringae* harpin-encoding gene hrpZm confers enhanced tolerance to *Phytophthora* root and stem rot in transgenic soybean. *Transgenic Res* 27: 277–288.

Du YT, Zhao MJ, Wang CT, Gao Y, Wang YX, Liu YW, Ma YZ (2018b) Identification and characterization of GmMYB118 responses to drought and salt stress. *BMC Plant Biol* 18: 1–18.

Endo M, Mikami M, Toki S (2016) Biallelic gene targeting in rice. *Plant Physiol* 170(2): 667–677.

Epinat JC, Arnould S, Chames P, Rochaix P, Desfontaines D, Puzin C, Patin A, Zanghellini A, Pâques F, Lacroix E (2003) A novel engineered meganuclease induces homologous recombination in yeast and mammalian cells. *Nucleic Acids Res* 31(11): 2952–2962.

Erdoğan İ, Cevher-Keskin B, Bilir Ö, Hong Y, Tör M (2023) Recent developments in CRISPR/Cas9 genome-editing technology related to plant disease resistance and abiotic stress tolerance. *Biology* 12(7): 1037.

Fauser F, Schiml S, Puchta H (2014) Both CRISPR/Cas-based nucleases and nickases can be used efficiently for genome engineering in *Arabidopsis thaliana*. *Plant J* 79(2): 348–359.

Food and Agriculture Organization FAO (2021) International Fund for Agricultural Development IFAD, World Food Programme WFP. *The state of food insecurity in the world 2021*. FAO.

Gaj T, Gersbach CA, Barbas CF (2013) ZFN, TALEN, and CRISPR/Cas-based methods for genome engineering. *Trends Biotechnol* 31(7): 397–405.

Gan P, Wong SM, Ding J (2022) Genome editing in crop improvement: current status and future prospects. *Plant Communications* 3(6):100373.

Gan WC, Ling AP (2022) CRISPR/Cas9 in plant biotechnology: applications and challenges. *Biotechnol Acta* 103(1): 81.

Gao C (2018) The future of CRISPR technologies in agriculture. *Nat Rev Mol Cell Biol* 19: 275–276.

Gaudelli NM, et al. (2017) Programmable base editing of A•T to G•C in genomic DNA without DNA cleavage. *Nature* 551: 464–471.

Ghebreyesus TA (2020) WHO on coronavirus pandemic: "The worst is yet to come".

Gilbertson L (2003) Cre–lox recombination: creative tools for plant biotechnology. *Trends Biotechnol* 21(12): 550–555.

Gong Z, Xiong L, Shi H, Yang S, Herrera-Estrella LR, Xu G, Zhu JK (2020) Plant abiotic stress response and nutrient use efficiency. *Sci China Life Sci* 63: 635–674.

Gupta RM, Musunuru K (2014) Expanding the genetic editing tool kit: ZFNs, TALENs, and CRISPR-Cas9. *J Clin Invest* 124(10): 4154–4161.

Haque E, Taniguchi H, Hassan MM, Bhowmik P, Karim MR, Smiech M, Zhao K, Rahman M, Islam T (2018) Application of CRISPR/Cas9 genome editing technology for the improvement of crops cultivated in tropical climates: recent progress, prospects, and challenges. *Front Plant Sci* 9: 617.

Haun W, Coffman A, Clasen BM, Demorest ZL, Lowy A, Ray E, Retterath A, Stoddard T, Juillerat A, Cedrone F, Mathis L (2014) Improved soybean oil quality by targeted mutagenesis of the fatty acid desaturase 2 gene family. *Plant Biotechnol J* 12(7): 934–940.

Hillary VE, Ceasar SA (2023) A review on the mechanism and applications of CRISPR/Cas9/Cas12/Cas13/Cas14 proteins utilized for genome engineering. *Mol Biotechnol* 65(3): 311–325.

Hsu PD, Lander ES, Zhang F (2014) Development and applications of CRISPR-Cas9 for genome engineering. *Cell* 157: 1262–1278.

Hua K, Peijin H, Jian-Kang Z (2022) Improvement of base editors and prime editors advances precision genome engineering in plants. *Plant Physiol* 188(4): 1795–1810.

Huang Y, Xuan H, Yang C, Guo N, Wang H, Zhao J, Xing H (2019) GmHsp90A2 is involved in soybean heat stress as a positive regulator. *Plant Sci* 285: 26–33.

Hyten DL, Song Q, Zhu Y, Choi IY, Nelson RL, Costa JM, et al. (2006) Impact of genetic bottlenecks on soybean genome diversity. *Proc Natl Acad Sci* 103: 16666–16671.

Jaganathan D, Ramasamy K, Sellamuthu G, Jayabalan S, Venkataraman G (2018) CRISPR for crop improvement: an update review. *Front Plant Sci* 9: 985.
Jiang F, Doudna JA (2017) CRISPR–Cas9 structures and mechanisms. *Annu Rev Biophys* 46(1): 505–529.
Jiang W, Bikard D, Cox D, et al. (2013) RNA-guided editing of bacterial genomes using CRISPR-Cas systems. *Nature Biotechnology* 31:233–239. https://doi.org/10.1038/nbt.2508
Jianing G, Yuhong G, Yijun G, Rasheed A, Qian Z, Zhiming X, Mahmood A, Shuheng Z, Zhuo Z, Xiaoxue W, Jian W (2022) Improvement of heat stress tolerance in soybean (*Glycine max* L) by using conventional and molecular tools. *Front Plant Sci* 13: 993189.
Jinek M, Chylinski K, Fonfara I, Hauer M, Doudna JA, Charpentier EA (2012) Programmable dual-RNA–guided DNA endonuclease in adaptive bacterial immunity. *Science* 337: 816–821.
Jones H (2015) Regulatory uncertainty over genome editing. *Nat Plants* 1: 14011.
Jung C, Till B (2021) Mutagenesis and genome editing in crop improvement: perspectives for the global regulatory landscape. *Trends Plant Sci* 26: 1258–1269.
Kantor A, McClements ME, MacLaren RE (2020) CRISPR-Cas9 DNA base-editing and prime-editing. *Int J Mol Sci* 21: 6240.
Katam R, Lin C, Grant K, Katam CS, Chen S (2022) Advances in plant metabolomics and its applications in stress and single-cell biology. *Int J Mol Sci* 23: 6985.
Kaul T, Easwaran M, Thangaraj A, Meyyazhagan A, Nehra M, Raman NM, Verma R, Sony SK, Motelb KFA, Bharti J, Gayacharan, Badapanda C, Balasubramanian B (2022) De novo genome assembly of rice bean (*Vigna umbellata*) – a nominated nutritionally rich future crop reveals novel insights into flowering potential, habit, and palatability centric traits for efficient domestication. *Front Plant Sci* 13: 739654.
Khorsandi P (2020) WFP chief warns of 'hunger pandemic' as global food crises report launched. *World Food Programme Insight*, 22 April 2020.
Kim H, Kim ST, Ryu J, Kang BC, Kim JS, Kim SG (2017) CRISPR/Cpf1-mediated DNA-free plant genome editing. *Nat Commun* 8: 1–7.
Kumar M, Prusty MR, Pandey MK, Singh PK, Bohra A, Guo B, Varshney RK (2023) Application of CRISPR/Cas9-mediated gene editing for abiotic stress management in crop plants. *Front Plant Sci* 14: 1157678.
Kumar S (2020) Abiotic stresses and their effects on plant growth, yield and nutritional quality of agricultural produce. *Int J Food Sci Agric* 4(4): 367–378.
Ladha JK, Pathak H, Krupnik TJ, Six J, Vankessel C (2005) Efficiency of fertilizer nitrogen in cereal production: retrospects and prospects. *Adv Agron* 87: 85–156.
Li C, Nguyen V, Liu J, Fu W, Chen C, Yu K, et al. (2019) Mutagenesis of seed storage protein genes in soybean using CRISPR/Cas9. *BMC Res Notes* 12: 1–7.
Li T, Liu B, Chen CY, Yang B (2016) TALEN-mediated homologous recombination produces site-directed DNA base change and herbicide-resistant rice. *J Genet Genomics* 43(5): 297–305.
Li T, Yang Y, Qi H, et al. (2023) CRISPR/Cas9 therapeutics: progress and prospects. *Sig Transduct Target Ther* 8: 36.
Li Z, Xing A, Moon BP, McCardell RP, Mills K, Falco SC (2009) Site-specific integration of transgenes in soybean via recombinase-mediated DNA cassette exchange. *Plant Physiol* 151(3): 1087–1095.
Liu M, Rehman S, Tang X, Gu K, Fan Q, Chen D, Ma W (2019) Methodologies for improving HDR efficiency. *Front Genet* 9: 691.
Liu Q, Yang F, Zhang J, Liu H, Rahman S, Islam S, Ma W (2021) Application of CRISPR/Cas9 in crop quality improvement. *Int J Mol Sci* 22(8): 4206.
Liu S, Jaouannet M, Dempsey DMA, Imani J, Coustau C, Kogel KH (2020) RNA-based technologies for insect control in plant production. *Biotechnol Adv* 39: 107463.

Luo SS, Sun YN, Zhou X, Zhu T, Zhu LS, Arfan M, Lin HH (2016) *Medicago truncatula* genotypes Jemalong A17 and R108 show contrasting variations under drought stress. *Plant Physiol Biochem* 109: 190–198.

Luo T, Ma C, Fan Y, Qiu Z, Li M, Tian Y, Shang Y, Liu C, Cao Q, Peng Y, Zhang S (2024) CRISPR-Cas9-mediated editing of GmARM improves resistance to multiple stresses in soybean. *Plant Sci* 112147.

Lynch JP (2019) Root phenotypes for improved nutrient capture: an underexploited opportunity for global agriculture. *New Phytol* 223(2): 548–564.

Manavalan LP, Guttikonda SK, Phan Tran LS, Nguyen HT (2009) Physiological and molecular approaches to improve drought resistance in soybean. *Plant Cell Physiol* 50: 1260–1276.

Marintcheva B (2018) Viral tools for genome manipulations in vivo. https://doi.org/10.1016/B978-0-12-810514-6.00003-9

Mishra R, Joshi RK, Zhao K (2020) Base editing in crops: current advances, limitations and future implications. *Plant Biotechnol J* 18(1): 20–31.

Mohamed HI, Khan A, Basit A (2024) CRISPR-Cas9 system mediated genome editing technology: an ultimate tool to enhance abiotic stress in crop plants. *J Soil Sci Plant Nutr* 24: 1799–1822.

O'Geen H, Abigail SY, Segal DJ (2015) How specific is CRISPR/Cas9 really? *Curr Opin Chem Biol* 29: 72–78.

Pan C, Sretenovic S, Qi Y (2021) CRISPR/dCas-mediated transcriptional and epigenetic regulation in plants. *Curr Opin Plant Biol* 60: 101980.

Pandey P, Irulappan V, Bagavathiannan MV, Senthil-Kumar M (2017) Impact of combined abiotic and biotic stresses on plant growth and avenues for crop improvement by exploiting physio-morphological traits. *Front Plant Sci* 8(8): 537.

Paul B, Montoya G (2020) CRISPR-Cas12a: functional overview and applications. *Biomed J* 43(1): 8–17.

Pazhamala LT, Kudapa H, Weckwerth W, Millar AH, Varshney RK (2021) Systems biology for crop improvement. *Plant Genome* 14: e20098.

Peoples MB, Brockwell J, Herridge DF, et al. (2009) The contributions of nitrogen-fixing crop legumes to the productivity of agricultural systems. *Symbiosis* 77(2): 89–107.

Population Reference Bureau (2021) Available at: https://www.prb.org/

Puchta H, Dujon B, Hohn B (1993) Homologous recombination in plant cells is enhanced by in vivo induction of double strand breaks into DNA by a site-specific endonuclease. *Nucleic Acids Res* 21: 5034–5040.

Qi LS, et al. (2013) Repurposing CRISPR as an RNA-guided platform for sequence-specific control of gene expression. *Cell* 152: 1173–1183.

Ran FA, et al. (2013) Genome engineering using the CRISPR-Cas9 system. *Nat Proto* 8: 2281–2308.

Rao MR, Moon HS, Schenk TM, Becker D, Mazarei M, Stewart Jr CN (2010) FLP/FRT recombination from yeast: application of a two gene cassette scheme as an inducible system in plants. *Sensors* 10(9): 8526–8535.

Ray DK, Mueller ND, West PC, Foley JA (2013) Yield trends are insufficient to double global crop production by 2050. *PLoS One* 8: 1–8.

Razzaq MK, Akhter M, Ahmad RM, Cheema KL, Hina A, Karikari B, Raza G, Xing G, Gai J, Khurshid M (2022) CRISPR-Cas9 based stress tolerance: new hope for abiotic stress tolerance in chickpea (*Cicer arietinum*). *Mol Biol Rep* 49(9): 8977–8985.

Sander J, Joung JK (2014) CRISPR-Cas systems for editing, regulating and targeting genomes. *Nature Biotechnology* 32:347–355. https://doi.org/10.1038/nbt.2842

Sander JD, Joung JK (2014) CRISPR-Cas systems for editing, regulating and targeting genomes. *Nat Biotechnol* 32(4): 347–350.

Saxena KB, Bohra A, Choudhary AK, Sultana R, Pazhamala L, Saxena RK (2021) The alternative breeding approaches for improving yield gains and stress response in pigeonpea [*Cajanus cajan* (L.) Millsp.]. *Plant Breed* 140: 76–86.

Saxena KB, Choudhary A, Srivastava R, Bohra A, Saxena R, Varshney RK (2019) Origin of early maturing pigeonpea germplasm and its impact on adaptation and cropping systems. *Plant Breed* 138: 243–251.

Silva G, Poirot L, Galetto R, Smith J, Montoya G, Duchateau P, Pâques F (2011) Meganucleases and other tools for targeted genome engineering: perspectives and challenges for gene therapy. *Curr Gene Ther* 11(1): 11–27.

Şimşek Ö, Isak MA, Dönmez D, Dalda Şekerci A, İzgü T, Kaçar YA (2023) Advanced biotechnological interventions in mitigating drought stress in plants. *Plants* 13(5): 717.

Sita K, Sehgal A, HanumanthaRao B, Nair RM, Vara Prasad PV, Kumar S, Gaur PM, Farooq M, M Siddique, Varshney RK, Nayyar H (2017) Food legumes and rising temperatures: effects, adaptive functional mechanisms specific to reproductive growth stage and strategies to improve heat tolerance. *Front Plant Sci* 8: 1658.

Song Q, Hyten DL, Jia G, Quigley CV, Fickus EW, Nelson RL, et al. (2015) Fingerprinting soybean germplasm and its utility in genomic research. *G3 Genes Genomes Genet* 5: 1999–2006.

Sun Y, Li J, Xia L (2016a) Precise genome modification via sequence-specific nucleases-mediated gene targeting for crop improvement. *Front Plant Sci* 7: 1928.

Sun Y, Zhang X, Wu C, He Y, Ma Y, Hou H, Guo X, Du W, Zhao Y, Xia L (2016b) Engineering herbicide-resistant rice plants through CRISPR/Cas9-mediated homologous recombination of acetolactate synthase. *Mol Plant* 9(4): 628–631.

Tachikawa M, Matsuo M (2024) Global regulatory trends of genome editing technology in agriculture and food. *Breed Sci* 74(1): 3–10.

Tilman D, Balzer C, Hill J, Befort BL (2011) Global food demand and the sustainable intensification of agriculture. *Proceedings of the National Academy of Sciences of the United States of America* 108(50):20260–20264. https://doi.org/10.1073/pnas.1116437108

Tingting L, Jinpeng Z, Xi Y, Kejian W, Yuchun R, Chun W (2023) Development and application of prime editing in plants. *Rice Sci* 30(6): 509–522.

Tsanova T, Stefanova L, Topalova L, Atanasov A, Pantchev I (2021) DNA-free gene editing in plants: a brief overview. *Biotechnol Biotechnol Equip* 35(1): 131–138.

Usman M, Bokhari SAM, Fatima B, Rashid B, Nadeem F, Sarwar MB, Nawaz-ul-Rehman MS, Shahid M, Ayub CM (2022) Drought stress mitigating morphological, physiological, biochemical, and molecular responses of guava (*Psidium guajava* L.) cultivars. *Front Plant Sci* 13: 878616.

Varshney R, Chen W, Li Y, et al. (2012) Draft genome sequence of pigeonpea (*Cajanus cajan*), an orphan legume crop of resource-poor farmers. *Nat Biotechnol* 30: 83–89.

Varshney RK (2021) The Plant Genome special issue: advances in genomic selection and application of machine learning in genomic prediction for crop improvement. *Plant Genome* 14: e20178.

Varshney RK, Thudi M, Pandey MK, Tardieu F, Ojiewo C, Vadez V, et al. (2018) Accelerating genetic gains in legumes for the development of prosperous smallholder agriculture: integrating genomics, phenotyping, systems modelling and agronomy. *J Exp Bot* 69(13): 3293–3312.

Verdoy D, Coba de la Peña T, Redondo FJ, Lucas MM, Pueyo JJ (2006) Transgenic *Medicago truncatula* plants that accumulate proline display nitrogen-fixing activity with enhanced tolerance to osmotic stress. *Plant Cell Environ* 29(10): 1913–1923.

Verma R, Bharti J, Kaul T, Sony SK, Fathy K, Thangraj A, Nehra M, Sony SK, Eswaran M, Kaul R (2022) An improved protocol for efficient regeneration, transformation and gene-editing using CRISPR/Cas9 in pigeonpea (*Cajanus cajan* (L.) Millsp.). *Research Square*. https://doi.org/10.21203/rs.3.rs-1997719/v1

Waltz E (2022) GABA-enriched tomatoes approved in Japan as first CRISPR-edited food crop. *Nature Biotechnology* 40:9–10.

Wang M, Wang S, Liang Z, Shi W, Gao C, Xia G (2018) From genetic stock to genome editing: gene exploitation in wheat. *Trends Biotechnol* 36(2): 160–172.

Wang T, Xun H, Wang W, Ding X, Tian H, Hussain S, et al. (2021) Mutation of GmAITR genes by CRISPR/Cas9 genome editing results in enhanced salinity stress tolerance in soybean. *Front Plant Sci* 12: 779598.

Wang W, Vinocur B, Altman A (2003) Plant responses to drought, salinity and extreme temperatures: towards genetic engineering for stress tolerance. *Planta* 218: 1–14.

Weeks DP, Spalding MH, Yang B (2016) Use of designer nucleases for targeted gene and genome editing in plants. *Plant Biotechnol J* 14(2): 483–495.

Wolt JD, Wang K, Yang B (2016) The regulatory status of genome-edited crops. *Plant Biotechnol J* 14(2): 510–518.

Wu X, Kriz AJ, Sharp PA (2014) Target specificity of the CRISPR-Cas9 system. *Quant Biol* 2: 59–70.

Wu Y, et al. (2015) Correction of a genetic disease by CRISPR-Cas9-mediated gene editing in mouse spermatogonial stem cells. *Cell Res* 25: 67–79.

Xu Y, Li Z (2020) CRISPR-Cas systems: overview, innovations and applications in human disease research and gene therapy. *Comput Struct Biotechnol* 18: 2401–2415.

Yadav RK, Tripathi MK, Tiwari S, Tripathi N, Asati R, Chauhan S, Tiwari PN, Payasi DK (2023) Genome editing and improvement of abiotic stress tolerance in crop plants. *Life* 13(7): 1456.

Yang Z, Du H, Xing X, Li W, Kong Y, Li X, Zhang C (2022) A small heat shock protein, GmHSP17.9, from nodule confers symbiotic nitrogen fixation and seed yield in soybean. *Plant Biotechnol J* 20(1): 103–115.

Yao D, Zhou J, Zhang A, Wang J, Liu Y, Wang L, Pi W, Li Z, Yue W, Cai J, Liu H, Hao W, Qu X (2023) Advances in CRISPR/Cas9-based research related to soybean [*Glycine max* (Linn.) Merr] molecular breeding. *Front Plant Sci* 14: 1247707.

Zhang F, Maeder ML, Unger-Wallace E, Hoshaw JP, Reyon D, Christian M, Li X, Pierick CJ, Dobbs D, Peterson T, Joung JK, Voytas DF (2010) High frequency targeted mutagenesis in *Arabidopsis thaliana* using zinc finger nucleases. *Proceedings of the National Academy of Sciences of the United States of America* 107(26):12028–12033. https://doi.org/10.1073/pnas.0914991107

Zhang JY, Cruz De Carvalho MH, Torres-Jerez I, Kang YUN, Allen SN, Huhman DV, Udvardi MK (2014) Global reprogramming of transcription and metabolism in *Medicago truncatula* during progressive drought and after rewatering. *Plant Cell Environ* 37(11): 2553–2576.

Zhang M, Cao J, Zhang T, Xu T, Yang L, Li X, Ji F, Gao Y, Ali S, Zhang Q, Zhu J, Xie L (2022) A putative plasma membrane Na+/H+ antiporter GmSOS1 is critical for salt stress tolerance in *Glycine max*. *Front Plant Sci* 13: 870695.

Zhang P, Du H, Wang J, Pu Y, Yang C, Yan R, Yang H, Cheng H, Yu D (2020) Multiplex CRISPR/Cas9-mediated metabolic engineering increases soybean isoflavone content and resistance to soybean mosaic virus. *Plant Biotechnol J* 18(6): 1384–1395.

Zhang X, Sun Y, Qiu X, Lu H, Hwang I, Wang T (2022) Tolerant mechanism of model legume plant *Medicago truncatula* to drought, salt, and cold stresses. *Front Plant Sci* 13: 847166.

Zhong X, Hong W, Shu Y, Li J, Liu L, Chen X, Islam F, Zhou W, Tang G (2022) CRISPR/Cas9 mediated gene-editing of GmHdz4 transcription factor enhances drought tolerance in soybean (*Glycine max* [L.] Merr.). *Front Plant Sci* 13: 988505.

Index

Pages in *italics* refer to figures and pages in **bold** refer to tables.

S

T

U

V

W

X

Z

For Product Safety Concerns and Information please contact our EU
representative GPSR@taylorandfrancis.com
Taylor & Francis Verlag GmbH, Kaufingerstraße 24, 80331 München, Germany

www.ingramcontent.com/pod-product-compliance
Ingram Content Group UK Ltd.
Pitfield, Milton Keynes, MK11 3LW, UK
UKHW022217190926
475702UK00003B/407
* 9 7 8 1 0 3 2 9 9 2 4 0 2 *